This volume of proceedings gives an up to date picture of the current status of studies of molecular clouds in our Galaxy and in external galaxies. The contributors review the current observational situation and consider a variety of theoretical questions concerning the structure of clouds, their dynamics and energetics and the physical processes that determine their properties.

Molecular Clouds

Molecular Clouds

*The proceedings of a conference at the Department of Astronomy,
University of Manchester, 26–30 March 1990*

Edited by

R. A. JAMES

Department of Astronomy, University of Manchester

and

T. J. MILLAR

Department of Mathematics, UMIST

CAMBRIDGE UNIVERSITY PRESS

Cambridge

New York   Port Chester

Melbourne   Sydney

Published by the Press Syndicate of the University of Cambridge
The Pitt Building, Trumpington Street, Cambridge CB2 1RP
40 West 20th Street, New York, NY 10011-4211, USA
10 Stamford Road, Oakleigh, Melbourne 3166, Australia

© Cambridge University Press 1991

First published 1991

Printed in Great Britain at the University Press, Cambridge

British Library cataloguing in publication data available

Library of Congress cataloguing in publication data available

ISBN 0 521 39543 7 hardback

CONTENTS

Preface   *xiii*

Names and addresses of participants   *xv*

Large-scale CO observations of our Galaxy and its nearest neighbours
P. Thaddeus   1

Orion-A molecular cloud: a multi-transition study of CO
A. Dutrey, A. Castets, G. Duvert, J. Bally, W.D. Langer and R.W. Wilson   13

First detection of ^{13}CO J = 6 → 5: large amounts of warm molecular gas
J. Stutzki, U.U. Graf, R. Genzel, A.I. Harris, R.E. Hills and A.P.G. Russell   17

Molecular clouds at the edge of the Galaxy
J. Brand and J.G.A. Wouterloot   21

Search for star formation in the Southern Coalsack
L.-Å. Nyman, L. Bronfman and P. Thaddeus   23

KOSMA submillimetre observations of outflows in L1228 and L1551
L. Haikala, M. Miller, K. Gierens and G. Winnewisser   25

Modelling the S140 molecular cloud: A multitransitional study of CO isotopomers
E.L.O. Bakes and G.J. White   27

Gamma-ray and far-infrared constraints on the N_{H_2}/W_{CO} conversion factor of molecular clouds
Hans Bloemen   29

The mass of molecular gas in the Galaxy
A.W. Wolfendale   41

High latitude molecular clouds
Leo Blitz   49

Galactic cirrus clouds in an HI loop
H. Meyerdierks and V. Großman   69

The high-latitude cloud towards HD210121
R. Gredel, E.F. van Dishoeck, C.P. de Vries and J.H. Black   73

Structure and energy balance of molecular clouds
R. Genzel 75

Millimetre-wave observations of interstellar molecules
M. Guélin, C. Rist and J. Cernicharo 97

Newly discovered CO outflows in OMC-1
T.L. Wilson, J. Schmid-Burgk, R. Mauersberger,
R. Güsten, A. Schulz and L.M. Ziurys 107

Detailed structure of the NGC 7538 molecular cloud
O. Kameya, N. Hirano, R. Kawabe and B. Campbell 111

Structure and physical properties of the molecular outflow in B335
N. Hirano, O. Kameya, T. Kasuga, T. Hasegawa,
S.S. Hayashi and T. Umemoto 115

Status of the James Clerk Maxwell telescope
G.D. Watt 119

Further CS observations of selected star-forming regions previously mapped in ammonia
G. Anglada, J. Buj, R. Estalella, R. López,
J. Pastor and P. Planesas 123

A 257–273 GHz spectral survey of the OMC-1 cloud core
J.S. Greaves and G.J. White 125

Cloud collapse and low mass star formation in Orion-KL
V. Migenes, K.J. Johnston, T.A. Pauls
and T.L. Wilson 127

The TMC-1 mapped in HC_5N
C. Codella, G. Bruni, G. Comoretto, G. Maccaferri,
G.G.C. Palumbo and F. Scappini 129

Water vapour in the Orion molecular cloud
R.F. Knacke and H.P. Larson 131

Magnetic fields in interstellar clouds
P.C. Myers 133

Star formation in dark globules
S.M. Scarrott 141

The relationship of optical to molecular outflows
T.P. Ray, R. Poetzel and R. Mundt 145

Cha T1 luminosity function P.R. Wesselius, T. Prusti, D.C.B. Whittet and R. Assendorp	149
Modelling the 2.2 micrometre polarization from the reflection nebula GL 2591 N.R. Minchin and J.H. Hough	153
Molecular clouds in the optical: dark clouds J.A. Stüwe	155
Extragalactic molecules R.S. Booth	157
CO and magnetic fields in spiral galaxies R. Beck, E.M. Berkhuijsen and E. Bajaja	179
Molecular absorption lines towards the nucleus of Centaurus A A. Eckart, M. Cameron, R. Genzel, J.M. Jackson, H. Rothermel, J. Stutzki, G. Rydbeck and T. Wiklind	183
OB associations and HII regions in the SMC Paolo Battinelli	187
Interstellar masers R.J. Cohen	189
$^{15}NH_3$ masers toward NGC 7538-IRS 1 Peter Schilke	199
Properties of H_2O masers from the Arcetri atlas G. Comoretto, J. Brand, M. Catarzi, R. Cesaroni, C. Giovanardi, M. Felli, M. Massi, F. Palagi, F. Palla and G. Tofani	203
Survey of H_2O masers associated with compact molecular clouds and ultracompact HII regions F. Palla, J. Brand, R. Cesaroni, G. Comoretto and M. Felli	207
Interstellar chemistry E. Herbst and T.J. Millar	209
CRESUS studies of ion-molecule reactions J.-B. Marquette	239

Molecular hydrogen line emission from C-shocks
M.D. Smith, P.W.J.L. Brand and A. Moorhouse   243

Hydromagnetic wave propagation, ambipolar diffusion and chemical fractionation in cold dark clouds
S.B. Charnley and W.G. Roberge   247

Fluorescent HD emission from photon-dominated regions
A. Sternberg   249

Models of the gas-grain interaction in dark clouds
P.D. Brown and S.B. Charnley   253

The chemistry of the early universe at the epoch of recombination
W.B. Latter and J.H. Black   255

Molecular processes in SN 1987A
J.M.C. Rawlings and D.A. Williams   257

Monte Carlo simulation of molecular cloud fragmentation
Marek Wolf   259

Clump formation in a non-uniform ultraviolet radiation field
J.-P. Chièze and C. de Boisanger   263

Interstellar gas cycling powered by star formation
A. Lioure and J.-P. Chièze   267

A self-regulated state for the warm, cool neutral and molecular Galactic gas
A. Parravano and J. Mantilla Ch.   269

Stabilization of the ISM by fluctuations
A. Just, B.M. Deiss and W.H. Kegel   271

Comets and molecular clouds: the sink and the source
M.E. Bailey   273

Molecular cloud isotopic chemistry and comets
V. Vanysek   291

Dust in molecular clouds
D.A. Williams   295

Solid CO and CO_2 in grain mantles
D.C.B. Whittet and H.J. Walker   309

Far-IR and submm studies of dust in molecular clouds
       D. Ward-Thompson and E.I. Robson 313

Critical study of the assignment of PAHs as carriers of the diffuse interstellar bands
       C. Cossart-Magos and S. Leach 317

Ice in Barnard 5 IRS2
       S.B. Charnley, D.C.B. Whittet and D.A. Williams 321

Does dust mean gas?
       S.J. Chapman and A.P. Whitworth 323

AROME, a balloon borne experiment: detection of the 3.3 micrometre feature
       N. Sales, M. Giard, E. Caux, J.M. Lamarre,
       F. Pajot and G. Serra 325

Author index 327

Object index 329

Subject index 333

PREFACE

This volume contains the proceedings of a conference on 'Molecular Clouds' which was held at the University of Manchester from March 26 – 30, 1990. The conference, the seventh in the Manchester Astronomical Conference series, brought together an international group of about 140 people, including many of the leading investigators in observational and theoretical studies of molecular clouds.

Twelve invited reviews were scheduled for the meeting, although owing to unforeseen circumstances. John Scalo's review on 'Hydrodynamics of star-forming regions' is missing from the proceedings. Due to illness Hans Bloemen was unable to attend, but we are very grateful to him for providing his review for inclusion in this volume. In his absence, and at very short notice, Arnold Wolfendale, in the words of Alex Dalgarno, 'gave a balanced account of the differences between himself and Bloemen and concluded that Wolfendale was correct'. We are particularly grateful to Professor Wolfendale both for this contribution and for a lively and entertaining speech following the conference dinner. We are also grateful to Professor Dalgarno for providing the conference summary which, as it has appeared in Newsletter No. 14 of the Collaborative Computational Project No. 7 (CCP7), is not included in these proceedings.

The remainder of the conference was taken up by a mixture of 27 contributed talks, of which 24 are contained in this volume and about 40 poster papers, over half of which appear here.

We would in particular like to thank Mrs E.B. Carling and Miss M. Thomas, of the Department of Astronomy for the care, efficiency and courtesy with which they looked after the organisational and administrative work of the meeting. Professor F.D. Kahn, Professor D. A. Williams and Dr. J. E. Dyson provided valuable input in planning the scientific content of the meeting. Financial support was provided by the Cambridge University Press, CCP7 and the SERC and we are pleased to thank them for their assistance. Finally, we would like to express our appreciation to all those who contributed to the meeting. In particular the invited speakers greatly simplified our organisational work by their willingness to commit themselves well in advance of the event.

December 1990

R.A. James
Department of Astronomy
University of Manchester

T.J. Millar
Department of Mathematics
UMIST

NAMES AND ADDRESSES OF PARTICIPANTS

Anglada, G. Department de Fisica de l'Atmosfera, Astronomia i Astrofisica, Universitat de Barcelona, 08028 Barcelona, Spain.

Bailey, M.E. Department of Astronomy, University of Manchester, Manchester M13 9PL.

Bakes, Emma L.O. Physics Department, Queen Mary and Westfield College, University of London, Mile End Road, London E1 4NS.

Balm, S.P. School of Chemistry and Molecular Sciences, University of Sussex, Falmer, Brighton BN1 9QJ.

Battinelli, P. Osservatorio Astronomico di Roma, V.le del Parco Mellini 84, I-00136 Roma, Italy.

Beck, R. Max-Planck-Institüt für Radioastronomie, Auf dern Hügel 69, D-5300 Bonn 1, F.R.Germany

Bell, K.L. Department of Applied Mathematics and Theoretical Physics, The Queen's University of Belfast, Belfast, Northern Ireland.

Blitz, L. Astronomy Program, University of Maryland, College Park, MD 20742, U.S.A.

Bloemen, J.B.G.M. Leiden Observatory, P.O. Box 9504, 2300 RA Leiden, The Netherlands.

de Boisanger, Constance C.E.A., Service de Physique et Technique Nucléaires, Centre d'Etudes de Bruyères-le-Châtel, Service PTN, F-91680 Bruyères-le-Châtel BP12, France.

Booth, R. Onsala Space Observatory, Chalmers University of Technology, S-43900 Onsala, Sweden.

Brand, J. Osservatorio Astrofisico di Arcetri, L.go Enrico Fermi 5, 50125 Firenze, Italy.

Brand, P.W.J.L. Department of Astronomy, University of Edinburgh, Royal Observatory, Edinburgh EH9 3HJ.

Brebner, Genevieve C. Department of Space Research, University of Birmingham, Birmingham B15 2TT.

Brooks, A. Department of Astronomy, University of Manchester, Manchester M13 9PL.

Brown, P.D. Canadian Institute for Theoretical Astrophysics, University of Toronto, Toronto, Ontario, M5S 1A1, Canada.

Bryce, Myfanwy Department of Astronomy, University of Manchester, Manchester M13 9PL.

Names of participants

Castets, A.	Groupe d'Astrophysique de Grenoble, B.P.53X 38 041 Grenoble, France.
Chapman, S.J.	Physics Department, University of Wales College of Cardiff, P.O. Box 913, Cardiff CF1 3TH, Wales.
Charnley, S.B.	Physics Department, Rensselaer Polytechnic Institute, Troy, New York 12180, U.S.A.
Chièze, J.-P.	C.E.A., Service de Physique et Technique Nucléaires, Centre d'Etudes de Bruyères-le-Châtel, Service PTN, F-91680 Bruyères-le-Châtel BP12, France.
Chrysostomou, A.	Department of Astronomy, University of Edinburgh, Edinburgh EH9 3HJ.
Codella, C.	Istituto di Spettroscopia Molecolare del CNR, 40126 Bologna, Italy.
Cohen, R.J.	Nuffield Radio Astronomy Laboratories, Jodrell Bank, Macclesfield, Cheshire SK11 9DL.
Comoretto, G.	Osservatorio Astrofisico di Arcetri, L.go E.Fermi 5, 50125 Firenze, Italy.
Cooper, D.L.	Department of Chemistry, University of Liverpool, Liverpool L69 3BX.
Dalgarno, A.	Harvard-Smithsonian Center for Astrophysics, 60 Garden Street, Cambridge, MA 02138, U.S.A.
Davis, C.J.	Department of Astronomy, University of Edinburgh, Edinburgh EH9 3HJ.
Dutrey, Anne	Groupe d'Astrophysique de Grenoble, B.P.53X 38 041 Grenoble, France.
Dyson, J.E.	Department of Astronomy, University of Manchester, Manchester M13 9PL.
Eckart, A.	Max-Planck-Institut für Physik und Astrophysik, Institut für Extraterrestrische Physik, 8046 Garching, F.R.Germany
Falgarone, Edith	Radioastronomie, Laboratoire de Physique ENS, 75231 Paris Cedex 05, France.
Feldman, P.A.	Herzberg Institute of Astrophysics, N.R.C.of Canada, Ottawa, Canada K1A 0R6.
Fernandes, A.J.L.	Department of Astronomy, University of Edinburgh, Edinburgh EH9 3HJ.
Flickinger, G.	Department of Physics, University of Alabama at Birmingham, Birmingham AL35255, U.S.A.

Flower, D.R.	Physics Department, University of Durham, Durham DH1 3LE.
Genzel, R.	Max-Planck-Institut für Extraterrestrische Physik, D-8046 Garching, F.R.Germany.
de Geus, E.J.	Astronomy Program, University of Maryland, College Park, MD 20742, U.S.A.
Gleghorn, J.	Department of Chemistry, University of Lancaster, Bailrigg, Lancaster LA1 4YA.
Greaves, Jane	Department of Physics, Queen Mary and Westfield College, University of London, Mile End Road, London E1 4NS.
Gredel, R.	European Southern Observatory, Casilla 19001, Santiago 19, Chile.
Großman, V.	Radioastronomisches Institut, Auf dem Hügel 71, D-5300 Bonn, F.R.Germany.
Guélin, M.	IRAM, 300 rue de la Piscine, 38406 St. Martin d'Hères, France.
Hagan, K.	Selwyn College, Cambridge CB3 9DQ.
Hahn, G.	Department of Astronomy, University of Manchester, Manchester M13 9PL.
Haikala, L.	I.Physikalisches Institut, Zülpicherstrasse 77, 5000 Köln, F.R.Germany.
Hare, J.	School of Chemistry and Molecular Sciences, University of Sussex, Brighton BN1 9QJ.
Harris, A.	Max-Planck-Institut für Extraterrestrische Physik, D-8046 Garching, F.R.Germany.
Henney, W.J.	Department of Astronomy, University of Manchester, Manchester M13 9PL.
Herbertz, R.	I.Physikalisches Institut, Zülpicherstrasse 77, 5000 Koln, F.R.Germany.
Herbst, E.	Department of Physics, Duke University, Durham NC 27706, U.S.A.
Hirano, N.	Nobeyama Radio Observatory, Nobeyama, Minamisaku, Nagano 384-13, Japan.
Hobson, M.P.	Radio Astronomy Group, Rutherford Building, Cavendish Laboratory, Madingley Road, Cambridge CB3 0HE.
Howe, D.A.	Department of Mathematics, UMIST, Manchester.
James, R.A.	Department of Astronomy, University of Manchester, Manchester M13 9PL.

Jeffrey, C.S.	University Observatory, Buchanan Gardens, St.Andrews, Fife KY16 9LZ.
Jones, D.K.	13 Gainford Gardens, Moston, Manchester M10 9GH.
Just, A.	Institut für Theoretische Physik, Robert Mayer-Strasse 10, Postfach 11 19 32, D-600 Frankfurt am Main 11, F.R.Germany.
Kameya, O.	Nobeyama Radio Observatory, Nobeyama, Minamisaku, Nagano 384-13, Japan.
Kemp, S.	Department of Astronomy, University of Manchester, Manchester M13 9PL.
Knacke, R.	ES63, Space Science Laboratory, NASA Marshall Space Flight Center, Huntsville, AL 35812, U.S.A.
Kramer, C.	I.Physikalisches Institut, Zülpicherstrasse 77, 5000 Köln 41, F.R.Germany.
Latter, W.B.	CITA, University of Toronto, McLennen Laboratories, Toronto, Ontario M5S 1A1, Canada.
Leach, S.	DAMAp (URA 812), Observatoire de Paris-Meudon, 92195 Meudon, France.
Lioure, A.	C.E.A., Service P.T.N., Centre d'Etudes de Bruyères-le-Châtel, F-91680 Bruyères-le-Châtel BP12, France.
Little, L.T.	Electronic Engineering Laboratories, University of Kent, Canterbury, Kent CT2 7NT.
Lloyd, H.W.	Department of Astronomy, University of Manchester, Manchester M13 9PL.
López, Rosario	Department de Fisica de la Atmosfera, Astronomia i Astrofisica, University of Barcelona, 08028 Barcelona, Spain.
Lynas-Grey, A.E.	Department of Physics and Astronomy, University College London, Gower Street, London WC1E 6BT.
Marquette, J.-B.	Laboratoire d'Aérothermique du CNRS, 4ter Route des Gardes, 92190 Meudon, France.
Martin-Pintado, J.	Centro Astronomico de Yebes, 19080 Guadalajara, Spain.
Masheder, M.R.W.	Department of Physics, University of Bristol, Bristol BS8 1TL.
McKeith, C.D.	Physics Department, Queen's University Belfast, Belfast BT7 1NN.
Meyerdierks, H.	Radioastronomisches Institut, Auf dem Hügel 71, D-5300 Bonn 1, F.R.Germany.

Names of participants

Migenes, V.	Nuffield Radio Astronomy Laboratories, University of Manchester, Jodrell Bank, Cheshire SK11 9DL.
Millar, T.J.	Department of Mathematics, UMIST, Manchester M60 1QD.
Miller, S.	Department of Physics and Astronomy, University College London, Gower Street, London WC1E 6BT.
Minchin, N.	Physics Division, School of Natural Sciences, Hatfield Polytechnic, Hatfield, Herts AL10 9AB.
Mirabel, I.F.	Service d'Astrophysique, CEA-CE N, Saclay, 91191 Gif-sur-Yvette Cedex, France.
Monteiro, Tania	Department of Mathematics, Royal Holloway College, University of London, Egham, Surrey TW20 0EX.
Moorhouse, A.	Department of Mathematics, UMIST, Manchester M60 1QD.
Moy, S.	Electronics Laboratory, The University, Canterbury, Kent CT2 7NT.
Myers, P.C.	Harvard-Smithsonian Center for Astrophysics, 60 Garden Street, Cambridge MA 02138, U.S.A.
Nejad, Lida A.M.	Department of Mathematics and Physics, Manchester Polytechnic, Manchester M1 5GD.
Nelson, A.H.	Physics Department, University of Wales College of Cardiff, P.O. Box 913, Cardiff CF1 3TH, Wales.
Noguchi, M.	Physics Department, University of Wales College of Cardiff, P.O. Box 913, Cardiff CF1 3TH, Wales.
Nyman, L.-Å.	ESO/La Silla, Casilla 19001, Santiago 19, Chile.
Parravano, A.	Royal Observatory, Blackford Hill, Edinburgh EH9 3HJ.
Persi, P.	Istituto Astrofisica Spaciale, CNR, 00044 Frascati, Italy.
Phillips, J. P.	Institut de Astrofisica de Canarias, La Laguna, Tenerife, Spain.
Pineau des Forets, G.	DAMAP, Observatoire de Paris, 92195 Meudon Principal, France.
Pongracic, Helen	Department of Physics, University of Wales, College of Cardiff, Cardiff CF1 8LD.
Ramsay, Suzanne K.	Royal Observatory, Blackford Hill, Edinburgh EH9 3HJ.
Rawlings, J.	Department of Mathematics, UMIST, Manchester M60 1QD.
Ray, T.P.	Dublin Institute for Advanced Studies, School of cosmic Physics, Dublin 2, Ireland.

Richer, J.	Cavendish Laboratory, Madingley Road, Cambridge CB3 0HE.
Roueff, Evelyne	DAMAp, Observatoire de Meudon, 92195 Meudon, France.
Sales, Nathalie	C.E.S.R., 9 Av. du Colonel Roche, BP 4346, 31029 Toulouse Cedex, France.
Sanderson, C.A.	Department of Mathematics, UMIST, Manchester M60 1QD.
Scalo, J.	Astronomy Department, University of Texas at Austin, Austin TX 78712-1083, U.S.A.
Scappini, F.	Istituto di Spettroscopia Molecolare del CNR, 40126 Bologna, Italy.
Scarrott, S.M.	Physics Department, University of Durham, Durham DH1 3LE.
Schilke, P.	Max-Planck-Institut fur Radioastronomie, Auf dem Hügel 69, D-5300 Bonn 1, F.R.Germany.
Skinner, C.	Nuffield Radio Astronomy Laboratories, University of Manchester, Jodrell Bank, Cheshire SK11 9DL.
Smith, M.D.	International School for Advanced Studies, Strada Costiere 11, I-34014 Trieste-Miramare, Italy.
Sternberg, A.	Max-Planck-Institut für Extraterrestrische Physik, D-8046 Garching bei Munchen, F.R.Germany.
Stutzki, J.	Max-Planck-Institut für Extraterrestrische Physik, D-8046 Garching bei Munchen, F.R.Germany.
Stüwe, J.A.	Astronomisches Institut der Ruhr Universität Bochum, Postfach 10 21 48, D-4630 Bochum, F.R.Germany.
Takeo, H.	National Chemical Laboratory for Industry, Isukuba, Obaraki 305, Japan.
Talbi, Dahbia	Laboratoire d'Astrochimie Quantique-Radioastronomie, E.N.S., 75005 Paris, France.
Taylor, S.	Department of Mathematics, UMIST, Manchester M60 1QD.
Thaddeus, P.	Harvard-Smithsonian Center for Astrophysics, 60 Garden Street, Cambridge MA 02138, U.S.A.
Tofani, G.	Osservatorio Astrofisico de Arcetri, L.go E.Fermi 5, 50125 Firenze, Italy.
Tringham, N.R.	Department of Astronomy, University of Manchester, Manchester M13 9PL.
Vanysek, V.	Remeis-Sternwarte, 8600 Bamberg, F.R.Germany and Department of Astronomy and Astrophysics, Charles University, 150 00 Praha, Švédská 8, Czechoslovakia.

Names of participants

Wagenblast, R.	Department of Mathematics, UMIST, Manchester M60 1QD.
Walker, R.N.F.	Physics Department, University of Bristol, Bristol BS8 1TL.
Ward-Thompson, D.	School of Physics and Astronomy, Lancashire Polytechnic, Preston PR1 2TQ.
Watt, G.D.	Joint Astronomy Centre, 665 Komohana Street, Hilo, Hawaii 96720, U.S.A.
Webster, A.	Royal Observatory, Blackford Hill, Edinburgh EH9 3HJ.
Wesselius, P.R.	Kapteyn Institute, P.O. Box 800, 9700 AV Groningen, The Netherlands.
Whittet, D.C.B.	School of Physics and Astronomy, Lancashire Polytechnic, Preston, PR1 2TQ.
Whitworth, A.	Physics Department, University of Wales College of Cardiff, P.O. Box 913, Cardiff CF1 3TH, Wales.
Williams, D.A.	Department of Mathematics, UMIST, Manchester M60 1QD.
Wilson, T.L.	Max-Planck-Institut fur Radioastronomie, Auf dem Hügel 69, D-5300 Bonn 1, F.R.Germany.
Wolf, M.	Department of Astronomy and Astrophysics, Charles University, 150 00 Praha, Švédská 8, Czechoslovakia.
Wolfendale, A.W.	Physics Department, University of Durham, Durham DH1 3LE.
Wolstencroft, R.D.	Royal Observatory, Blackford Hill, Edinburgh EH9 3HJ.
Wouterloot, J.G.A.	I.Physikalisches Institut, Zülpicherstrasse 77, 5000 Köln 41, F.R.Germany.
Yamagata, T.	National Astronomical Observatory, Mitaka, Tokyo 181, Japan.

Large-scale CO observations of our Galaxy and its nearest neighbours

P. THADDEUS

HARVARD-SMITHSONIAN CENTER FOR ASTROPHYSICS, 60 GARDEN STREET, CAMBRIDGE, MASSACHUSETTS 02138, U.S.A.

A fairly complete inventory of molecular clouds in the Galaxy and its three nearest neighbors, the two Magellanic Clouds and the Andromeda Nebula (M31), is now at hand from large-scale CO surveys we have made with a small millimeter-wave telescope in the United States and an almost identical instrument in Chile. It is useful at the beginning of a general meeting on molecular clouds to exhibit briefly some of this very large collection of data, to set the stage for the papers on specific objects to follow, and to obtain a notion of the great richness of the molecular content of spiral and irregular galaxies.

Nearly all our surveys have been of the CO line from the fundamental $(J = 1 \to 0)$ rotational transition of the dominant isotopic species at 115 GHz, the line which, with current instrumentation, is the most useful general tool for the study of interstellar molecular gas. All the data shown here are of this line, which nearly everywhere is observed only in emission, given no strong background continuum sources exist at 115 GHz. Figure 1, the entire Milky Way in molecular clouds (Dame et al., 1987), is a composite of a number of individual northern and southern surveys – five unbiased surveys along the Galactic plane and 11 of particular regions, such as the well known star forming areas in Orion and in Taurus (figure 3). To survey in a reasonable time the huge area shown in figure 1 – more than 7700 square degrees, or nearly 20% of the celestial sphere – observations of most of it were made at an angular resolution of 0.5° – 3.4 times larger than the primary beam of our 1.2m telescopes, which have a beamwidth at half-maximum of 8.8'. This was accomplished not by stopping down or under-illuminating the telescopes, but simply by stepping through a 4x4 array of positions during an integration, a procedure that acquires information at essentially the same rate as observing with a larger beam but requires no hardware modifications. For uniformity in patching together the individual surveys in Figure 1, those done at higher angular resolution, such as that of Orion and Monoceros (Maddalena et al., 1986), have been numerically smoothed to an angular resolution of 0.5°.

Even at such a low resolution, the main molecular features of the Galaxy stand out clearly, as a comparison of Figure 1 with the schematic diagram of Figure 2 illustrates. Most of the molecular gas in our system is evidently within 6 kpc of the Galactic Center (for $R_\odot = 8.5$ kpc), because the average intensity of CO emission falls off quite rapidly more than 45° from the Center in either direction along the plane. In the ℓ, v diagram of Figure 1, a thin circularly rotating ring within the solar circle is transformed into a straight diagonal line passing through zero velocity at $\ell = 0°$, with a slope that increases as the size

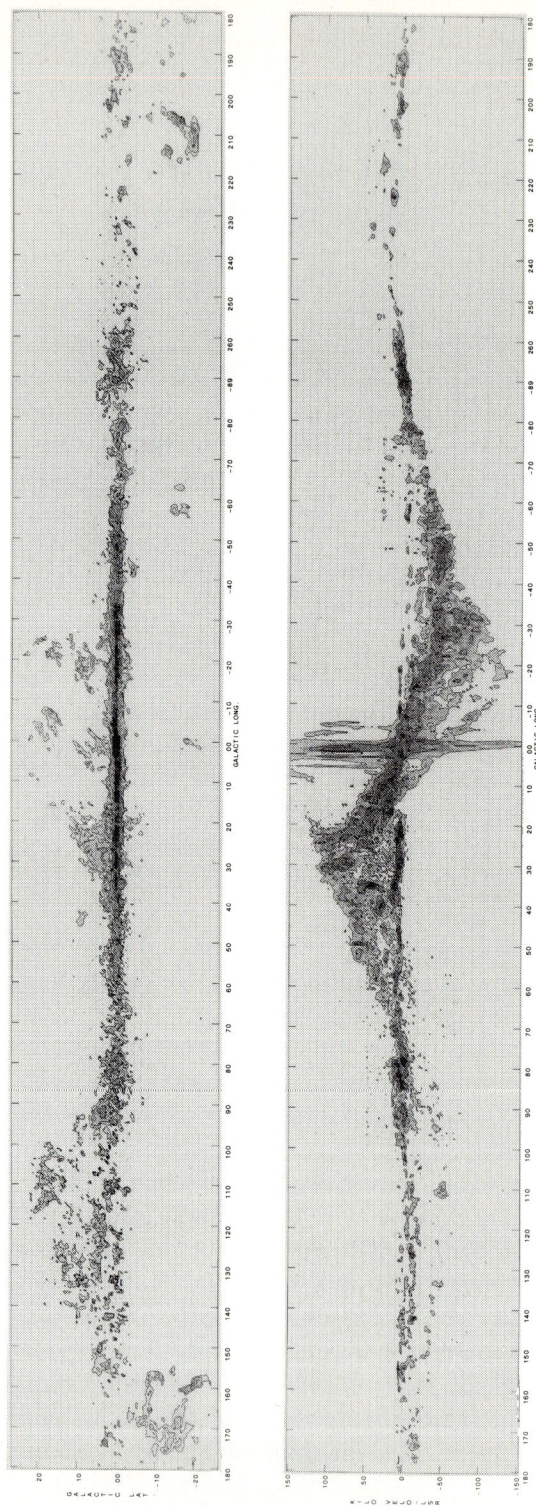

Figure 1: The Milky Way in molecular clouds. *Top*: velocity integrated CO emission at an angular resolution of 0.5°. *Bottom*: longitude-velocity diagram for material within 3.25° of the Galactic equator (thus omitting high-latitude local molecular gas such as that in Ophiuchus, Taurus and Orion in the map above). For contour levels and other details, see Dame *et al.* (1987).

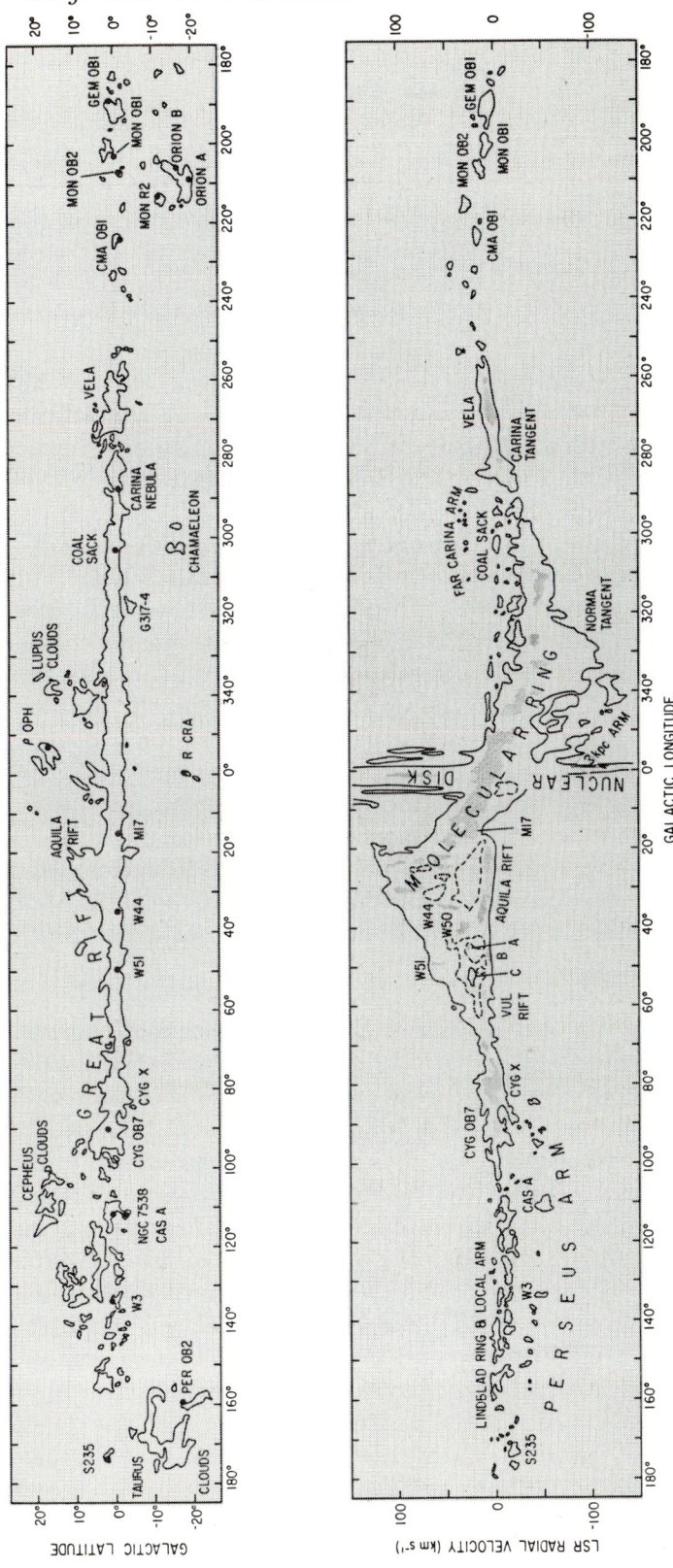

Figure 2: Identification of the main features in Figure 1.

of the ring is reduced. The intense, nearly vertical lane at $\ell = 0°$, for example, is mainly emission from the nuclear disk: dense, rapidly rotating molecular gas within a few 100 pc of the Center. The thick diagonal bar of emission extending about 30° on either side of the Center is the so-called molecular ring, the peak in the distribution of molecular clouds about halfway between the solar circle and the Center. The thinner diagonal lane just below is molecular gas in the 3 kpc Expanding Arm first observed in HI: an arm on the near side of the Center, displaced to negative velocity at $\ell = 0°$ because it is not only rotating about the Center but also expanding outward toward us at a speed of 53 km s^{-1} (Oort, 1977).

Finally, the thin, rather spotty horizontal line of emission at zero velocity across much of the diagram is material very near the solar circle – mainly rather modest molecular clouds within a few 100 pc of the Sun associated with dark optical rifts in the Milky Way. Notice the large gaps in this local lane of emission in the Carina region of the southern hemisphere between $\ell = 290°$ and $330°$, in contrast to the intense emission found at the same distance on the other side of the Galactic Center, produced by the large dark clouds in the northern Aquila, Vulpecula and Cygnus Rifts (Dame et al., 1987). These large windows in local obscuration are one of the reasons why optical Galactic astronomers are so enamored of the southern hemisphere; at certain directions in Carina it is possible to see for distances of up to 6 kpc squarely in the Galactic plane with little obscuration.

The large gaps in the CO distribution between the molecular ring and the nuclear region (partially masked at negative longitudes by the 3 kpc Expanding Arm) are an enigmatic feature of the molecular distribution, produced by an almost complete absence of molecular clouds and other Population I between $R = 1$ and 4 kpc. The agency responsible for removing gas from this vast region, or keeping it out in the first place, is unknown. An annular gap of this sort in the distribution of molecular clouds is present in some external spirals, but absent in others (Young and Scoville, 1991).

In addition to the wide-latitude surveys summarized in Figure 1, we have undertaken deep, closely sampled surveys at the full resolution of our telescopes along most of the inner Galactic plane (i.e., the first and fourth Galactic quadrants), which are omitted here for reasons of space. From this data, Bronfman (1988) made a detailed study of the density and thickness of the molecular disk in both hemispheres and showed that it is slightly but significantly displaced from the Galactic equator defined by the IAU. He confirms, as we found from our first deep northern survey (Cohen et al., 1980), that the molecular ring is a quite asymmetrical structure, considerably broader in the southern hemisphere than in the northern, with large gaps and holes – a poor ring, in other words. Even in the ℓ,v diagram in Figure 1, the broadening of the ring in the southern hemisphere at negative longitudes is fairly evident.

Presumably, the molecular ring is largely – perhaps even entirely – the sum of molecular clouds in the crowded inner arms of the Galaxy, and if so it ought to be possible to partition the clouds into the various arms and derive a convincing model of the inner molecular Galaxy. That task has turned out, however, to be difficult – as 21 cm astronomers have known for a long time – largely owing to the kinematic distance ambiguity within the solar circle and the existence of noncircular motion. No generally accepted model of the arrangement of the molecular arms in the inner Galaxy has yet been constructed.

Farther from the Galactic Center, however, the evidence for large-scale spiral structure is unmistakable in CO and easily discernible in Figure 1. The classical Perseus arm beyond the solar circle, for example, is the long arc in the ℓ,v diagram at negative velocity between $\ell \approx 80°$ and $170°$, and the Carina arm is the loop crossing the plane at $\ell \approx 280°$ (Grabelsky et al., 1988). Most of the emission in both these arms is apparently concentrated in the largest concentrations, such as the giant molecular cloud (GMC) at $\ell = 111°$, associated with the Cas A supernova remnant at 3 kpc; the sprinkling of clouds in the upper branch of the Carina loop are GMCs on the far side of the Galaxy, so large that even at a resolution of 0.5° they are discernible to distances as great as 20 kpc. All the analyses that we and others have done with higher resolution data confirm this general impression, finding that the mass spectrum of Galactic molecular clouds is rather flat, with most of the mass near the top of the mass spectrum in the GMCs with masses of $\sim 1 \times 10^6 \, M_\odot$ and larger.

It is clear from a comparison of Figure 1 with other large scale surveys of Population I – e.g. dark nebulae, the IRAS 100 μm all sky survey and the COS-B survey of diffuse Galactic γ-rays (Figure 4) – that little of the molecular mass of the Galaxy lies beyond the boundaries of our wide-angle CO survey, but it is also true that over 75% of the celestial sphere remains almost entirely unsurveyed in CO. A very extensive high-latitude molecular cloud toward Polaris recently discovered with our telescope in Cambridge is shown in Figure 5 (Heithausen and Thaddeus, 1990). This object, which we call the Polaris Flare, is so faint in CO that had it been included in Figure 1 it would have largely fallen beneath the lowest contour; it is undoubtedly quite close to the Sun, to judge from the extinction vs. distance observed against a number of circumpolar stars near Polaris. The number of similar local objects at high latitude, and the fraction of the sky covered by them, is largely unknown. Blitz and co-workers (e.g. Magnani, Blitz and Mundy, 1985) have made spot CO observations and limited maps of 100 μm cirrus clouds and diffuse reflection nebulae at high latitude, but aside from our observations of the Polaris Flare and a previous study by de Vries, Heithausen and Thaddeus (1987) of a nearby high latitude cloud almost no systematic surveys have been done.

The Magellanic Clouds are ideal objects for a small telescope, because they subtend too large a solid angle to be surveyed with large single-dish instruments or interferometers. Star formation in the LMC, for example, occurs over a region of order $6° \times 6°$, and with present instrumentation fully sampling such an area with a beam smaller than about $5'$ is impractical. In fact, the LMC turned out to be an ambitious project even at our lower resolution, because the molecular clouds there are quite faint in CO relative to those in the Galaxy. To achieve the map shown in Figure 6, the result of a two-year survey from Chile, an integration time of nearly an hour per point was required to bring out the weak CO emission.

A remarkable property of the molecular clouds in the LMC is their enormous size, possibly the result of the rigid-body rotation curve and the absence of

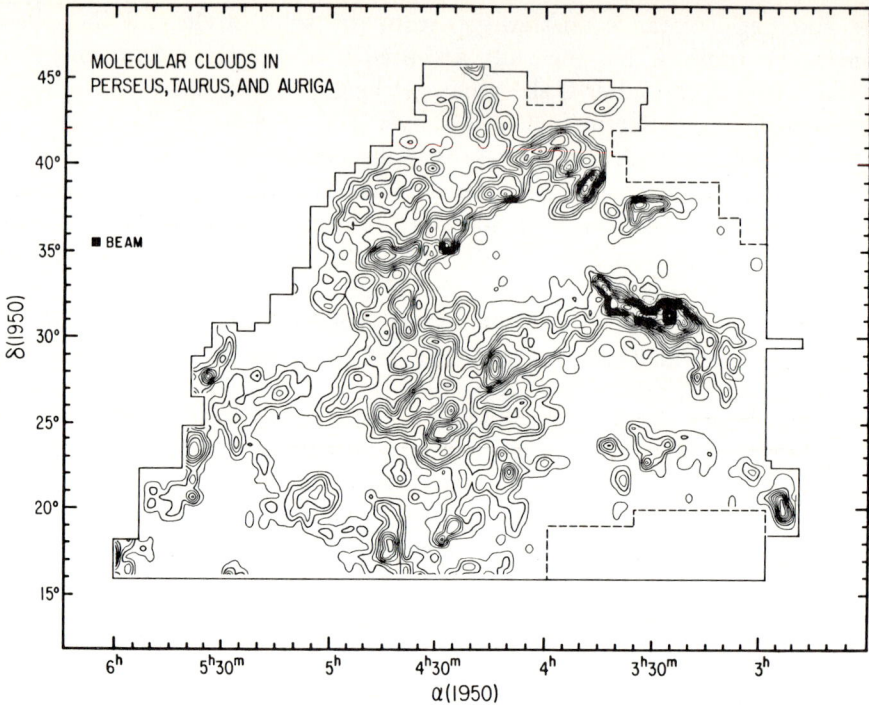

Figure 3. CO survey of the Taurus clouds and adjacent dark nebulae at a resolution of 0.5° – one of the component surveys of Figure 1 (Ungerechts and Thaddeus, 1987). Contours of velocity-integrated intensity of the CO $1 \to 0$ transition are from 0.5 by 1.5 K km s^{-1}.

differential rotation within the inner few kpc of the system. The object stretching south of 30 Doradus and Supernova 1987a, for example, is nearly 3 kpc long and its mass is estimated to be in the vicinity of $6 \times 10^7 M_\odot$. Apart from our nuclear disk, no molecular concentration with such a large mass exists in the Milky Way, and even the nuclear disk is dwarfed in size by the 30 Dor Cloud. As in the Galaxy, molecular clouds in the LMC are apparently fractal objects, with structure on many scales – even at our modest resolution the 30 Dor Cloud evidently contains a number of component clouds or clumps on the scale of 100–300 pc, and with the new 15m SEST telescope its structure is now being observed on a scale still finer by an order of magnitude.

Toward the 30 Dor Cloud and elsewhere in the LMC, the correlation of molecular clouds with supernova remnants is surprisingly high. Supernova 1987a and over 80% of the SNRs in the LMC cataloged by Mathewson *et al.* (1985) lie toward a molecular cloud, or so close to one that an association is plausible.

On calculating the virial masses of the molecular clouds in the LMC, it soon became apparent to us that the clouds there are deficient in CO, or at any rate that the CO lines there are abnormally weak relative to the Galaxy. We find that the ratio of the H_2 column density to the velocity-integrated CO line intensity, $N(H_2)/W(CO)$, is about 6 times higher in the LMC than here, and on the basis of this mass calibration we have estimated the masses of the molecular clouds

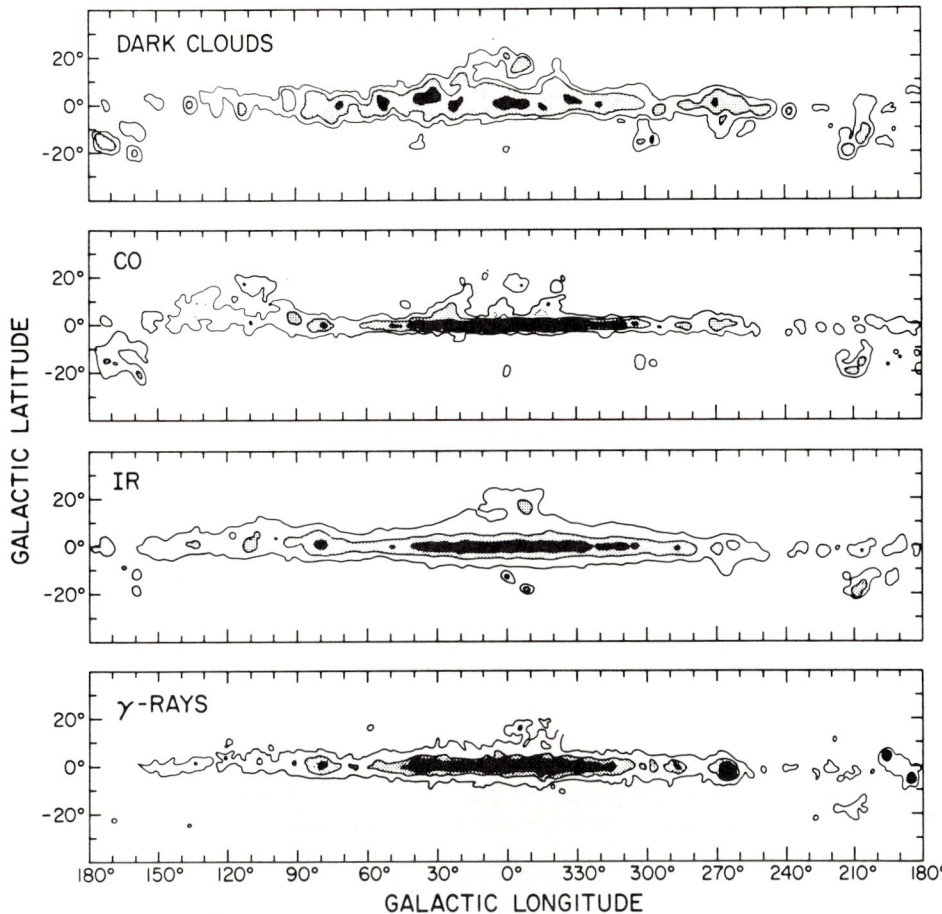

Figure 4: The Galactic distribution of dark nebulae, CO and IRAS 100 μm emission, smoothed to the 2.5° resolution of the COS-B high energy diffuse γ-ray survey – Figure 9 from Dame et al. (1987).

shown in Figure 6. The lower metallicity in the LMC – 4 to 6 times lower than in the Galaxy – suggests an important generalization: $N(H_2)/W(CO)$ scales as the inverse first power of the metallicity, or perhaps a little more steeply. This is potentially an important result, because nearly all molecular mass calibrations of external galaxies have been based upon the Galactic calibration, irrespective of metallicity.

The beamwidth of our 1.2m telescopes at 115 GHz, 8.8′, is the same as that of the 100m Bonn telescope at 21 cm. When I first saw Cram, Roberts and Whitehurst's (1980) rather intricate map of the 21 cm line in M31 done with the 100m telescope, I was impressed at how much detail was visible on such an angular scale, and I resolved that when we had a sufficiently sensitive receiver we would attempt to survey M31 in CO. Such a receiver, based on an SIS mixer developed by A.R. Kerr and S.-K. Pan, was installed by E.S. Palmer on our Cambridge telescope several years ago; it has now been refined to the point that

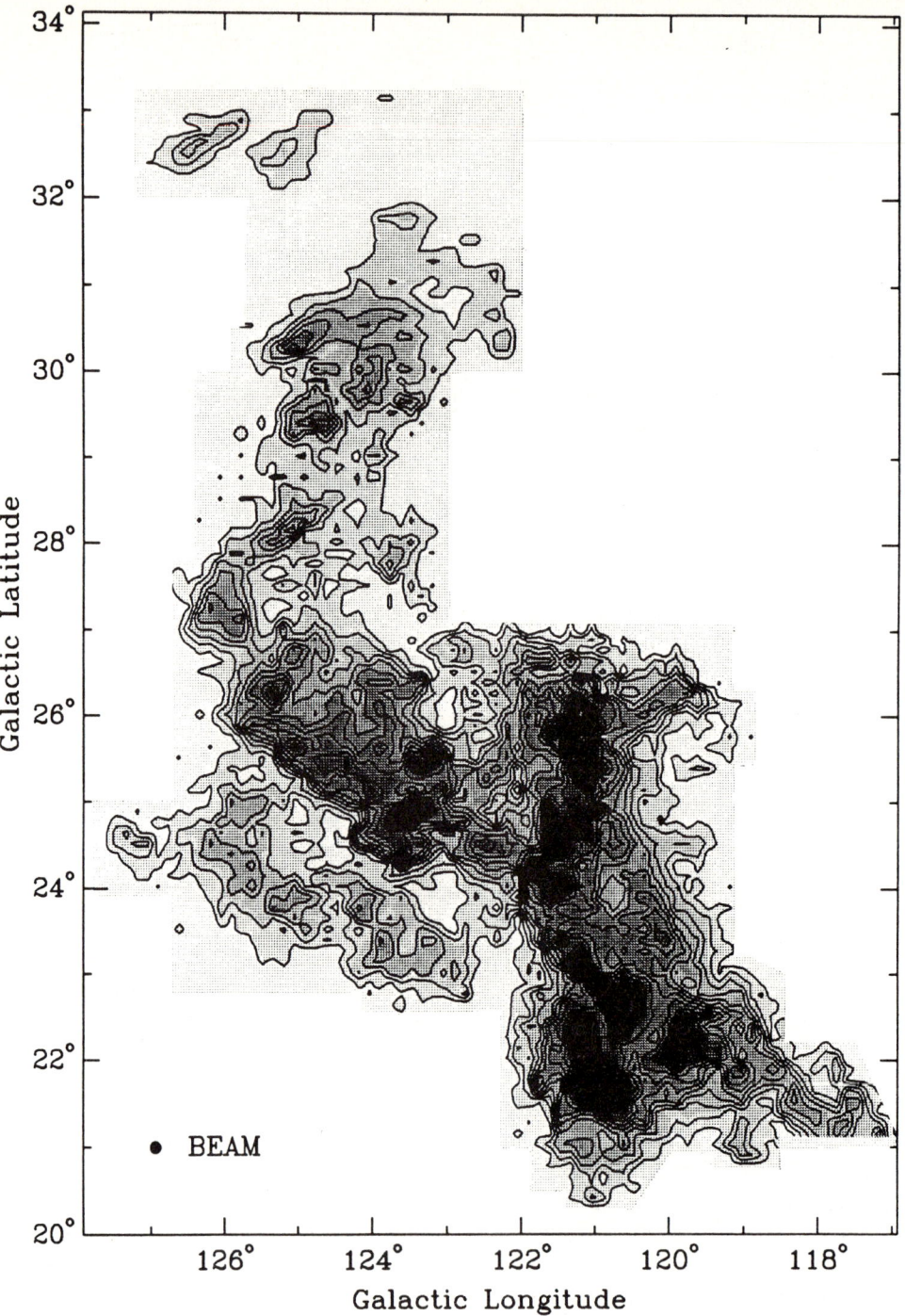

Figure 5: The Polaris Flare, a high-latitude molecular cloud above the boundaries of the composite survey in Figure 1. Contours of velocity-integrated CO intensity range from 0.4 to 13 K km s^{-1} in steps of 0.8 K km s^{-1}. From Heithausen and Thaddeus (1990).

Large-scale CO *observations*

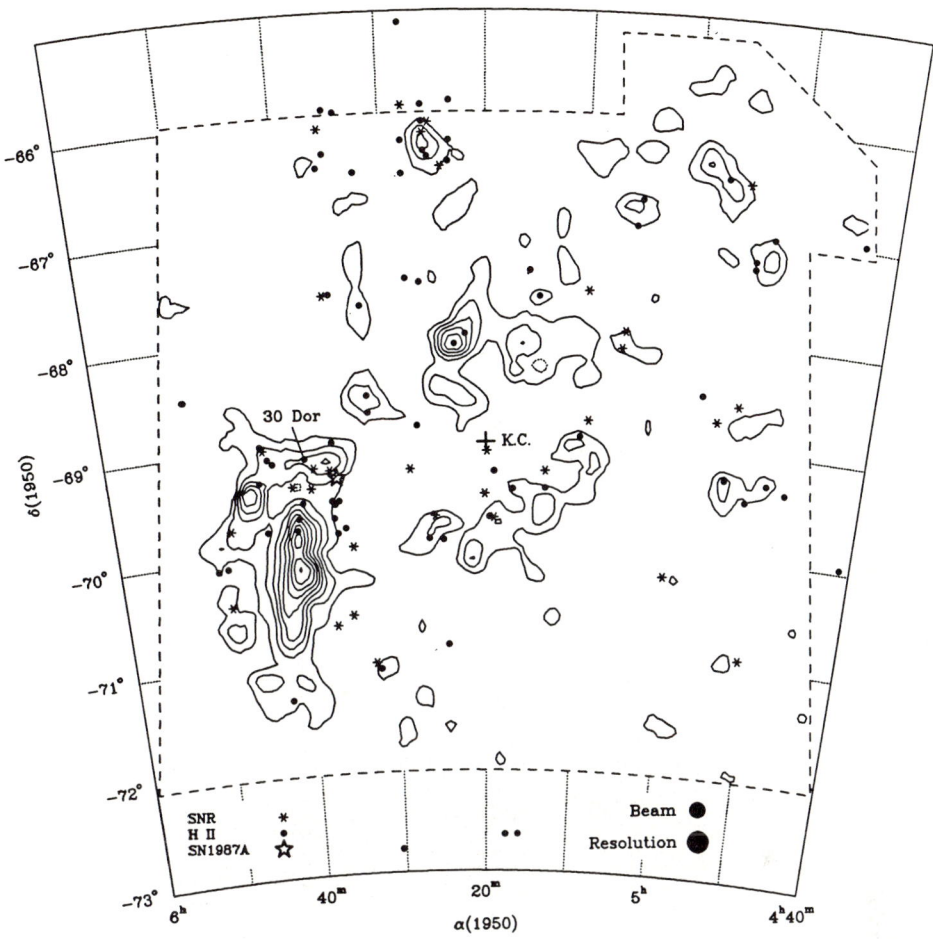

Figure 6: Map of velocity-integrated CO intensity in the LMC, smoothed to an angular resolution of 12′ to bring out the extensive weak emission – from Cohen *et al.* (1988). The contour interval is $0.38\,\mathrm{K\,km\,s^{-1}}$, which is about the $1.6\,\sigma$ noise level. Also shown are supernova remnants from the catalog of Mathewson *et al.* (1985), the location of SN 1987a, and HII regions (6 cm continuum sources not identified as SNRs). K.C. marks the kinematic center of the LMC according to de Vaucouleurs and Freeman (1973).

it is one of the most sensitive spectral-line receivers in operation at 115 GHz, with a single-sideband temperature of only 70 K at that frequency. Using this receiver, the first half of a two-year survey of M31 has just been completed by Enrico Koper, a graduate student from Leiden, and Tom Dame. A summary of the work to date is shown in Figures 7 and 8.

M31 is the most challenging survey we have yet undertaken, because averaged over our 9′ beam the CO emission is never very strong – the linear resolution at M31 is 1700 pc, considerably larger than the dimensions of the largest molecular complexes, to judge from high-resolution studies of particular regions done by others with the Bell, Kitt Peak and IRAM telescopes. At the most intense position the peak line temperature is less than 0.03 K. Even with our very sensitive

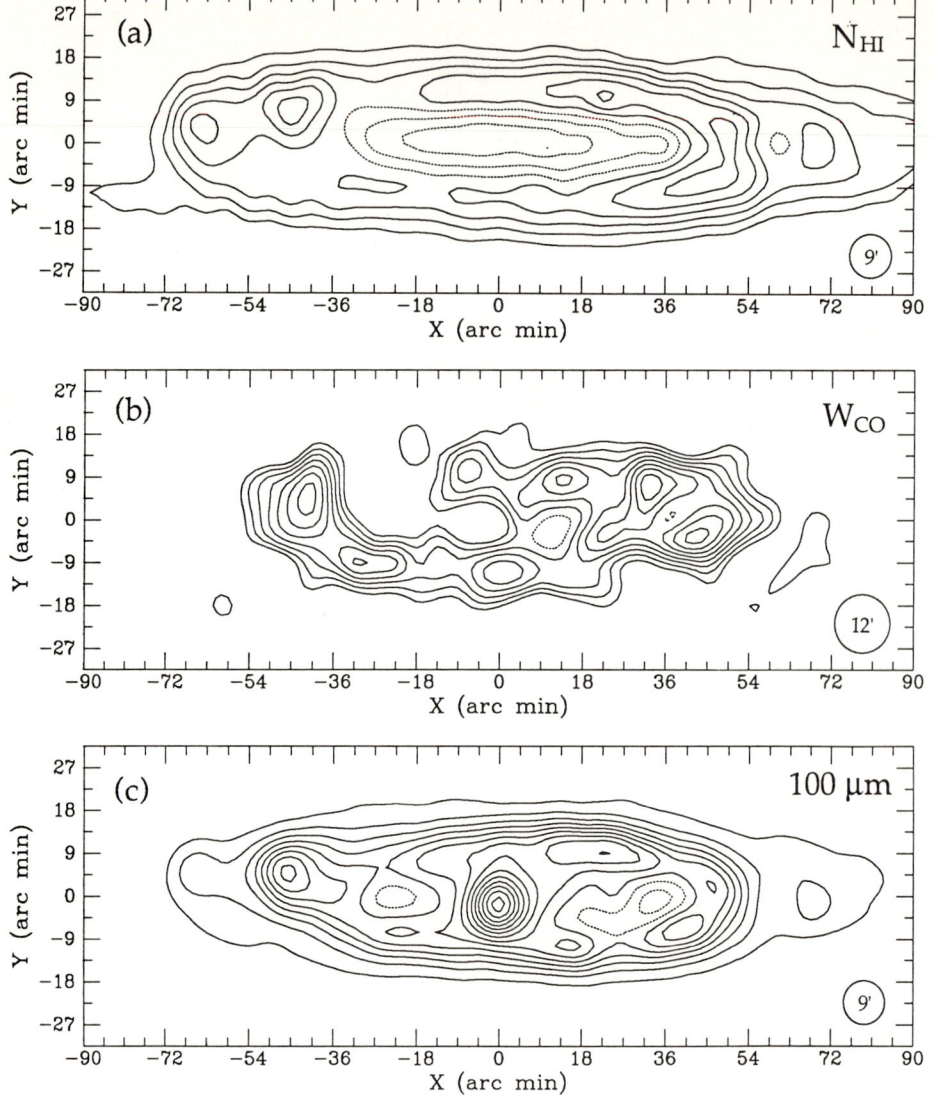

Figure 7: HI, CO and far-IR surveys of M31 (from Dame et al., 1990). X and Y are the standard Baade-Arp offset coordinates. (a) HI column density from the 21 cm survey of Cram et al. (1980). Contours are from 5.0 by 2.5×10^{20} cm^{-2}. (b) Velocity-integrated intensity of the CO $1 \rightarrow 0$ line, spatially smoothed to an angular resolution of 12′. Contours are from 0.3 by 0.1 K km s^{-1}, the lowest at a signal-to-noise level of about 3σ. (c) IRAS 100 μm emission smoothed to a resolution of 9′. Contours are from 1.8 by 1.8 MJy ster^{-1}.

receiver, very long integrations have been necessary, as well as extraordinary care to suppress baseline ripple. Our strategy has been to observe on the same Baade-Arp grid with 4.5′ spacing adopted by Cram, Roberts and Whitehurst, spending about 1 hour per position during the first year and covering the entire galaxy out to a radius of at least 15 kpc – the entire area where strong 100 μm emission has

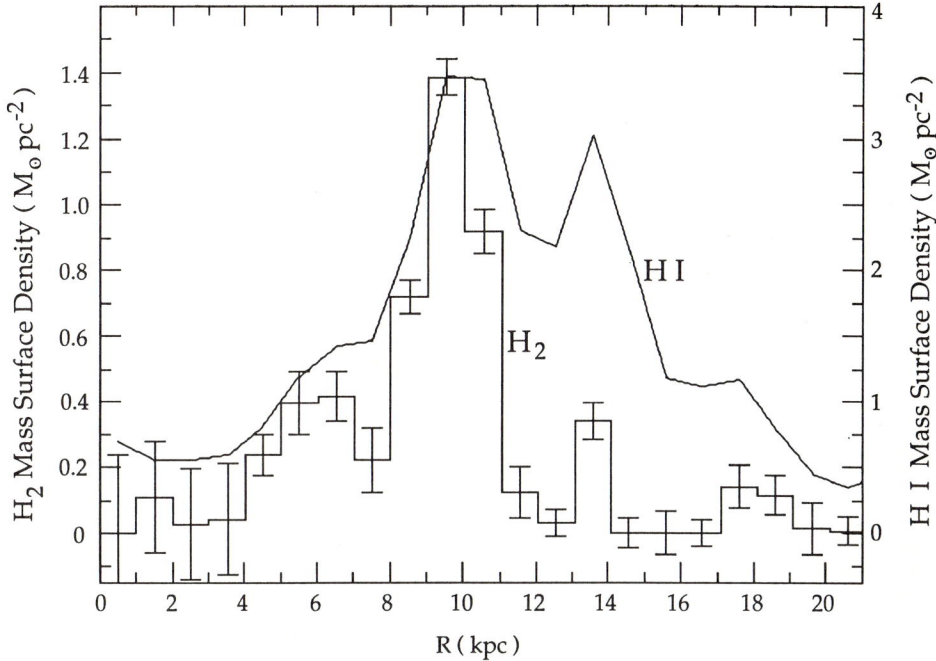

Figure 8: Radial distribution of the surface density of atomic and molecular gas in M31, the result of a fit of axisymmetric models to the HI and CO survey data (Dame *et al.*, 1990).

been detected by IRAS. Even with the long integrations employed – equivalent to nearly 4 hours per beamwidth – smoothing the survey to a resolution of about 12′, as shown in Figure 7, has been necessary to attain adequate signal-to-noise. By essentially repeating the same net of observations during the coming year and summing the two data sets, it will be possible to avoid such smoothing, achieving a somewhat sharper map than the preliminary one in Figure 7.

Even the first year's data are extremely interesting when analysed. It is evident from Figure 7 that the large ring with a radius of roughly 10 kpc in the distribution of both 21 cm and 100 μm emission (and conspicuous as well in HII regions and other Population I) is present also in CO. The CO emission appears much more clumpy on the scale of 2–3 kpc than that of these other two gas tracers, and in Figure 7 a large gap is conspicuous to the upper left in the CO ring whose counterpart at 21 cm and 100 μm is weak. It will be interesting to see whether this anomaly in the molecular cloud distribution persists as a feature of such high-contrast as data is added to the survey during the coming observing season (fall, winter and spring 1990-91).

Koper and Dame have fit axially symmetric models of the gas in M31 to the Bonn 21 cm data and their own CO observations, to obtain the average radial distribution of HI and CO shown in Figure 8. Both distributions exhibit a steady increase in mass surface density with galactocentric distance R to a peak at about 10 kpc, as one might predict from Figure 7, and a general decrease beyond, the molecular surface density falling off more rapidly than the atomic, just as in the Milky Way. In both distributions there is weak evidence for a secondary peak at $R \approx 13\,\mathrm{kpc}$, possibly attributable to one or more of the outer arms of the system.

Adopting the Galactic $N(H_2)/W(CO)$ ratio to calculate the molecular mass of M31, in lieu of a specific determination of this ratio, Koper and Dame derive an average molecular surface density at the peak of the M31 ring of $\sim 1\,\mathrm{M_\odot\,pc^{-2}}$, a factor of about 2.5 less than the atomic surface density. The total H_2 mass they find for M31 is $\sim 3 \times 10^8\,\mathrm{M_\odot}$. For the Galaxy, in contrast, Bronfman et al. (1988) find that the molecular surface density at the peak of the molecular ring at about 4 kpc is somewhat greater than the atomic, and the total molecular mass (excluding the dense molecular region of the Galactic Center, which as Figure 7 shows has no counterpart in M31) is $1.2 \times 10^9\,\mathrm{M_\odot}$. Relative to the Galaxy, M31 therefore appears to be deficient in molecular gas by a factor of about 4, a surprising finding because the amount of HI in the two systems has been determined to be about the same (Cram et al., 1980; Henderson et al., 1982). Whether this molecular deficiency is the result of lower metallicity, or some other factor, remains to be determined.

References

Bronfman, L., Cohen, R.S., Alvarez, H., May, J. and Thaddeus, P., 1988, Astrophys. J., **324**, 248.
Cohen, R.S., Cong, H.-I., Dame, T.M. and Thaddeus, P., 1980, Astrophys. J., **239**, L53.
Cohen, R.S., Dame, T.M., Gary, G., Montani, J., Rubio, M. and Thaddeus, P., 1988, Astrophys. J., **331**, L95.
Cram, T.R., Roberts, M.S. and Whitehurst, R.N., 1980, Astr. Astrophys. Suppl., **40**, 215.
Dame, T.M., Ungerechts, H., Cohen, R.S., de Geus, E.J., Grenier, I.A., May, J., Murphy, D.C., Nyman, L.-Å. and Thaddeus, P., 1987, Astrophys. J., **322**, 706.
Dame, T.M., Koper, E., Israel, F.P. and Thaddeus, P., 1990, in *Dynamics of Galaxies and Molecular Cloud Distribution*, Proceedings of IAU Symposium No. 146, (in press).
de Vaucouleurs, G. and Freeman, K.G., 1973, Vistas Astr., **14**, 163.
Grabelsky, D.A., Cohen, R.S., Bronfman, L. and Thaddeus, P., 1988, Astrophys. J., **331**, 181.
Heithausen, A. and Thaddeus, P., 1990, Astrophys. J., **353**, L49.
Henderson, A.P., Jackson, P.D. and Kerr, F.J., 1982, Astrophys. J., **263**, 116.
Maddalena, R.J., Morris, M., Moscowitz, J. and Thaddeus, P., 1986, Astrophys. J., **303**, 375.
Magnani, L., Blitz, L. and Mundy, L., 1985, Astrophys. J., **295**, 402.
Mathewson, D.S., Ford, V.I., Tuohy, I.R., Mills, B.Y., Turtle, A.J. and Helfand, D.J., 1985, Astrophys. J. Suppl., **58**, 197.
Oort, J.H., 1977, Ann. Rev. Astron. Astrophys., **15**, 295.
Ungerechts, H. and Thaddeus, P., 1987, Astrophys. J. Suppl., **63**, 645.
de Vries, H.W., Heithausen, A. and Thaddeus, P., 1987, Astrophys. J., **319**, 723.
Young, J.S. and Scoville, N.Z., 1991, Ann. Rev. Astron. Astrophys., **29**, (in press).

Orion-A molecular cloud: a multi-transition study of CO

A. DUTREY [1], A. CASTETS[1], G.DUVERT[1], J. BALLY [2], W.D. LANGER[2] and R.W. WILSON[2]

Abstract Applying an LVG model on a multi-isotopes study of CO (^{13}CO and $C^{18}O$), we have derived the H_2 density and the ^{13}CO column density maps of the Orion A region. Our analysis shows that the use of the peak ^{12}CO antenna temperature as a gas kinetic temperature tracer through the cloud gives unphysical results. Temperature gradients must exist in the cloud along most lines-of-sight, which is consistent with models where the cloud surface is heated by UV photons coming from young stars. Several low column density regions exhibit unusually high gas density resulting from shocks expanding between the HII region and the molecular complex.

1 Introduction

To analyse the excitation conditions of Orion A, we combined ^{12}CO, ^{13}CO and $C^{18}O$ observations in the transitions $1 \rightarrow 0$ and $2 \rightarrow 1$, obtained at the AT&T Bell Laboratories 7-meter telescope (USA) and at the "Groupe d'Astrophysique de Grenoble" radiotelescope (POM-2, Plateau de Bure, France) respectively. We used an LVG code to derive the H_2 density and the ^{13}CO column density. The results presented here concern a map covering 1° in declination by 0.5° in right ascension and centered on the BN-KL nebula.

2 Results

We obtained physically reasonable densities, only when the kinetic temperature is modeled to have large gradient along the line of sight. The ^{12}CO, optically thicker than ^{13}CO, is only representative of the outer layers of the cloud, photo-heated by starlight. Then, we cannot use ^{12}CO as kinetic indicator without making an important mistake. To pass round this problem we assumed a constant kinetic temperature of 25 K (deduced from NH_3 data) in the cores (given by a ^{13}CO column density greater than 5×10^{15} cm^{-2}) and 40 K in the envelope (^{13}CO column density lower than 5×10^{15} cm^{-2}).

The figures 1 for $n(H_2)$ and 2 for $N(^{13}CO)$ summarize our main results. In the south (from $Dec = +2.5'$ to $Dec = -30'$) the ^{13}CO reveals several components with different excitation temperature and line opacities (clearly separated in the $C^{18}O$ data). In order to estimate the excitation conditions we analysed the southern part of the cloud with respect to these different components. The

---

[1] GROUPE D'ASTROPHYSIQUE DE GRENOBLE, B.P.53X 38 041 GRENOBLE, FRANCE.
[2] AT&T BELL LABORATORIES, BOX 400 HOLMDEL, NEW JERSEY 07, USA.

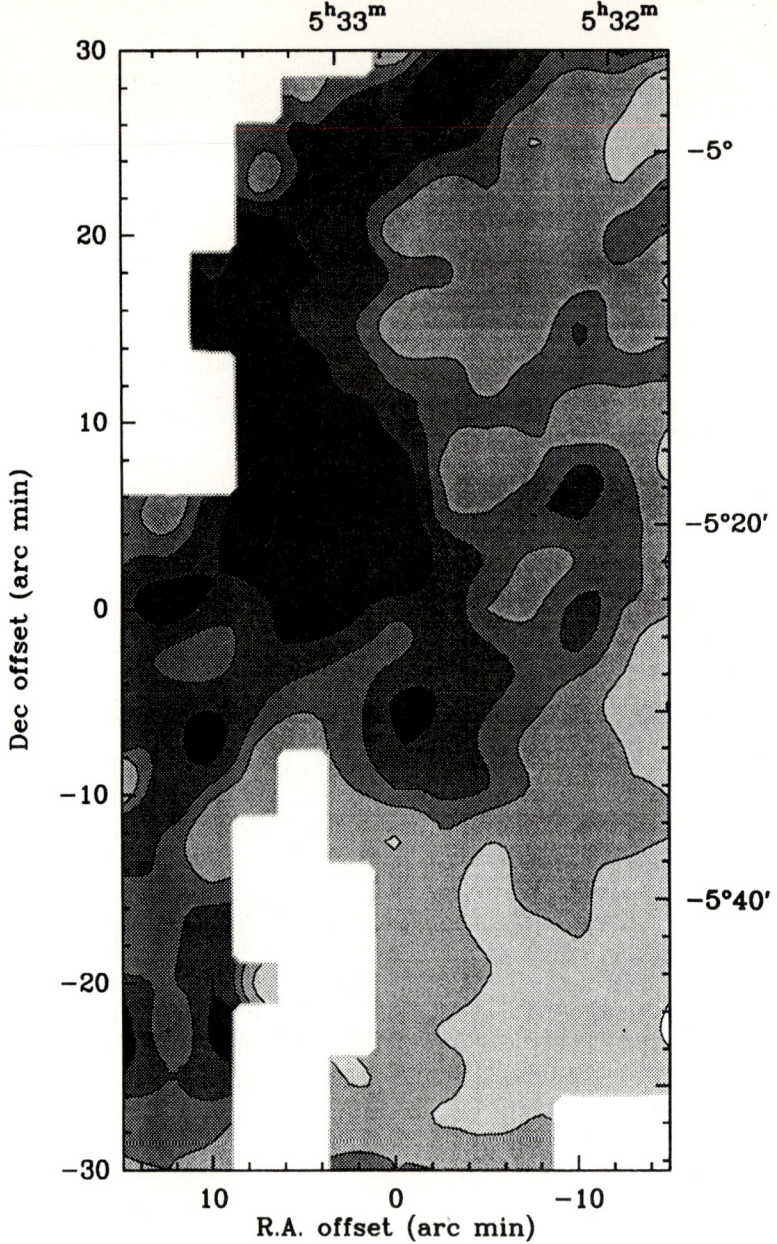

Figure 1: $n(H_2)$ density map derived from LVG analysis. Contour levels are 300, 1000, 1800, 2500, 4000, 6000 cm^{-3}.

figures refer to the "main component" of the cloud (i.e. which corresponds to the main filamentary structure).

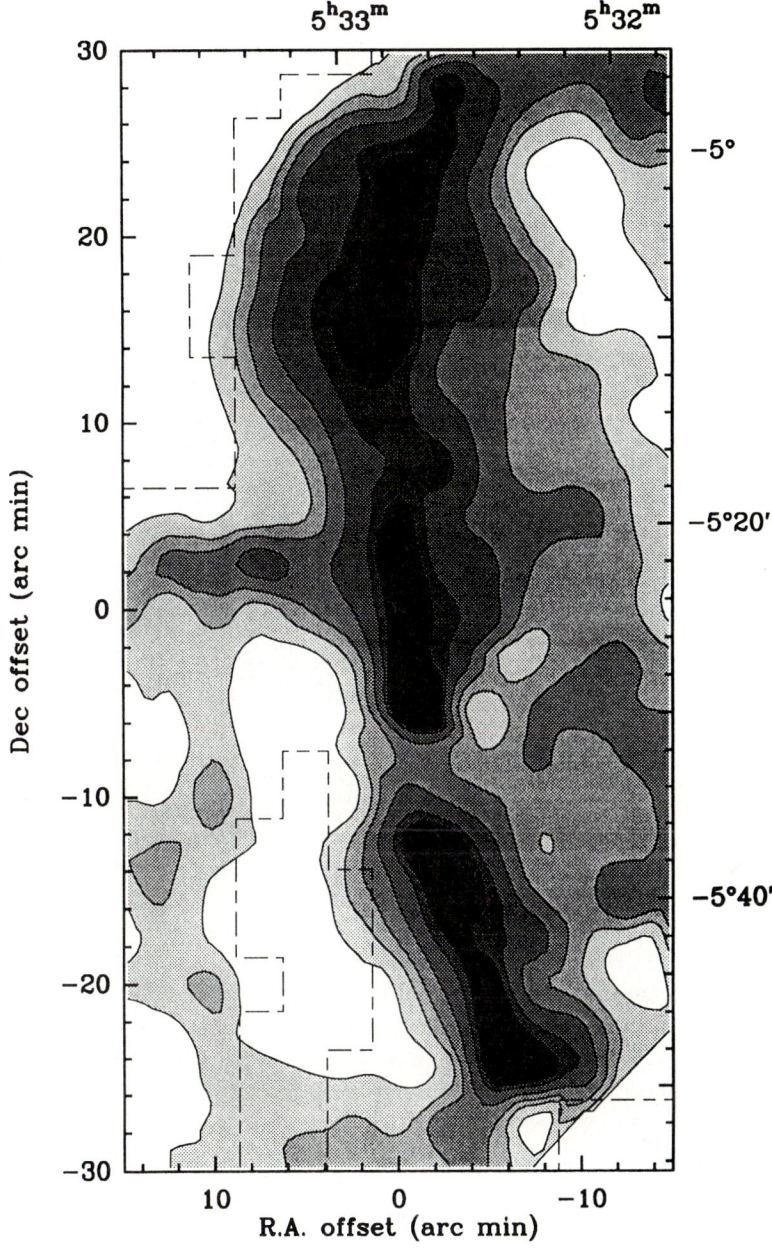

Figure 2: ^{13}CO column density map derived from LVG analysis. Contour levels are 5×10^{15}, 10^{16}, 1.6×10^{16}, 2.5×10^{16}, 4×10^{16}, 6.5×10^{16} cm^{-2}.

The analysis of our maps leads to the following important points:

1 The N(^{13}CO) map exhibits an overall appearance which is very similar to the integrated area map (see the map in Castets *et al.*, 1990).

2 In the OMC1 core we found a factor 2.5 difference in the column density computed by LVG and LTE method. The LTE method uses the peak

^{12}CO for the excitation temperature of the entire ^{13}CO emitting region while the LVG method derives the excitation temperature from the ^{13}CO emission. If the estimate of the kinetic temperature from the ^{12}CO peak is not representative of the ^{13}CO emitting region, the LTE method can give unreliable results. The LTE method is more sensitive than LVG to errors in the excitation temperature in the emitting region.

The H$_2$ density map exhibits several high density features in region of low column density while the dense cores seen in high density tracer as CS or high column density tracer as C^{18}O are not visible. In these regions, the ^{13}CO fails to probe the dense cores. It probes only the outer denser regions located in layers where the density is not sufficiently high to obtain CS emission (these features are not present in our CS (2 → 1) or in C^{18}O (2 → 1) integrated area maps, (to be published).

a) On the southeastern edge, a low column density, high density region (about 3000 cm^{-3}) is associated with the expansion of the HII region M42 which drives a shock into the neutral material.

b) In the north-west to the "Dark Bay" (clearly seen in the east-north-east of M42 in N(^{13}CO) map as a distinct feature) a second high density feature, corresponding to the north to the HII region M43 suggests the existence of a partial shell of shocked compressed gas.

c) Near NGC 1977, the N(^{13}CO) exhibits a curvature to the west while the n(H$_2$) map presents a local maximum.

3   Conclusion

In order to better understand the excitation conditions in GMC, we have realised in Orion A, the first multi-transition and multi-isotopes analysis of CO. We conclude that multi-transitions and multi-isotopes study can give a more realistic picture of clouds. We found that a temperature gradient must exist on most lines-of-sight in the cloud and that ^{12}CO is not useful as a kinetic temperature indicator. Due to opacity problems, the ^{12}CO emission and ^{13}CO emission do not probe the same layers in the cloud. Furthermore as they do not show the dense cores, we need some high density tracers as CS (details appear in Castets et al., 1990) .

Reference

Castets, A., Duvert, G., Dutrey, A., Bally, J., Langer, W.D., Wilson, R.W., 1990, Astron. Astrophys., **234**, 469.

First Detection of ^{13}CO J = 6 → 5: Large Amounts of Warm Molecular Gas

J. STUTZKI[1] U.U. GRAF[1], R. GENZEL[1], A.I. HARRIS[1], R.E. HILLS[2] AND A.P.G. RUSSELL[1,3]

Introduction. Submillimeter and far-infrared observations of carbon monoxide (CO) (Jaffe, Harris, and Genzel, 1987, Harris *et al.*, 1987a, Genzel, Poglitsch, and Stacey, 1988, Schmid-Burgk *et al.*, 1989, Boreiko, Betz, and Zmuidzinas, 1989) have indicated the presence of warm, dense molecular gas near regions of recent star forming activity. Estimates based on the intensity ratio between mid-J and high-J ^{12}CO lines in M17 and S106 (Harris *et al.*, 1987a) gave a lower limit of $\approx 10^{18}$ cm^{-2} ($\tau(^{12}$CO $7 \to 6) \approx 1$) to the CO column density of quiescent gas at temperatures of at least 100 K and H$_2$ densities of 10^4 to 10^6 cm^{-3}, corresponding to between 5% and 20% of the gas mass in these regions. In order to obtain a better estimate of the column densities, it is thus of great interest to observe isotopic mid-J CO lines, which are likely to be optically thin.

2. Observations and Results. We detected the ^{13}CO J = 6 → 5 line on 1989 November 27 with the MPE cooled Schottky submillimeter heterodyne receiver (Harris *et al.*, 1987b) mounted at the JCMT, on Mauna Kea, Hawaii. Details on the observational and calibration procedure are discussed by Graf *et al.*(1990a). Line temperatures listed below are on a Rayleigh-Jeans main beam brightness temperature scale.

Strong ^{13}CO J =6 → 5 emission was detected from a number of positions towards Orion IRc2 (Figure 1a), H1C, the Orion Bar, and NGC 2024 (Figure 1b). Towards NGC 2023 no strong ^{13}CO J = 6 → 5 emission was detected.

3. Warm Gas Column Densities. The ^{13}CO J = 6 level lies 111 K above ground state. The J = 6 → 5 transition has an A-coefficient of 1.9×10^{-5} s^{-1}, and in optically thin gas the critical density is $\approx 2 \times 10^5$ cm^{-3}. The high brightness temperature seen in this transition is an immediate proof of the presence of large amounts of warm, dense gas. In the optically thin limit a typical integrated line intensity of 120 K km s^{-1} implies a ^{13}CO column density in the J = 6 level of 5×10^{15} cm^{-2}. Assuming LTE conditions, the maximum fractional population of the J = 6 level is $\approx 11\%$. This constrains the total column density in warm ^{13}CO to $\geq 5 \times 10^{16}$ cm^{-2}.

Tighter constraints are obtained by comparing the observed source parameters with the results of escape probability, radiative transfer calculations shown in Figure 2. The optically thick ^{12}CO J = 6 → 5 line profiles differ substantially from the isotopic 6 → 5 line (Figure 1). We therefore base our analysis on the ^{13}CO lines only, since these are likely to have only moderate optical depths and

---

[1] MAX-PLANCK-INSTITUT FÜR EXTRATERRESTRISCHE PHYSIK, 8046 GARCHING, F.R.G.
[2] MULLARD RADIO ASTRONOMY OBSERVATORY, CAMBRIDGE, U.K.
[3] JOINT ASTRONOMY CENTRE, HILO, HAWAII, U.S.A.

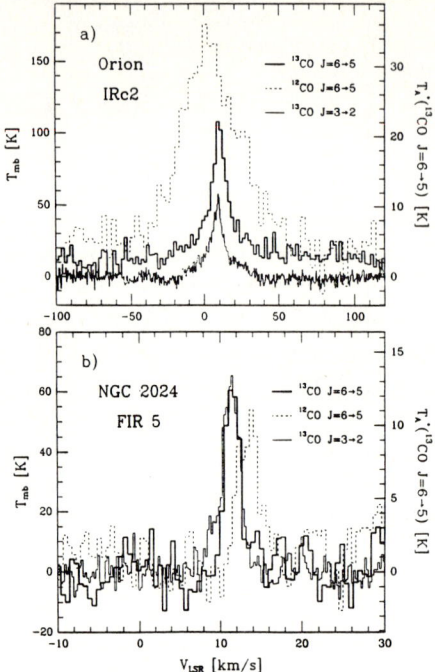

Figure 1: Overlay of CO lines detected towards a) Orion IRc2 ($5^h\,32^m\,47.0^s$, $-5°\,24'\,23''$ (1950)) and b) NGC 2024 (FIR 5, $5^h\,39^m\,12.8^s$, $-1°\,57'\,04''$(1950). Mezger et al. 1988): ^{13}CO $J = 6 \to 5$ (heavy solid line), ^{12}CO $J = 6 \to 5$ (dashed line), and ^{13}CO $J = 3 \to 2$ (light solid line).

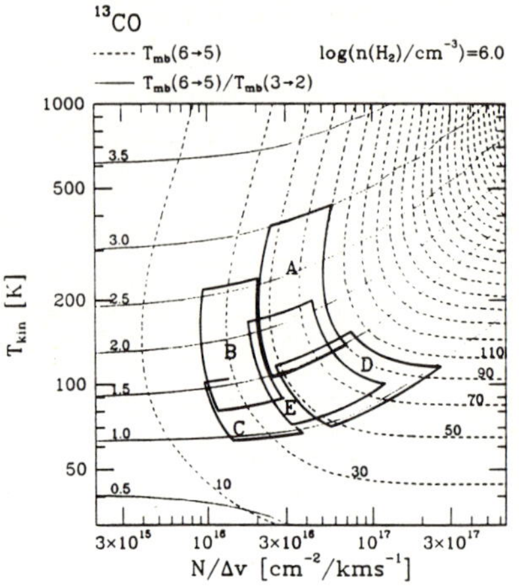

Figure 2: Radiative transfer results compared with the observed source parameters. We present the results in the kinetic gas temperature/^{13}CO column density plane at a density of 10^6 cm^{-3}. Dashed lines are curves of constant ^{13}CO $J = 6 \to 5$ brightness temperatures, dotted lines represent constant ratios of $(6 \to 5)/(3 \to 2)$ brightness temperatures. The heavy boxes outline the range of parameters consistent with the observed values for the individual sources: A: Orion IRc2 (spike), B: Orion IRc2 (plateau), C: Orion Bar (lower limits), D: NGC 2024 (0″,0″), E: NGC 2024 (-15″,15″).

thus trace a similar volume of gas. The similarity of the ^{13}CO $6 \to 5$ and $3 \to 2$ line profiles supports this assumption.

The column density is constrained by the intensity of the ^{13}CO $J = 6 \to 5$ emission to N(^{13}CO) = $4\times10^{16} - 5\times10^{17}$ cm^{-2}. The $(6 \to 5)/(3 \to 2)$ intensity ratio constrains the temperature to $\geq$ 100 K. A certain fraction of the $3 \to 2$ emission could come from cooler gas not traced by the higher energy $6 \to 5$ transition. The temperatures given above thus have to be considered as lower limits. Over this range of parameters the ^{13}CO $J = 6 \to 5$ line is still optically thin ($\tau \leq 0.6$, except for NGC 2024 (0,0): $\tau \approx 1.5$).

4. Orion B (NGC 2024): Two Gas Components. The observations show an

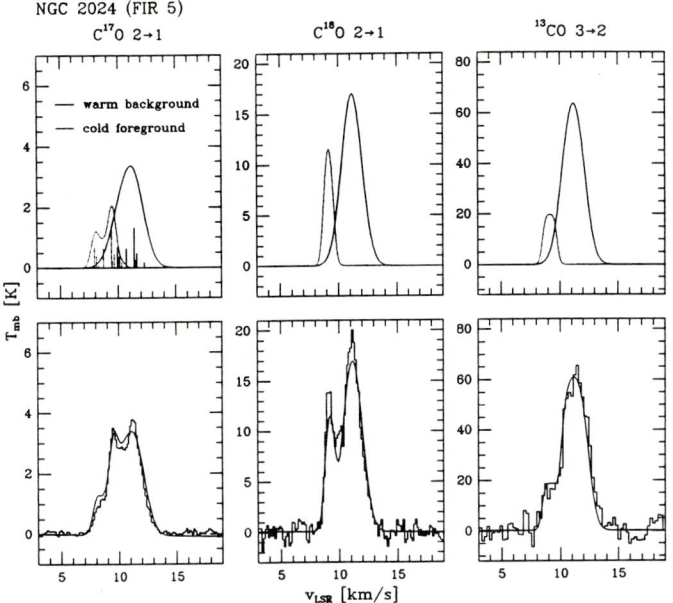

Figure 3: Low-J, rare isotopic lines observed towards NGC 2024 (histogram) and the best fit by a two component model, showing a warm background (solid line) and cold foreground (dotted line) component with the parameters discussed in the text. The upper panels also include the nuclear quadrupole hyperfine components for the $C^{17}O$ $J = 2 \rightarrow 1$ transition. Note in particular that the cold component has enough column density and low enough temperature to explain the apparent shift in the ^{12}CO $J = 6 \rightarrow 5$ line center as being due to self absorption of an intrinsically very bright line centered at 11 km s^{-1} by the cold ($J_\nu(T_{ex})$ = a few K), high opacity (τ about 40 for the above excitation parameters) foreground component. Due to the low temperatures the opacity of the cold component drops rapidly with increasing rotational quantum number and indeed the observed ^{12}CO $J = 7 \rightarrow 6$ line (Graf et al., 1990b) is much brighter, slightly wider, and less shifted than the $J = 6 \rightarrow 5$ line.

apparent velocity shift between the ^{12}CO and ^{13}CO $J = 6 \rightarrow 5$ emission from NGC 2024. In addition, low-J, rare isotopic CO transitions show rather complicated profiles (Figure 3). $C^{17}O$ $J = 2 \rightarrow 1$ is additionally affected by nuclear quadrupole hyperfine structure. We simultaneously fitted the three lines shown in Figure 3 with two velocity components $T(v) = T_{ex} \times \{1 - \exp(-\tau \times \sum s_j \phi(v - v_j))\}$ along the line of sight. The foreground component can absorb the background component. The relative strengths s_j are given by the abundance ratios of the different isotopes (plus LTE excitation factors), or in case of the $C^{17}O$ hyperfine components by the spectroscopic factors.

A two velocity component model fits the data very well. The fit identifies a cold foreground component ($v_0 = 9.3$ km s^{-1}, $\Delta v = 0.8$ km s^{-1}, $T_{ex} = 18$ K, $N(H_2) = 4 \times 10^{22}$ cm^{-2}) with parameters very similar to the ones for the OH absorption component along the dust bar (Barnes et al., 1989). The warm background component ($v_0 = 11.2$ km s^{-1}, $\Delta v = 2$ km s^{-1}, $T_{ex} = 70$ K, $N(H_2) = 2.2 \times 10^{23}$ cm^{-2}) agrees in velocity, velocity width, and column density very well with the bulk emission from the molecular ridge (e.g. Moore et al., 1989). The column density is about twice that derived above for the warmer, ^{13}CO J = 6 → 5 emitting gas component. A temperature gradient in the warm component, with about 2/3 of the column density at temperatures around 50 K and 1/3 at temperatures of around 200 K gives a consistent picture for all observed CO transitions.

5. *What Heats the Gas?* Shocks are the dominant heating mechanism in the Orion plateau source (Draine and Roberge, 1984, Chernoff, Hollenbach and McKee, 1982). However, due to the narrow lines observed shocks are ruled out as a heating source in the other cases.

Collisional heating of the gas by warm dust is ruled out by the high gas temperatures (> 100 K) inferred from the observations. Note however, that in reverse the large column densities of warm molecular gas can easily explain similarly large column densities of warm dust being heated to about 50 K dust temperature by collisions with the gas.

The observations seem to indicate a correlation between the column density of warm molecular gas and the UV luminosity of the sources (e.g. the non detection of ^{13}CO J = 6 → 5 from the low UV-luminosity source NGC 2023). However, all present photodissociation region models (e.g. Burton, Hollenbach, and Tielens, 1989, Sternberg, 1990), including more recent extensions to much higher densities, fail by about an order of magnitude to explain the observed ^{13}CO mid-J emission. It is essentially impossible to penetrate $Av \approx 100$ mag of molecular material by the UV radiation in order to heat it. This also holds for cases of extreme clumpiness.

We thus conclude that at present we do not understand the heating mechanism for the large column densities of quiescent warm molecular gas observed.

References

Barnes, P.J., Crutcher, R.M., Bieging, J.H., Storey, J.W.V. and Willner, S.P., 1989, Astrophys. J., **342**, 883.
Boreiko, R.T., Betz, A.L. and Zmuidzinas, J., 1989, Astrophys. J., **337**, 332.
Burton, M., Hollenbach, D. and Tielens, A.G.G.M., 1989, 22nd ESLAB Symposium, *Infrared Spectroscopy in Astronomy*, (ESA SP-290), p. 141.
Chernoff, D.F., Hollenbach, D.J. and McKee, C.F., 1982, Astrophys. J., **259**, L97.
Draine, B.T. and Roberge, W.G., 1984, Astrophys. J., **282**, 491.
Genzel, R., Poglitsch, A. and Stacey, G.J., 1988, Astrophys. J., **333**, L59.
Graf, U.U., Genzel, R., Harris, A.I., Hills, R.E., Russell, A.P.G. and Stutzki, J., 1990a, Astrophys. J., **358**, L49.
Graf, U.U., Genzel, R., Harris, A.I., Hills, R.E., Russell, A.P.G. and Stutzki, J., 1990b, (in preparation).
Harris, A.W., Stutzki, J., Genzel, R., Lugten, J.B., Stacey, G.J. and Jaffe, D.t¿, 1987a, Astrophys. J. **322**, L49.
Harris et al. 1987b, Int.J.Infrared Millimeter Waves, **8**, 857.
Jaffe, D.T., Harris, A.I. and Genzel, R., 1987, Astrophys. J. **361**, 231.
Moore, T.J.T., Chandler, C.J., Gear, W.K. and Mountain, C.M., 1989, Mon. Not. Roy. Astron. Soc., **237**, 1p.
Schmid-Burgk et al. 1989, Astron. Astrophys., **215**, 150.
Sternberg, A., 1990, (in preparation).

Molecular clouds at the edge of the Galaxy[†]

J. BRAND[1] and J. G. A. WOUTERLOOT[2]

Introduction and Observations. From studies of the outer Galaxy (Fich and Blitz, 1984, Brand, 1986) it has become clear that the Galaxy contains HII region/molecular cloud complexes at distances R from the galactic centre of up to 20 kpc. We have derived, from CO measurements towards IRAS sources having the colours of star forming regions, the distribution of CO clouds (or equivalently H_2) between $R = 8.5$ and 20 kpc (Wouterloot and Brand, 1989, Wouterloot et al., 1990).

The results of Wouterloot and Brand (1989) can be used to select those clouds at the largest (kinematic) distances from the galactic centre (that are located at the edge of the molecular disk of the Galaxy) to obtain the properties (sizes and masses) of these clouds.

Doing this, we have studied 56 clouds with $16\,\mathrm{kpc} < R < 20\,\mathrm{kpc}$ using the 3m KOSMA telescope for the northern hemisphere clouds and the 15m SEST for the southern ones. Their distance from the Sun is about 10 kpc, which translates to linear beamsizes of the two telescopes at 115 GHz of about 12 pc (KOSMA) and 2.5 pc (SEST).

At Gornergrat we have observed 32 clouds in ^{12}CO (J = 1 → 0) and for 5 clouds also observed ^{13}CO (J = 1 → 0) at the IRAS position. With a velocity resolution of $0.4\,\mathrm{km\,s^{-1}}$, the sensitivity is 0.05 – 0.10 K for ^{12}CO (J = 1 → 0) and 0.01 – 0.03 K for ^{13}CO (J = 1 → 0). Thirty clouds were detected. Subsequently we have mapped 13 of these in ^{12}CO (J = 1 → 0) on a 4′ grid, tracing the clouds until the signal disappeared.

With the SEST we have mapped 14 clouds in ^{12}CO(J = 1 → 0) on a 40″ or 80″ grid. With a velocity resolution of $0.11\,\mathrm{km\,s^{-1}}$, the rms noise in the spectra is about 0.5 K. Some positions have been observed in ^{13}CO (J = 1 → 0) with an rms noise ≈ 0.1 K. In addition to these maps, ^{12}CO and ^{13}CO were observed at the IRAS position in 11 clouds.

Results. The 13 molecular clouds mapped at Gornergrat were detected at between 2 and 22 positions. The largest clouds are more extended in one direction, reaching lengths of 80 –100 pc. All clouds have a maximum in T_A^* at the IRAS position. We derive virial masses assuming spherical, homogeneous clouds. The size of the cloud is approximated from the square root of the surface area where emission is detected. It ranges from 14 to 55 pc with a median value of about 25 pc. For the line width we take the value at the IRAS position. Masses then range from $7.6 \times 10^3\,M_\odot$ to $8.4 \times 10^4\,M_\odot$.

The size of the clouds mapped with the SEST ranges between 3 and 50 pc. Virial masses were derived in the same way as for the northern clouds. They range from $1.6 \times 10^3\,M_\odot$ to $8.3 \times 10^4\,M_\odot$, which is comparable to the num-

---

[†] Partly based on observations collected at the European Southern Observatory, La Silla
[1] OSS. ASTROFISICO DI ARCETRI, FIRENZE, ITALY.
[2] I. PHYSIKALISCHES INSTITUT, KÖLN, GERMANY

bers obtained for the northern clouds. The sum of the masses of the clouds is $4.1 \times 10^5\,M_\odot$ in the north and $3.0 \times 10^5\,M_\odot$ in the south for about equal galactic longitude intervals investigated. We do not find clouds above $10^5\,M_\odot$, which are often found in the inner Galaxy.

We found no difference in $T_A^*(^{12}CO)/T_A^*(^{13}CO)$ between clouds in the outer Galaxy (where the ratio was determined at the IRAS position only), and those in the inner parts, but the data are too scarce to draw any firm conclusions yet.

A more extensive account of this work can be found in Brand and Wouterloot (1990).

References

Brand, J., 1986, *Ph.D Thesis, University of Leiden.*
Brand, J. and Wouterloot, J.G.A., 1990, in *The Interstellar Disk-Halo Connection in Galaxies*, ed. J.B.G.M. Bloemen (in press).
Fich, M. and Blitz, L., 1984, Astrophys. J. **279**, 125
Wouterloot, J.G.A. and Brand, J., 1989, Astron. Astrophys. Suppl., **80**, 149
Wouterloot, J.G.A., Brand, J., Burton, W.B. and Kwee, K.K., 1990, Astron. Astrophys., **230**, 21

Search for star formation in the Southern Coalsack

L.-Å. NYMAN[1], L. BRONFMAN[2] and P. THADDEUS[3]

The Southern Coalsack is the most prominent dark cloud in the southern Milky Way. It is situated at a distance of about 180 pc, and its mass has been estimated to be about $3500\,M_\odot$ from a CO(1 → 0) survey made by Nyman et al. (1989). A cloud of this mass would be expected to contain young stars, but so far none has been found in the searches made for T Tauri stars, flare stars, and HH objects. A literature review of the Coalsack is given in Nyman (1990).

The Coalsack contains many globules and there are many IRAS sources with the color–color characteristics of embedded young stars or dense molecular cores situated in its direction. We have used the 15m Swedish ESO Sub-mm Telescope (SEST) on La Silla in Chile (see Booth et al., 1989 for a description of the telescope) to observe the CO, ^{13}CO, $C^{18}O$ (1 → 0), and CS (2 → 1) transitions toward 23 globules and 20 IRAS sources to search for molecular outflows and dense cores. Because the Coalsack is situated in the Galactic plane, there are several massive background molecular clouds in the line of sight. These clouds are situated in the near and the far side of the Sagittarius–Carina spiral arm at distances of about 2 kpc and 14 kpc respectively. The emission lines of the background clouds are easily distinguished from the emission lines of the Coalsack because of their different radial velocities.

We observed 23 globules from the list in Bowers et al. (1980). 12 of them were detected in $C^{18}O$ emission, which implies molecular hydrogen column densities larger than $5 \times 10^{21}\,{\rm cm}^{-2}$. The lines are narrow ($\sim 0.6\,{\rm km\,s}^{-1}$) and weak ($T_{mb}$ between 0.3 K and 1.9 K). Only four globules were found to have CS (2 → 1) emission. Two of them (numbers 1 and 2 in the list of Bowers et al.) are situated in the densest part of the cloud, and one of them, Tapia 2, is described below. The other two (numbers 9 and 10) are dense, compact globules at the edge of the main cloud. No evidence for molecular outflows were found in the CO data.

The IRAS sources were selected using the criteria $F_{60}/F_{25} > 1$ and $F_{25}/F_{12} > 1$ (if the source was detected at 12 μm). These criteria include T Tauri stars, dense cores, and compact HII regions. Only sources situated within the $2\,{\rm K\,km\,s}^{-1}$ contour of the large scale CO survey of Nyman et al. (1989) were selected, giving a total of 20 objects. None of the IRAS sources are associated with a globule. $C^{18}O$ emission at the velocities of the Coalsack were detected toward three of the sources, two of which are situated in the dense western part of the cloud (see below) and one is near a globule. CS emission at the velocities of the Coalsack was not detected toward any of the IRAS sources, and there is no

---

[1] SEST PROJECT, ESO, CASILLA 19001, SANTIAGO 19, CHILE.
[2] UNIVERSIDAD DE CHILE, CASILLA 36-D, SANTIAGO, CHILE.
[3] CENTER FOR ASTROPHYSICS, 60 GARDEN STREET, CAMBRIDGE, MA 02138, U.S.A.

indication of molecular outflows in the CO data. However, eight of the sources have wide $C^{18}O$ and CS emission lines at velocities corresponding to those of the background cloud at the near side of the Sagittarius–Carina arm, and five have emission lines at velocities corresponding to those of the clouds at the far side, indicating that these IRAS sources are associated with the background clouds and not the Coalsack itself.

The densest globule in the Coalsack is Tapia 2 (Tapia, 1973). It is situated in the western part of the cloud together with two other globules. All of them are embedded in a medium of lower density. We mapped the globule in the CO, ^{13}CO, $C^{18}O$ (1 → 0), and CS (2 → 1) transitions with a spacing of 1.5' (*i.e.* every second beamwidth) in 12' long strips in the N–S, E–W, NE–SW, and NW–SE directions. The inner 1.5' were mapped every beamwidth. The CO and ^{13}CO intensities are fairly constant over the region indicating that the lines become optically thick already in the lower density medium surrounding the globules. The CO excitation temperature is about 12 K. The $C^{18}O$ data shows a peaked distribution toward the center of the globule with a FWHM of 4.5', and there is a velocity gradient from east to west, in agreement with the general velocity gradient in the region (Nyman *et al.*, 1989). The linewidths are typically $0.6 \,\mathrm{km\,s^{-1}}$, perhaps with a slight increase toward the center. Even at distances larger than 6' from the globule $C^{18}O$ emission is still visible and probably originates in the lower density medium. The molecular hydrogen mass of the globule, as estimated from the $C^{18}O$ data, is about $10 \, M_\odot$, if the contribution from the lower density medium is excluded. The CS (2 → 1) intensity distribution and linewidths are similar to the $C^{18}O$ data, although the CS line intensity is much less peaked toward the center of the globule. The CS intensities imply densities larger than $10^5 \,\mathrm{cm^{-3}}$ in the central parts of the globule.

Conclusion

From our molecular line observations of globules and IRAS sources we do not find any evidence for star formation presently taking place in the Coalsack. It is not clear why there is no star formation, and further searches for young stars should be performed at different wavelengths.

References

Booth, R.S., Delgado, G., Hagström, H., Johansson, L.E.B., Murphy, D.C., Olberg, M., Whyborn, N.D., Greve, A., Hansson, B., Lindström, C.O. and Rydberg, A., 1989, Astron. Astrophys., **216**, 315.
Bowers, P.F., Kerr, F.J. and Hawarden, T.G., 1980, Astrophys. J., **241**, 183.
Nyman, L.-Å., Bronfman, L. and Thaddeus, P., 1989, Astron. Astrophys., **216**, 185.
Nyman, L.-Å., 1990, in *Low Mass Star Formation in Southern Molecular Clouds*, ed. B. Reipurth, (ESO, in press).
Tapia, S., 1973, in *Interstellar Dust and Related Topics*, IAU Symp. 52, ed. J.M. Greenberg and J.M. Van der Hulst, (Reidel, Dordrecht), p. 43.

KOSMA submillimeter observations of outflow in L1228 and L1551

L. HAIKALA[1,2], M. MILLER[1], K. GIERENS[1] and G. WINNEWISSER[1]

L1228

L1228 is a high latitude (111.°6, 20.°1) dark cloud (the distance is estimated to be 100 to 200 pc). ^{12}CO J = 1→0 observations of the cloud and of the large (18' by 9') bipolar outflow associated with the IRAS point source 20582+7724 are described in Haikala and Laureijs (1989). The IR luminosity of the outflow central source is very low (0.4 L$_\odot$ at 100 pc). Besides the large outflow L1228 is remarkable for the large extent of strong ^{12}CO self absorption which is observed over a one by a half degree area in the cloud.

The outflow was observed in ^{12}CO J = 3→2 using a 2' grid. The observations confirm the large extent of the outflow and show it to be well collimated. The self absorption is pronounced in the ^{12}CO J = 3→2 spectra. The molecular outflow in L1228 and a ^{12}CO J = 3→2 spectrum in the blue outflow lobe are shown in Figure 1.

L1551

We have mapped a 5' × 5' area in the red lobe of L1551. Two of the observed ^{12}CO J = 3→2 spectra are shown in Figure 2. ^{12}CO J = 2→1 spectra given by Levreault (1988) in the same positions are also shown. Within the noise the spectra are identical both in intensity and and in velocity range (by a coincidence the spatial resolution of these observations are the same). The similarity of the CO J = 3→2 and CO J = 2→1 shows that the *outflowing matter must be optically thick* at these transitions, it *must be clumped* and it *must be dense* (N(H$_2$) ≥ 10^4). This conclusion is confirmed by the ^{12}CO J = 1→0 spectra obtained with the BELL LABS 7m telescope (Bally, 1990), in the whole area observed in ^{12}CO J = 3→2. The line shapes in intensity are identical also in these two transitions.

---

[1] I. PHYSIKALISCHES INSTITUT, ZÜLPICHERSTRASSE 77, 5000 KÖLN, F.R.G.
[2] OBSERVATORY AND ASTROPHYSICS LABORATORY, TÄHTITORNINMÄKI, SF-00130 HELSINKI 13, FINLAND.

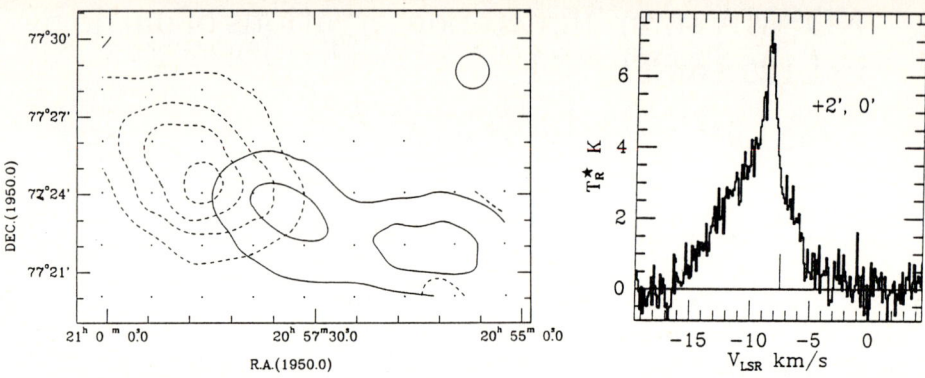

Figure 1. (a) ^{12}CO J = 3→2 blue shifted (dashed line) and redshifted (continuous line) emission in L1228. (b) A ^{12}CO J = 3→2 spectrum in L1228. The offset given in the spectrum is from IRAS 20582+7724 and the tickline is at the velocity of the parent cloud.

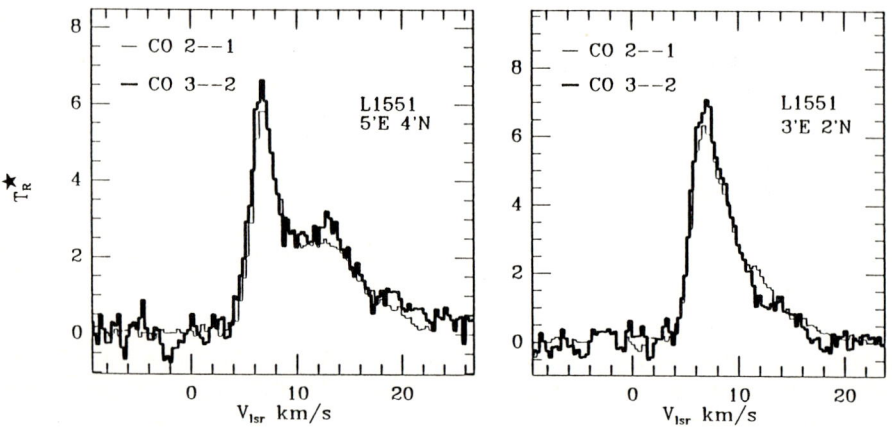

Figure 2. Two ^{12}CO J = 3→2 spectra in L1551. Offsets are from L1551 IRS 5. ^{12}CO J = 2→1 spectra from Levreault (1988) in the same positions are also shown.

References

Bally, J, 1990, private communication.
Haikala, L.K. and Laureijs, R.J., 1989, Astron. and Astrophys., **223**, 287.
Levreault, R.M., 1988, Astrophys. Journ. Suppl., **67**, 283.

Modelling the S140 molecular cloud: A multitransition study of CO isotopomers

EMMA L.O. BAKES AND GLENN J. WHITE

PHYSICS DEPT., QUEEN MARY AND WESTFIELD COLLEGE, UNIVERSITY OF LONDON, MILE END ROAD, LONDON E1 4NS, U.K.

Introduction. The S140 molecular cloud consists of a dense core containing a CO hotspot where massive, young stars are formed. Within the hotspot are three mid-infrared sources IRS1, 2 and 3 with deep silicate absorption features (Beichman, Becklin and Wynn-Williams, 1979), embedded in an extended dark cloud CO envelope L1204.

There is an H_2O maser coincident with IRS1 (White and Little, 1975) and an asymmetric, intrinsically collimated bipolar outflow oriented NW to SE. The probable driving force of the outflow is IRS1 (Hayashi *et al.*, 1985).

External to the cloud, there is an HII region excited by two B stars. The cloud is bordered by an $H\alpha$ rim (Sharpless 1959), excited primarily by one of the B stars, HD211880 (Blair *et al.*, 1978) and clearly is a good example of an edge-on ionisation front.

Results. JCMT observations of S140 using CO $J = 1 \rightarrow 0$, $2 \rightarrow 1$, $3 \rightarrow 2$, $4 \rightarrow 3$ transitions plus the isotopomers ^{13}CO and $C^{18}O$ have yielded the following results:

1) A large velocity gradient model (Goldreich and Kwan, 1974) provided a satisfactory fit for transitions of CO isotopes up to $J = 3 \rightarrow 2$. However, once the $J = 4 \rightarrow 3$ transition was included, no suitable fit could be found, bringing into question the applicability of a simple, single component model of uniform kinetic temperature and density. A hot ridge of gas was observed for the ^{12}CO isotope close to the ionization front of S140, suggesting heating by the external UV field. Qualitative analysis yielded a simple model of a clumpy cloud, where FUV photons from the B stars external to the cloud are absorbed by dust grains. As a result, electrons are ejected from the surface of the grains in a grain-photoelectric heating process. The electrons then collisionally deposit kinetic energy to the gas.

2) Masses calculated for the cloud via several methods were found to depend significantly on the value of $n(H_2)$ used, which in turn depends on the molecule used to analyse the cloud. Our estimates were lower than those previously calculated – 180 $M_\odot$ in a region of radius 1' compared with 1000 $M_\odot$ obtained by Blair *et al.* (1978), where we have used $n(H_2) = 10^4 \, cm^{-3}$ rather than the $n(H_2) = 10^5 \, cm^{-3}$ estimated from molecules of higher dipole moment. Our assumptions were based on a theoretical calculation of the critical $n(H_2)$ at which collisional and radiative excitation rates were equal for a given transition. The calculation assumed optically thin gas – radiative trapping modifies the critical density, which scales with escape probability. Using the data, the escape probability was calculated and $n(H_2)$ was scaled accordingly.

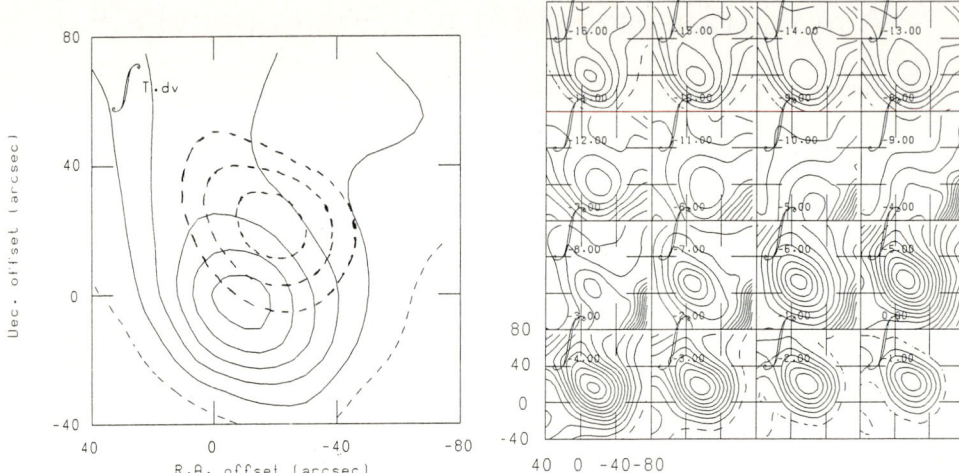

Figure 1: Map of integrated emission in the blue (solid contours) and red (dashed contours) wings. The blue wing is integrated from -37.0 to -12.0 km s^{-1} and the red wing from -4.0 to 6.0 km s^{-1}. Map centre RA (1950) (22^h 17^m 40.88^s), Dec (1950) (63° 03' 44.98").

Figure 2: Maps of antenna temperature integrated in velocity channels of width 1.0 km s^{-1} from -16.0 to 0.0 km s^{-1} with 16 temperature contours ranging from 2.0 to 32.0 K. The axes are labelled with right ascension and declination offsets in arcseconds.

3) The ^{12}CO spectral lines showed a position dependent asymmetry, suggesting a compact, bipolar outflow oriented NW to SE, with the centres of its red and blue lobes separated by $\sim 35''$ (Figure 1), similar to that measured by Hayashi *et al.* (1987). Further proof of bipolarity is given by velocity channel maps (Figure 2). The peak of the T_a^* contours shifts across the map as the velocity changes. This is especially apparent via comparison of the maps integrated from -16 to -15 km s^{-1} and -1 to 0 km s^{-1}.

4) A feature was resolved at Dec (63° 05' 00") (1950) whose RA (1950) varied from (22^h 17^m 50.88^s) to (22^h 17^m 10.88^s) when integrating T_a^* over the range of the blue and red-shifted ^{12}CO velocity wings respectively. This could be an additional feature of the outflow or an interaction zone with material ejected from the outflow (Schwartz, 1989).

References

Beichman, C.A., Becklin, E.E. and Wynn-Williams, C.G., 1979, Astrophys. J., **232**, L47.
Blair, G.N., Evans, N.J., Vanden Bout, P.A. and Peters, W.L., 1978, Astrophys. J., **219**, 896.
Goldreich, P. and Kwan, J., 1974, Astrophys. J., **189**, 441.
Hayashi, M., Hasegawa, T., Omodaka, T., Hayashi, S. and Miyawaki, T., 1987, Astrophys. J., **312**, 327.
Schwartz, P.R., 1989, Astrophys. J., **338**, L25.
Sharpless, S., 1959, Astrophys. J. Suppl., **4**, 257.
White, G.J. and Little, L.T., 1975, Astrophys. Lett., **16**, 151.

Gamma-ray and far-infrared constraints on the N_{H_2}/W_{CO} conversion factor of molecular clouds

HANS BLOEMEN

SPACE RESEARCH LEIDEN AND LEIDEN OBSERVATORY, P.O. BOX 9504,

2300 RA LEIDEN, THE NETHERLANDS

1 Introduction

The $J = 1 \to 0$ transition of the ^{12}CO molecule, next most abundant after H_2, is relatively easy to observe and therefore frequently used as a probe of molecular gas. This is a welcome alternative to direct H_2 observations, which are restricted to either far-ultraviolet wavelengths where cold H_2 appears in absorption, but the number of suitable background stars is small (*e.g.* Spitzer and Jenkins, 1973, Savage *et al.*, 1977), or near-infrared wavelengths where excited H_2 appears in emission, but this possibility is limited to regions where H_2 is heated by shocks or an intense ultraviolet radiation field (*e.g.* Shull and Beckwith, 1982, Black and van Dishoeck, 1987). It is evidently essential to know how to convert from the (integrated) CO brightness temperature W_{CO} to a column density of H_2, a subject of much discussion in several respects (see *e.g.* Maloney and Black, 1988, Bloemen, 1989). This W_{CO}-to-N_{H_2} conversion factor (hereafter referred to as X) may vary from cloud to cloud owing to temperature and abundance variations. In fact, because of its optical thickness for most molecular clouds, it is not evident that the $J = 1 \to 0$ transition can be used at all to obtain H_2 column densities. On a Galactic scale, when averaging over a large number of clouds, CO appears to be a useful tracer of N_{H_2}. Evidence has come particularly from studies of the COS-B γ-ray observations (Lebrun *et al.*, 1983, Bloemen *et al.*, 1986), which have also provided an estimate of the average Galactic X value and given some insight into a possible large-scale (radial) variation of X throughout the Galaxy. Such a gradient may exist because abundance gradients have been found, although X may not be a strong function of CO abundance, and a temperature gradient may be present. Because of their limited sensitivity, the γ-ray observations available at the moment are not suitable for the mass calibration of individual clouds, except for a few massive complexes.

New insight into the mass calibration of CO surveys has recently been obtained from the IRAS observations. The particular importance of the infrared calibration method is that it can be applied, in principle, to large-scale CO surveys of the Galactic disk as well as to CO observations of individual local clouds. This paper describes both the γ-ray and infrared calibration methods and the resulting constraints on X, with emphasis on Galactic large-scale analyses. Important information on X, not discussed in this paper, can also be obtained from extinction measurements of clouds at medium and high Galactic latitudes (*e.g.* Dickman, 1978, Frerking *et al.*, 1982, Magnani *et al.*, 1988).

2 Inferences from Gamma-Ray Observations

2.1 Decomposition of the γ-ray Milky Way

The diffuse Galactic γ-ray emission with energies between 10 MeV and 10 GeV originates probably primarily from the interactions of cosmic-ray (CR) particles with interstellar gas. The relevant processes are electron bremsstrahlung and the decay of π° mesons, produced by nuclear interactions of cosmic rays with nuclei of the interstellar gas. Basically, the observed γ-ray intensity in a certain direction traces the product of gas density and CR density, integrated along the line of sight. For instance, the γ-ray intensity is a measure of the total gas column density, practically irrespective of the composition or the physical state of the gas, if the CR density can be assumed to be constant along the line of sight: $I_\gamma = \varepsilon_\gamma N_H$, where ε_γ is the γ-ray emissivity per unit of mass. Under the assumption that HI and CO observations provide a good estimate of N_H, this relation can be written as $I_\gamma = \varepsilon_\gamma(N_{HI} + 2XW_{CO})$, where W_{CO} is the velocity-integrated CO brightness temperature and $X \equiv N_{H_2}/W_{CO}$ in units of molecules cm^{-2} K^{-1} km^{-1} s. Both ε_γ and X can be estimated from a correlation study of HI, CO, and γ-ray surveys if the spatial distributions of the atomic and molecular gas are significantly different.

If cosmic rays are of Galactic origin then a radial gradient of the CR intensity can be expected on a large scale in the Galaxy, simply because more potential CR sources are present in the inner regions. This large-scale variation of the CR density can be determined simultaneously in the correlation analysis described above, using the velocity information of HI and CO surveys as a distance indicator (in combination with the Galactic rotation curve). Bloemen et al. (1986) and Strong et al. (1988) have performed such an analysis. They constructed HI and CO maps of the Milky Way for 6 Galacto-centric distance intervals [†] (R = 2–4, 4–8, 8–10, 10–12, 12–15, and $>$ 15 kpc) and fitted the observed γ-ray intensity distribution by the relation

$$I_\gamma = \left\{ \sum_{i=1}^{6} \varepsilon_\gamma(R_i)(N_{HI,i} + 2Y_\gamma W_{CO,i}) \right\} + I_{b,\gamma}, \qquad (1)$$

where the sum is over the 6 rings and $\varepsilon_\gamma(R_i)$ ($i = 1, \cdots, 6$), Y_γ, and $I_{b,\gamma}$ are the free parameters, determined by a maximum-likelihood method. $I_{b,\gamma}$ is the isotropic γ-ray background level (including the instrumental background). The use of Y_γ instead of X is addressed below. The 6 wide intervals are dictated by the requirement that the gas distributions in the rings be distinctly different in order to be able to determine the contribution of each ring to the observed γ-ray emission and to separate the contributions from atomic and molecular gas. The region fitted was limited to the latitude range $|b| < 10°$ and two $\sim 20°$ wide longitude intervals, centered on the Galactic centre and anticentre directions

---

[†] $R_\odot = 10$ kpc

were excluded, because of the poor kinematic distance information there. The model fitted to the γ-ray data included other components (omitted from the above relation for clarity), such as an estimate of the γ-ray emission produced by inverse-Compton scattering of high-energy CR electrons on the interstellar photon field and some well-known γ-ray point sources. This analysis was done for each of the three γ-ray energy intervals generally studied by the COS-B group (70–150 MeV, 150–300 MeV, and 300 MeV – 5 GeV — these bands were chosen, among other reasons, because they have approximately equal counting statistics).

2.2 Constraints on N_{H_2}/W_{CO}

Equation 1 contains a parameter Y_γ (instead of X) to remind us of the fact that the CR density inside molecular clouds may differ from that in the ambient medium,

$$Y_\gamma = \frac{\varepsilon_{\gamma,H_2}}{\varepsilon_{\gamma,HI}} X, \qquad (2)$$

where $\varepsilon_{\gamma,HI} \equiv \varepsilon_\gamma$. If CR particles are not excluded from, or concentrated in, molecular clouds, then Y_γ equals X independent of γ-ray energy. Theoretical work indicates that only low-energy CR protons (estimates range from $E_p \lesssim 300$ MeV to $\lesssim 50$ MeV, Skilling and Strong, 1976, Cesarsky and Völk, 1978), which are irrelevant for the production of high-energy γ rays, fail to penetrate a dense cloud completely. The alternative, an enhanced CR density in clouds, particularly CR electrons, cannot be excluded (see *e.g.* review by Völk, 1983). This would imply that $X < Y_\gamma$. If only the density of CR electrons is enhanced, then Y_γ for the 70–150 MeV range should be larger than Y_γ for the 150–300 MeV and 300 MeV – 5 GeV intervals (because electron bremsstrahlung barely contributes to these high-energy intervals), so $X < Y_\gamma$ for low energies and $X = Y_\gamma$ for high energies. The parameter Y_γ may also be influenced by the possible presence of a population of unresolved γ-ray point sources. If such a population exists, with an angular distribution similar to that of CO, then X is overestimated (*i.e.*, again, $X < Y_\gamma$).

From the most complete study of this type (Strong *et al.*, 1988), using the final COS-B data base, the Columbia CO survey (Dame *et al.*, 1987), and several HI surveys, all covering the entire Galactic disk, the Y_γ value was found to be $Y_\gamma = (2.3 \pm 0.3) \times 10^{20}$ molecules cm^{-2} K^{-1} km^{-1} s (150 MeV – 5 GeV). In an earlier analysis (Bloemen *et al.*, 1986), dealing mainly with the 1st and 2nd Galactic quadrants, it was found that $Y_\gamma = (2.75 \pm 0.35) \times 10^{20}$ molecules cm^{-2} K^{-1} km^{-1} s (70 MeV – 5 GeV). The difference between these two estimates, although statistically insignificant, results mainly from the different γ-ray energy intervals used; particularly the work of Strong *et al.* showed evidence for a larger Y_γ value in the low-energy interval (70 – 150 MeV), namely $\sim 3.3 \times 10^{20}$. This may be due to an enhanced electron density inside clouds, as discussed above. Also, it may simply result from the fact that the angular resolution of COS-B is poor at low energies, which complicates the model fitting. Anyway, the Y_γ value obtained for these low energies is the least reliable indicator of X, so Strong et al. stated that the 150 MeV – 5 GeV range is to be preferred for estimating X.

Fitting a more extended model to the data, Bloemen et al. and Strong et al. found that the ratio $Y_\gamma(2-8\,\text{kpc})/Y_\gamma(>8\,\text{kpc})$ is not significantly different from unity. Hence there is no indication for a significant difference between the average X value for $R = 2 - 8$ kpc and that found for $R > 8$ kpc. Variations on smaller scales cannot be excluded. Particularly in the Galactic-centre region, X may be drastically different (Blitz et al., 1985). On the other hand, γ-ray analyses of some individual molecular cloud complexes in the solar neighbourhood — the Orion-Monoceros and Cepheus clouds — lead to consistent results (Bloemen et al., 1984, Grenier and Lebrun, 1990).

Strictly speaking, for the reasons given above, the best Y estimate of $\sim 2.3 \times 10^{20}$ should be regarded as an upper limit of X. Although the γ-ray data do not provide a stringent lower limit, there is no reason (from the γ-ray point of view) to believe that X is much smaller than 2.3×10^{20}. The main arguments supporting this statement are the good agreement between model and data and the consistency between the Y value obtained from the large-scale analysis and the values obtained from similar analyses of the Orion-Monoceros and Cepheus regions, where the source contribution is probably negligible.

The reliability of the γ-ray calibration method has been disputed because of some differences between the above-mentioned results obtained by the COS-B group and those obtained by the Durham group from parallel analyses of the COS-B data (Bhat et al., 1986, Wolfendale, 1988). They advocate an X value of 1.5×10^{20} in the solar vicinity and 1×10^{20} in the molecular ring at $4 - 7$ kpc from the Galactic centre. The reason for the differing results appears to be twofold (for details see Bloemen, 1989): (a) the radial γ-ray emissivity distribution adopted by the Durham group is steeper than the one derived by the COS-B group (a difference by a factor of ~ 1.8 between $R = 10$ and 5 kpc) and (b) the specific CO surface densities used by the Durham group (from Sanders et al., 1984) are relatively high. The latter point is discussed in detail by Bronfman et al. (1988), who showed that the high CO surface densities presented by Sanders et al. can be attributed to two facts. Firstly, the W_{CO} values of the survey used by Sanders et al. are simply $\sim 20\%$ higher than those of the Columbia survey. Secondly, there is some debate among CO observers on the radial-unfolding procedure to be applied; using the procedure advocated by Bronfman et al. would lead to an average CO surface density in the inner Galaxy that is $\sim 40\%$ lower than derived by Sanders et al. The method used by the COS-B group does not depend on such unfolding procedures. Altogether, these effects account for a correction factor of $1.8 \times 1.2 \times 1.4 \approx 3.0$ (although the factor of 1.4 may of course not be applicable). This renders the differences fully understandable.

The most important difference between the Durham approach and that of the COS-B group is that the Durham group does not determine simultaneously $\varepsilon_\gamma(R)$ and Y. They prefer to use independent information on $\varepsilon_\gamma(R)$, which is largely dictated by the fact that the Durham group does not use the large-scale Columbia survey, but CO observations restricted to regions within $\sim 1°$ from the Galactic plane. A complete correlation analysis as done by the COS-B group is therefore not feasible. The radial distribution of ε_γ used in their work is mainly based on a correlation analysis of γ-ray data and HI data for selected disk regions at $|b| \gtrsim 2.5°$ where the H_2 contribution to the gas column density was expected to be small (Issa et al., 1981). This method is necessarily restricted to Galacto-centric radii between $\sim 7\,\mathrm{kpc}$ and $\sim 13\,\mathrm{kpc}$, but the resulting emissivity distribution was argued to be similar to that of the surface density of supernova remnants, and the latter was therefore taken to be representative of $\varepsilon_\gamma(R)$ throughout the Galaxy.

3 Inferences from Infrared Observations

3.1 Decomposition of the FIR Milky Way

The IRAS surveys of the Galactic disk in the 60 μm and 100 μm bands can be unravelled by a procedure similar to that applied to the γ-ray data. This approach offers a variety of possibilities: determining the Galacto-centric distribution of the FIR emissivity, studying the FIR properties of the dust emission from different gas components separately (particularly atomic and molecular gas), and setting a constraint on the CO-to-H_2 conversion factor, on which we concentrate here. Such an analysis was performed by Bloemen et al. (1990). The HI and CO data sets used were the same as in the γ-ray analysis of Strong et al. (1988) and the selected rings were identical as well. As in the γ-ray analysis, it was assumed that the 60 μm and 100 μm maps can be represented as a linear combination of the HI and CO maps of the selected radial intervals — with the exclusion of those regions that have a strong contribution from HII regions, as discussed below:

$$I_\lambda = \left\{ \sum_{i=1}^{6} \varepsilon_\lambda(R_i)(N_{\mathrm{HI},i} + 2Y_\lambda W_{\mathrm{CO},i}) \right\} + I_{b,\lambda}, \qquad (3)$$

where $\varepsilon_\lambda(R_i)$ ($i = 1, \cdots, 6$), Y_λ, and $I_{b,\lambda}$ are the free parameters. $\varepsilon_\lambda(R_i)$ is the infrared emissivity [MJy sr^{-1} (H atom)$^{-1}$ cm^2] of the dust mixed with the atomic gas in the ith distance interval, $N_{\mathrm{HI},i}$ and $W_{\mathrm{CO},i}$ are the HI and CO maps for this interval, and $I_{b,\lambda}$ is an isotropic background. The parameter Y_λ is the product of X and the ratio of the emissivity of the dust associated with the molecular to that of the atomic gas, $\varepsilon_{\lambda,H_2}/\varepsilon_{\lambda,HI}$, so

$$Y_\lambda = \frac{\varepsilon_{\lambda,H_2}}{\varepsilon_{\lambda,HI}} X, \qquad (4)$$

where $\varepsilon_{\lambda,HI} \equiv \varepsilon_\lambda$. The parameter Y_λ includes effects related to different conditions in the atomic and molecular gas, such as different dust temperatures, dust-to-gas ratios, and dust properties. We return to this parameter in Section 3.3. A maximum-likelihood technique, as applied in the γ-ray work of the COS-B group, was used to determine the values of the fit parameters. Again, the fit region was limited to the latitude range $|b| < 10°$. Two 20° longitude intervals, centered on the Galactic centre and anti-centre directions, were excluded.

The infrared emission from warm dust associated with ionized gas is not included in the model, simply because this is very hard to accomplish. Of particular importance are regions of massive star formation (HII regions and their interfaces with molecular clouds), which are known to be intense FIR emitters (an OB association heats during the initial 30–40% of its lifetime mainly the dust in its immediate surroundings; Leisawitz and Hauser, 1988). However, their contribution to the infrared luminosity of the Galactic disk is currently believed to be rather small; estimates have decreased considerably from about 60–80% (Mezger et al., 1982, Cox et al., 1986) to about only 20% in recent work (Cox and Mezger, 1988). Boulanger and Pérault (1988) found that in the solar vicinity most of the infrared emission comes indeed from interstellar matter not associated with OB star formation. HII regions and their interfaces with molecular clouds appear in the IRAS data as sharp peaks, which Bloemen et al. partly eliminated by excluding from the fit those bins with $60\,\mu\mathrm{m}/100\,\mu\mathrm{m}$ intensity ratios exceeding 0.35 (corresponding to dust temperatures $T_d \gtrsim 27\,\mathrm{K}$, for a λ^{-2} emissivity law). All excluded bins coincide with the positions of known HII regions. Owing to the line-of-sight integration of cold and warm dust, this criterion rejects only regions most affected by the presence of HII, but it was verified that the results do not depend sensitively on the actual criterion used.

3.2 Large-scale FIR Properties

Different versions of the above model were tested, using the likelihood-ratio technique, in order to investigate whether all free parameters are indeed required. Each of the model versions has different constraints on the fit parameters, such as $Y_{60} = Y_{100}$ (equivalent to $\varepsilon_{60,\mathrm{H_2}}/\varepsilon_{100,\mathrm{H_2}} = \varepsilon_{60,\mathrm{HI}}/\varepsilon_{100,\mathrm{HI}}$ — see equation 4) and $\varepsilon_{60}(R)/\varepsilon_{100}(R) = $ constant. In summary, the basic fit and test results were the following:

> The infrared emissivity per nucleon, $\varepsilon_\lambda(R)$, decreases significantly with increasing Galactocentric radius (Figure 1).
>
> $Y_{60} = (0.65 \pm 0.11) \times 10^{20}$ molecules cm^{-2} K^{-1} km^{-1} s
> $Y_{100} = (0.84 \pm 0.20) \times 10^{20}$ molecules cm^{-2} K^{-1} km^{-1} s.
>
> The $60\,\mu\mathrm{m}/100\,\mu\mathrm{m}$ emissivity ratio is constant as a function of R within uncertainties (Figure 1), $\varepsilon_{60,\mathrm{HI}}/\varepsilon_{100,\mathrm{HI}} \equiv \varepsilon_{60}/\varepsilon_{100} = 0.27 \pm 0.03$ and $\varepsilon_{60,\mathrm{H_2}}/\varepsilon_{100,\mathrm{H_2}} \equiv (Y_{60}/Y_{100}) \cdot (\varepsilon_{60,\mathrm{HI}}/\varepsilon_{100,\mathrm{HI}}) = 0.20 \pm 0.06$, although the latter may have some radial dependence in view of the uncertainty.

Introducing an additional free parameter $r \equiv Y_\lambda(2\text{–}8\,\mathrm{kpc})/Y_\lambda(>8\,\mathrm{kpc})$ did not improve the fit; r was found to be consistent with unity.

Figure 2 shows longitude profiles of the observed and modelled $60\,\mu\mathrm{m}$ and $100\,\mu\mathrm{m}$ intensities. These figures indicate that there is generally good agreement, but some strong peaks in the data are not fitted by the model (particularly at

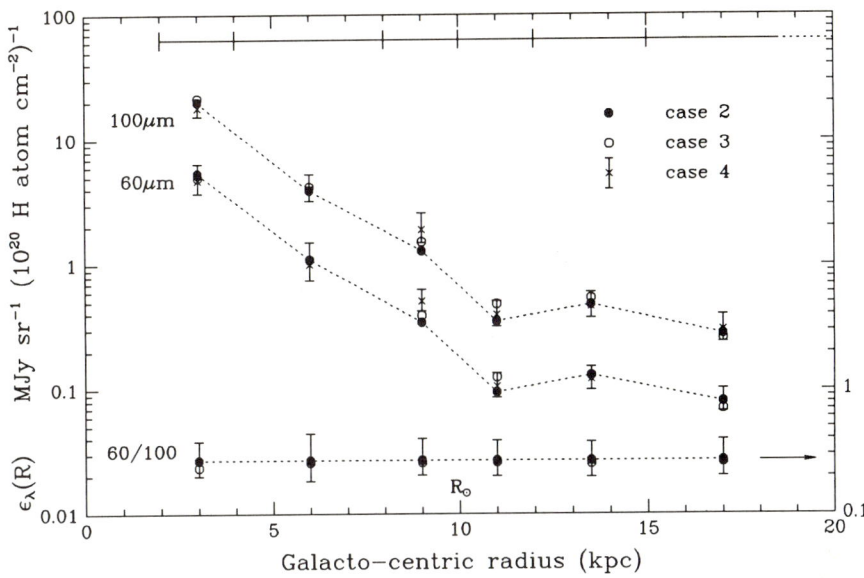

Figure 1: Galacto-centric distributions of the infrared emissivity of dust associated with atomic gas. Case 2 corresponds to the preferred model discussed in the text. Within uncertainties, the radial dependence of the emissivity distribution for dust mixed with molecular gas is identical, but the 60μm/100μm ratio is smaller (see text).

60 μm). These are probably mainly the HII regions that were not excluded by the criterion. The longitude profiles suggest that these remaining HII regions did not have a strong influence on the fitting process because most peaks appear to be excesses on top of the model predictions. This is gratifying and in fact not surprising because the peaks contribute to only a minor fraction of the total number of pixels fitted. The fits were repeated with none of the HII regions excluded and it was found that the parameter values are indeed not significantly different. This gives further assurance that the results do not depend sensitively on the criterion for exclusion of HII regions.

The strong radial gradient of the infrared emissivity distribution in the inner Galaxy, together with the constancy of the 60 μm/100 μm emissivity ratio suggests, at first glance, a strong radial gradient in the dus-to-gas ratio. Between $R_\odot$ and $R = 5$ kpc, the emissivity increases by a factor of ~ 8. The energy density of the Galactic interstellar radiation field (ISRF), however, is generally estimated to increase by roughly the same factor (Mathis et al., 1983). The radial gradient in the ISRF should therefore be the preferred interpretation of the radial gradient in the infrared emissivity distribution, but the constancy of the 60 μm/100 μm emissivity ratio is difficult to understand if the model of standard grains (Mathis et al., 1977) is assumed to be applicable. This finding can be understood, however, if the emission in the 60 μm band has a considerable contribution from small particles (or large molecules) that are not in thermal equilibrium (Léger and Puget, 1984, Draine and Anderson, 1985, Puget et al., 1985).

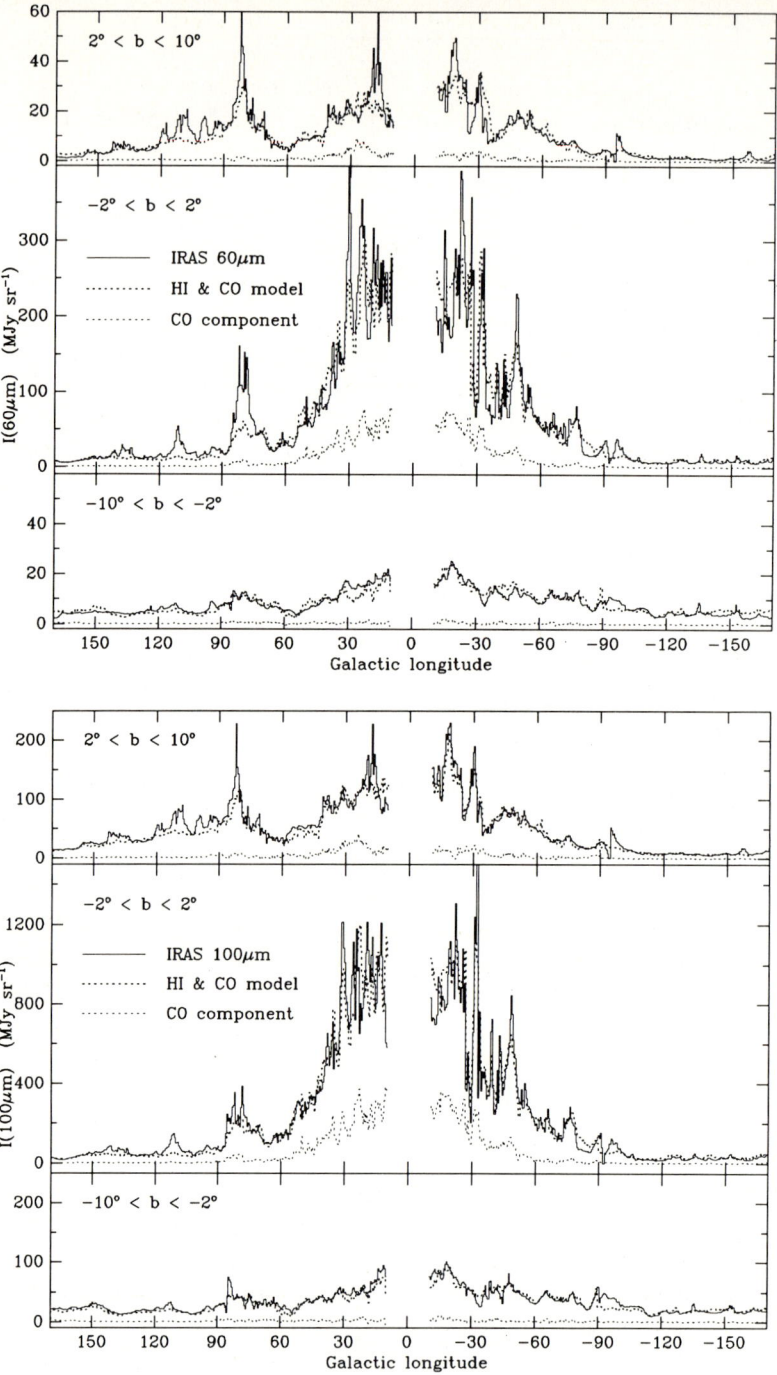

Figure 2: Longitude distributions of the observed and expected infrared intensities in the 60 μm and 100 μm bands. The regions excluded from the analysis because of the presence of intense HII regions, poor kinematic-distance information, and striping artifacts are excluded from these profiles as well. The peaks in the observed profile, not fitted by the model, result from HII regions not rejected by the exclusion criterion discussed in Section 3.1. From Bloemen *et al.*, 1990.

3.3 Constraints on N_{H_2}/W_{CO}

In order to derive X from the Y_λ values one needs to know the ratio between the infrared emissivities per nucleon inside and outside molecular clouds, that is, $\varepsilon_{\lambda,H_2}/\varepsilon_{\lambda,HI}$ (see equation 4). This ratio can be expected to vary among clouds of different categories (discussed below); we should focus on giant molecular clouds, which dominate the CO emission from the Galactic disk. In principle, both Y_{60} and Y_{100} can be used to derive X. They could be combined in an elegant way if the 60 μm/100 μm colour were a good indicator of the dust temperature, but owing to the probably significant contribution from small grains to the emission in the 60 μm band this approach is not feasible. Further, this procedure would be sensitive to the contamination from HII regions not rejected by the exclusion criterion. Because both the presence of small particles and the remaining HII regions mainly affect the parameter estimates for the 60 μm band, the most reliable constraint on X can be obtained from the 100 μm band alone, on which we concentrate in the following discussion.

The close resemblance between the observed 100 μm emission and the model (Figure 2), with a simple axisymmetric dust emissivity distribution, suggests that mainly the general ISRF, rather than the radiation from embedded and nearby stars, is responsible for heating the dust in molecular clouds. Indeed, as noted above, studies of nearby clouds associated with HII regions (Leisawitz and Hauser, 1988) show that only during the initial phase of its lifetime, an OB association heats its parent molecular cloud; stars in older sub-associations, away from the cloud, can be considered contributors to the general ISRF. The regions most affected by high-mass star formation are apparently mostly rejected by the exclusion criterion applied. Furthermore, approximately half of the total Galactic CO emission appears to originate from cool molecular clouds, not associated with (giant) HII regions (*e.g.*, Stark, 1979, Solomon *et al.*, 1985) and can therefore be expected to be heated predominantly by the general ISRF.

Let us proceed on the assumptions that the external ISRF is the dominant dust heating source for most of the molecular gas and that the dust properties and gas-to-dust ratio inside molecular clouds do not differ significantly from those in the ambient atomic medium. The average FIR emissivity from a cloud is then largely determined by its optical depth and geometry, since these determine the attenuation of the radiation field and therefore the average heating rate inside a cloud. For the simple case of a uniform, spherical cloud heated from outside by a uniform radiation field, Boulanger and Pérault (1988) derived explicit values of the ratio between the 100 μm emissivities inside and outside the cloud (that is, $\varepsilon_{100,H_2}/\varepsilon_{100,HI}$), from studies of radiative transfer and dust heating by Flannery *et al.* (1980) and Mathis *et al.* (1983). They found that for total visual extinctions, A_v, of 2, 5, and 10 mag through the centre of the cloud, $\varepsilon_{100,H_2}/\varepsilon_{100,HI} = 0.54$, 0.37, and 0.25, respectively. Although the actual geometry of a molecular cloud is undoubtedly important, their modelling indicates that the radiation field inside clouds is attenuated on average by a factor of at least 2. This attenuation implies, according to equation 4, that Y_{100} is a strict lower limit on the average

Table 1 *Values of Y_{100}, either given in the source or derived from quantities given there. These can be regarded as estimates of the N_{H_2}/W_{CO} ratio if the 100 μm emissivity (per nucleon) is the same inside and outside the molecular cloud(s).*

Source	Y_{100}[1]	Clouds
Weiland et al. (1986)	0.6[2]	HLC 16, 20, 30 (Magnani et al., 1985)
de Vries et al. (1987)	0.5	Ursa Major (Lynds 683)
de Geus (1988)	0.6	Ophiuchus
Heithausen & Mebold (1989)	0.6	Ursa Major (Lynds 629, 683, 691)
Boulanger et al. (1989)[3]	0.4-0.6	Auriga, Cepheus, Chamaeleon,
	0.9	Lupus, Mon R2, Taurus, Orion
Pérault et al. (1990)	0.8	Galactic disk
Sodroski et al. (1989)	0.7	"
Bloemen et al. (1990)	0.8	"

[1] Units: 10^{20} molecules cm^{-2} K^{-1} km^{-1} s.
[2] From I_{100}/W_{CO}, assuming $\varepsilon_{100,HI} = 1$ MJy sr^{-1} $(10^{20}$ H atom cm$^{-2})^{-1}$.
[3] The authors excluded regions with clear evidence for star formation, mainly important for Orion.

Galactic X value (i.e., $X > 0.8 \times 10^{20}$ molecules cm^{-2} K^{-1} km^{-1} s) and, more likely, $X \gtrsim 2Y_{100}$, i.e. $X \gtrsim 1.6 \times 10^{20}$.

Estimates of Y_{100} follow from two other IRAS studies of the Galactic disk (Sodroski et al., 1989, Pérault et al., 1990): $Y_{100} \simeq 0.7 \times 10^{20}$ and $Y_{100} \simeq 0.8 \times 10^{20}$ molecules cm^{-2} K^{-1} km^{-1} s, respectively. Both groups do not explicitly determine Y_{100} (they use the γ-ray estimate of X), but the values follow directly from further quantities given. The consistency of the three independent estimates is of particular interest because widely different methods of analysis were used, especially in the treatment of HII regions, illustrating, once more, that the specific way of dealing with this phase of the ISM has no important impact on parameter estimates. Another approach was followed by Broadbent et al. (1989). They performed a spectral decomposition, including the 150 μm and 250 μm data from Hauser et al. (1984), and derived an X value of 1.0×10^{20} for the inner Galaxy ($X = 1.2 \times 10^{20}$ if the Columbia data is used instead of the CO data from Sanders et al., 1984 — see Section 2.2). The importance of using these additional observations was emphasized by MacLaren and Wolfendale (1990). However, the small sky coverage of the 150 μm and 250 μm data used ($\ell = 15° - 52°$ and $|b| \lesssim 2°$) does not allow a radial decomposition, which, in view of the strong radial emissivity gradient (Figure 1), is a clear limitation in the approach of Broadbent et al. (1989).

Finally, let me note that correlation studies of CO, HI, and 100 μm observations of several local molecular clouds have lead to very similar Y_{100} values, which is quite remarkable (a compilation is given in Table 1). Most authors do not give explicitly the Y_{100} value, but it is readily derived from other quantities given. The clouds span a considerable range in optical depths, so, as discussed above, $\varepsilon_{100,H_2}/\varepsilon_{100,HI}$, and thus Y_{100}, is expected to vary if X is the same for all clouds. Note that the trend, if any, visible in Table 1 (larger Y_{100} values for giant molecular clouds) is quite opposite to this expectation. The small spread in Y_{100} values suggests that different X values apply to clouds of differing extinctions: X appears to be anticorrelated with $\varepsilon_{100,H_2}/\varepsilon_{100,HI}$. The underlying reason may be the inverse proportionality between X and the gas temperature (Kutner and Leung, 1985, van Dishoeck and Black, 1987, Maloney and Black, 1988).

Acknowledgements

I very much appreciate the invitation by the organizers to write this contribution, although I was not able to attend the meeting.

References

Bhat, C.L., Mayer, C.J. and Wolfendale, A.W., 1986, Philos. Trans. Roy. Soc. Lond. A, **319**, 249.
Black, J.H. and van Dishoeck, E.F., 1987, Astrophys. J., **322**, 412.
Blitz, L., Bloemen, J.B.G.M., Hermsen, W. and Bania, T.M., 1985, Astron. Astrophys., **143**, 267.
Bloemen, J.B.G.M., 1989, Ann. Rev. Astr. Astrophys., **27**, 469.
Bloemen, J.B.G.M., Caraveo, P.A., Hermsen, W., Lebrun, F., Maddalena, R.J., Strong, A.W. and Thaddeus, P., 1984, Astr. Astrophys., **139**, 37.
Bloemen, J.B.G.M., Deul, E.R. and Thaddeus, P., 1990, Astr. Astrophys, **233**, 437.
Bloemen, J.B.G.M., Strong, A.W., Blitz, L., Cohen, R.S., Dame, T., Grabelsky, D.A., Hermsen, W., Lebrun, F., Mayer-Hasselwander, H.A. and Thaddeus, P., 1986, Astr. Astrophys., **154**, 25.
Broadbent, A., MacLaren, I. and Wolfendale, A.W., 1989, Mon. Not. Roy. Astron. Soc., **237**, 1075.
Bronfman, L., Cohen, R.S., Alvarez, H., May, J. and Thaddeus, P., 1988, Astrophys. J., **324**, 248.
Boulanger, F., 1989, in *Physics and Chemistry of Interstellar Molecular Clouds*, ed. G. Winnewisser and J.T. Armstrong, (Springer, Berlin, Heidelberg, New York), p. 30.
Boulanger, F. and Pérault, M., 1988, Astrophys. J., **330**, 964.
Cesarsky, C.J. and Völk, H.J., 1978, Astron. Astrophys., **70**, 367.
Cox, P., Krügel, E. and Mezger, P.G., 1986, Astron. Astrophys., **380**, 396.
Cox, P. and Mezger, P.G., 1989, Astron. Astrophys. Review, **1**, 49.
Dame, T.M., Ungerechts, H., Cohen, R.S., de Geus, E., Grenier, I.A., May, J., Murphy, D.C., Nyman, L.-A. and Thaddeus, P., 1987, Astrophys. J., **322**, 706.
Dickman, R. L., 1978, Astrophys. J. Suppl, **37**, 407.
van Dishoeck, E.F. and Black, J.H., 1987, in *Physical Processes in Interstellar Clouds*, ed. G.E. Morfill and M. Scholer, (Reidel, Dordrecht), p. 241.
Draine, B.T. and Anderson, N., 1985, Astrophys. J., **292**, 494.
Flannery, B.P., Roberge, W. and Rybicki, G.B., 1980, Astrophys. J., **236**, 598.
Frerking, M.A., Langer, W.D. and Wilson, R.W., 1982, Astrophys. J., **262**, 590.
de Geus, E.J., 1988, PhD Thesis, University of Leiden.
Grenier, I.A. and Lebrun, F. 1990, Astrophys. J., **360**, 129.
Hauser, M.G., Silverberg, R.F., Stier, M.T., Kelsall, T., Gezari, D.Y, Dwek, E., Walser, D., Mather, J. and Cheung, L.H., 1984, Astrophys. J., **285**, 74.
Heithausen, A. and Mebold, U., 1989, Astron. Astrophys., **214**, 347.
Issa, M.R., Riley, P.A., Strong, A.W. and Wolfendale, A.W., 1981, J. Phys. G, **7**, 973.
Kutner, M.L. and Leung, C.M., 1985, Astrophys. J., **291**, 188.

Lebrun, F., Bennett, K., Bignami, G.F., Bloemen, J.B.G.M., Buccheri, R., Caraveo, P.A., Gottwald, M., Hermsen, W., Kanbach, G., Mayer-Hasselwander, H.A., Montmerle, T., Paul, J.A., Sacco, B., Strong, A.W., Wills, R.D., Dame, T.M., Cohen, R.S. and Thaddeus, P., 1983, Astrophys. J., **274**, 231.
Léger, A. and Puget, J.L., 1984, Astron. Astrophys., **137**, L5.
Leisawitz, D. and Hauser, M., 1988, Astrophys. J., **332**, 954.
MacLaren, I. and Wolfendale, A.W., 1990, J. Phys. G., (in press).
Magnani, L., Blitz, L. and Mundy, L., 1985, Astrophys. J., **295**, 402.
Magnani, L., Blitz, L. and Wouterloot, J.G.A., 1988, Astrophys. J., **326**, 909.
Maloney, P. and Black, J.H., 1988, Astrophys. J., **325**, 389.
Mathis, J.S., Mezger, P.G. and Panagia, N., 1983, Astron. Astrophys., **128**, 212.
Mathis, J.S., Rumpl, W. and Nordsieck, K.H., 1977, Astrophys. J., **217**, 425.
Mezger, P.G., Mathis, J.S. and Panagia, N., 1982, Astron. Astrophys., **105**, 372.
Pérault, M., Boulanger, F., Puget, J.L. and Falgarone, E., 1990, Astrophys. J., (submitted).
Puget, J.L., Léger, A. and Boulanger, F., 1985, Astron. Astrophys., **142**, L19.
Sanders, D.B., Solomon, P.M. and Scoville, N.Z., 1984, Astrophys. J., **276**, 182.
Savage, B.D., Bohlin, R.C., Drake, J.F. and Budich, W., 1977, Astrophys. J., **216**, 291.
Shull, J.M. and Beckwith, S., 1982, Ann. Rev. Astr. Astrophys., **20**, 163.
Skilling, J. and Strong, A.W., 1976, Astron. Astrophys., **53**, 253.
Sodroski, T.J., Dwek, E., Hauser, M.G. and Kerr, F.J., 1989, Astrophys. J., **336**, 762.
Solomon, P.M., Sanders, D.B. and Rivolo, R., 1985, Astrophys. J., **292**, L19.
Spitzer, L. and Jenkins, E.B., 1975, Ann. Rev. Astr. Astrophys., **13**, 133.
Stark, A.A., 1979, Ph. D. Thesis, University of Princeton.
Strong, A.W., Bloemen, J.B.G.M., Dame, T.M., Grenier, I.A., Hermsen, W., Lebrun, F., Nyman, L.-Å., Pollock, A.M.T. and Thaddeus, P., 1988, Astron. Astrophys., **207**, 1.
Völk, H.J., 1983, Space Sci. Rev., **36**, 3.
de Vries, H.W., Heithausen, A. and Thaddeus, P., 1987, Astrophys. J., **319**, 723.
Walterbos, R.A.M. and Schwering, P.B.W., 1987, Astron. Astrophys., **180**, 27.
Weiland, J.L., Blitz, L., Dwek, E., Hauser, M.G., Magnani, L. and Rickard, L., 1986, Astrophys. J., **306**, L101.
Wolfendale, A.W., 1988, in *Molecular Clouds in the Milky Way and External Galaxies*, ed. R. Dickman, R. Snell and J. Young, (Springer-Verlag, Heidelberg), p. 76.

The mass of molecular gas in the Galaxy

A.W. WOLFENDALE

PHYSICS DEPARTMENT, UNIVERSITY OF DURHAM, DURHAM DH1 3LE, U.K.

Abstract Attention is given to the question of the mass of molecular gas in the Galaxy, particularly within the solar circle. It is concluded that our previous estimate for the ratio of the masses of molecular and atomic hydrogen, 60%, is still valid. Arguments are put forward justifying this value in the face of other, higher, estimates.

1 Introduction

Determinations of the mass of molecular gas in the Galaxy, and its spatial variation, are important for a number of reasons. Firstly, the efficiency for star formation and the related chemical evolution of the ISM as a function of position in the Galaxy depends on this quantity. Secondly, the gas is intimately connected with dust in the ISM and thus with the whole area of interstellar chemistry. The bulk of the molecular gas is in the form of H_2 in the so-called Giant Molecular Clouds and these clouds have considerable dynamical interest; their collisions with stars are responsible, in part at least, for the dependence of stellar velocities on time from formation, they may well be the seat of acceleration of low energy cosmic rays and their collisions with the Oort cloud of comets has undoubted importance for the number of comets seen at earth (and impinging on it).

Equally important is the question of the mass of gas in other galaxies. The relevance of the mass of gas in our Galaxy is that it is usual to adopt the same conversion factor to go from the measured quantity, commonly $\int T(^{12}CO)dv$ for the $J = 1 \rightarrow 0$, ^{12}CO line, to the column density of H_2. Thus, a mistake in the Galactic value yields a mistake in them all. In fact, a determination of the correct conversion value for the Galaxy does not guarantee the correct value for other galaxies insofar as it almost certainly varies from galaxy to galaxy and from place to place within a galaxy. Thus, it is imperative to endeavour to determine the dependence of the conversion value on the parameters of the ISM at the place where the H_2-density is sought.

The history of the subject is interesting. Until about 15 years ago the presence of large quantities of H_2 in the Galaxy was not realised; it was presumed that the radiation fields were so strong that most molecules would be broken up. Molecular hydrogen, unlike atomic hydrogen, has no easily observed transitions and thus its presence is inferred from lines emitted by other molecules excited, in turn, by collisions with H_2. The discovery came from the observation of the line of ^{12}CO already mentioned. It is usual to write the conversion factor as $X = N(H_2)/W_{CO}$ in units of 10^{20} mols cm^{-2} K^{-1} km^{-1} s where $W_{CO} = \int T(^{12}CO)dv$. The early X-values were dramatically high (see Bronfman et al., 1988, for a summary) but more recently they have fallen. Nevertheless, in the author's view the majority are still too high.

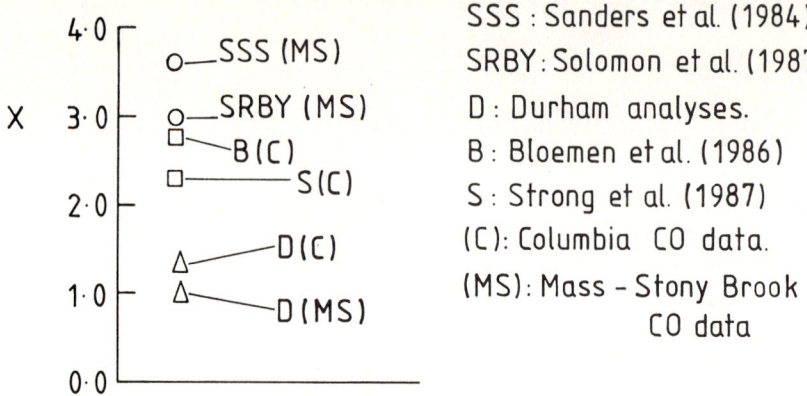

Figure 1: A summary of recent X-values for the Inner Galaxy. The Durham analyses comprise Bhat *et al.* (1985, 1986), Broadbent *et al.* (1988, 1989), MacLaren *et al.* (1989), MacLaren and Wolfendale (1990).

Figure 1 gives a summary of recent values; the purpose of the present paper is to convince the reader that the lowest value shown is nearest the truth.

2 Techniques

Many methods have been used to estimate X; each has its own advantages and its own shortcomings.

(i) Visual extinction. This method is applicable to local clouds where the extinction due to the gas-associated dust is related to the total column density of gas (HI and H_2) by the relation $N(H) = 1.9 \times 10^{20} A_v$ atoms cm^{-2} determined by Bohlin *et al.* (1978). The basic problem here is the extent to which the high A_v-regions of local clouds are representative of similar A_v- regions in the Inner Galaxy where most of the H_2 resides. We consider that they are not, because of the likely gradient of the dust to gas ratio with Galactocentric distance, R (dust/gas ratio rises as R falls; Issa *et al.*, 1990) and because the high A_v values for clouds at smaller R represent large extended regions of lower volume density than is found in the small, high density cores of the smaller local clouds. Simple application of the local A_v/N_H relation to Inner Galaxy clouds leads to an overestimate of X.

(ii) The virial theorem. Application of the virial theorem to GMC offers a deceptively simple way of determining their masses – and thereby their X-values. The mass of a virialised cloud of radius R is related to its velocity width σ_v by $M = kR\sigma_v^2$. Unfortunately, there are problems associated with each and every one of the parameters in the expression. k is a function of the density distribution in the cloud and this is not, in general, known, furthermore the usual assumption of radial symmetry: $\rho(r) \alpha \ r^{-n}$, is essentially untrue. R suffers from

a similar problem – where must the cloud boundary be placed and how must a correction be applied for material outside this boundary? The whole question of definition of discrete clouds in the presence of confusing other, unrelated, material along the line of sight is a most difficult one. Finally, σ_v is often taken to be the root mean square velocity dispersion for the easily measured ^{12}CO line. In fact, clouds are optically thick in ^{12}CO (although some alleviation comes from the presence of cloud substructure) and the results are very different when optically thin lines are used. There is also the problem of confusion here: material wrongly included in "the cloud" will often enhance σ_v considerably. Finally there is the whole question of the extent to which the GMC are virialised. It is inevitable that some clouds are just forming (by accretion of cloudlets?) and others are dispersing (due to tidal shears, SNR pressure, etc.). Neglect of these considerations leads to an overestimate of X.

(iii) X-ray absorption. X-ray sources near the Galactic Centre are seen through absorbing material along the line of sight. The presence of absorption, largely photo-electric effect by carbon and oxygen atoms, causes a downturn in the measured energy spectrum as one proceeds below about 2 keV. With the usual assumption that the energy spectrum of the photons on emission is a power law with constant exponent the column density of gas (weighted in a known fashion towards C and O) follows from the measured spectral shape. Problems with this method include uncertainty in the intrinsic spectrum (absorption close to the star itself causes a spectral downturn) and most particularly a lack of precise knowledge of the whereabouts of many of the X-ray emitters with respect to the absorbing gas. A further difficulty is that conversion from measured absorption to total column density of molecular gas involves a knowledge of the metallicity as a function of position along the line of sight. The effect of neglect of these factors appears to lead to a random error in X, some factors leading one way and some the other.

(iv) Infra-red absorption. A rather similar method involves a determination of the extinction in the infra-red to stars in the Galactic Centre region. The extinction is converted to A_v and, in turn, to N(H). The problems are similar to those under (iii): uncertainty in the location of the stars in question and lack of precise knowledge of the dust to gas ratio and its dependence on position. The effect of lack of consideration of these effects is probably rather similar to the situation in the previous section, although there may be bias effects towards missing Galactic Centre stars in the dense central regions which lead to significant under-estimates of X.

(v) Gamma rays. It is true to say that it was the introduction of this technique that led to a reappraisal of the early, dramatically high, gas mass estimates. The principle is simple: the gamma rays come from cosmic rays interacting with all the gas in the "target" cloud - atomic and molecular, alike. However, this method has its drawbacks, too. Thus, the cosmic ray intensity is not, in fact, known with any precision anywhere in the Galaxy other than just above our own atmosphere. Indeed, the author's entry into this field was due to a

wish to use measured gamma ray fluxes and "known" gas masses to estimate the cosmic ray intensity in remote regions. It is likely that the cosmic rays in question can penetrate the molecular gas clouds but insofar as GMC are the seat of considerable astrophysical activity: SNR, HII regions, turbulence ..., it would not be surprising if the cosmic ray intensity were higher in some clouds (the "active" clouds) and lower in others (the "inert" clouds). We do not yet know how to parameterise such activity. Another problem concerns the gamma ray fluxes. The contemporary spatial resolution of gamma ray detectors is insufficient to resolve most of the discrete gamma ray sources in the Inner Galaxy which account for an uncertain fraction of the flux. Neglect of the discrete source contribution leads to an overestimate of X as does the (common) assumption that all clouds are inert.

3 A summary of the determinations of X

3.1 General summary

Inspection of Figure 1, bearing in mind the strictures of Section 2, leads immediately to the case for preferring our low value. Table 1 endeavours to summarise our reasons for recommending reductions in the derived X- values.

A further complication concerns the differences between the W_{CO} values reported by different workers. Thus, to define "X" alone, without reference to the CO data to which it should be applied, leads to errors. There have been two major surveys of ^{12}CO, by the Massachusetts-Stony Brook group (Sanders et al., 1984, and universally available tapes) and by the Columbia group (Dame et al., 1986, Bronfman et al., 1988). The former survey was more restricted in ℓ and b than the latter but taken with a much bigger telescope. Both surveys have covered the Inner Galaxy, low latitude region, where most of the Galactic molecular gas resides but the local GMC have been the province of the Columbia group alone. In the common region of the Galaxy the differences in W_{CO} amount to some 20 – 30%, with the Columbia values being lower.

In what follows we denote estimates of X with respect to the Massachusetts-Stony Brook data by X_{MS} and, correspondingly X_C when the Columbia data are adopted. The final analysis, in which the total mass of molecular gas is estimated, is not sensitive to whichever X is used provided that consistency is observed.

3.2 The Durham Analyses

Our early work (Bhat et al., 1985, 1986) used all the techniques described in Section 2 and found, for the Inner Galaxy, $X_{MS} \simeq 1.0$. Later work has examined some of the techniques in greater depth. Concerning γ-rays, Richardson and Wolfendale (1988) studied local and Inner Galaxy GMC in some detail and

Table 1 *Summary of recent X-values for the Inner Galaxy, as quoted by the authors (and the technique) the correction recommended by the author, f, (and the reason) and the resultant value of X^1. (MS denotes Massachusetts-Stony Brook and C denotes Columbia).*

Author (technique)	X	f (reason)	X^1
Sanders et al. (1984) (A_v)	3.6	$\simeq 0.4$ (dust/gas and metallicity)	1.4 (MS)
Scoville et al. (1987) (virial theorem)	3.0	$\simeq 0.37$ (use σ_v from ^{13}CO)	1.1 (MS)
Strong et al. (1988) (cosmic γ-rays)	2.3	$\simeq 0.67$ (discrete γ- sources)	1.5 (C) 1.2 (MS)
Bloemen et al. (1989) (Far I.R.)	> 1.6	0.75 (smaller correction for absorption)	1.2 (C) 1.0 (MS)

Figure 2 gives a summary. It will be noted that X is apparently falling with increasing γ-ray energy! We attribute this feature, tentatively, to the generation of γ-rays in GMC by either embedded γ-ray sources or by extra fluxes of low energy cosmic rays generated within the clouds. It is the neglect of these sources that we have endeavoured to correct in the COS B analysis of Strong et al. (1987) (see Figure 1, Table 1, and Figure 2).

Turning to the FIR technique, our work has centred on using the IRAS surveys at 60 μm and 100 μm and those (albeit of low statistical precision) at 150 μm and 250 μm by Hauser et al. (1984). The analysis involves the assumption of an increase in dust to gas ratio as R diminishes, following the prescription of Cox et al. (1986). The results are shown in Figure 3, denoted D1 using both the MS CO data (denoted "Stony-Brook") and the Columbia CO data. Also shown are the results of Bloemen et al. (1990) who derived a lower limit of 0.8 and then corrected it upwards (see Table 1). Our view, described in some detail in MacLaren and Wolfendale (1990), is that the amount of cold dust (virtually unobservable at 100 μm) is smaller than commonly thought and that the correction factor to allow for it, or rather the H_2 associated with this cold dust, is smaller than adopted by Bloemen et al. (1990).

Finally, we consider application of the Virial Theorem. Our result (MacLaren

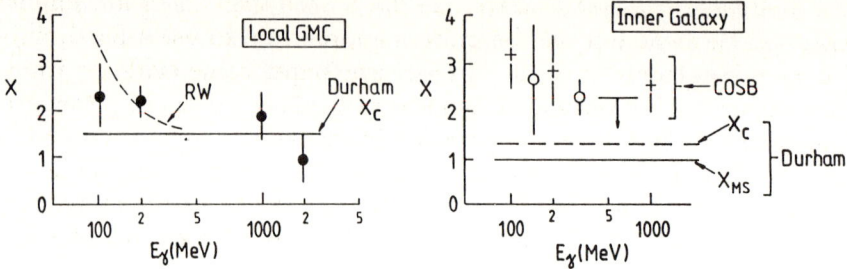

Figure 2: Summary of γ-ray determinations of X, after Richardson and Wolfendale (1988). RW : Riley and Wolfendale (1984). "COS B" : Strong et al. (1987) - these values relate to the Columbia CO- survey. The "Durham" value of $X_{MS} \sim 1.0$ is increased to 1.3 for comparison with the COS B values. The COS B values should be reduced to allow for γ-ray sources in the Inner Galaxy. Our own estimate of the reduction factor is given in Table 1.

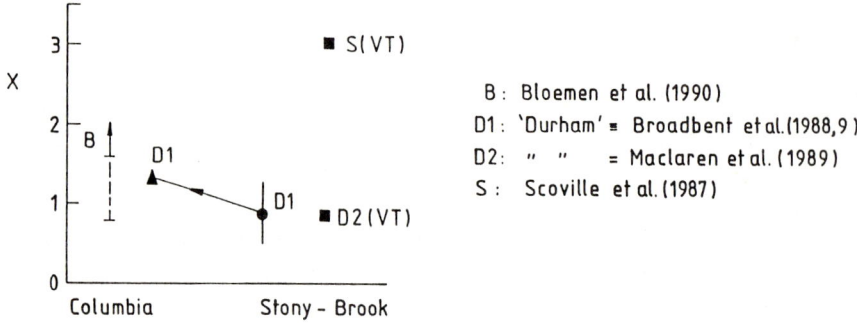

Figure 3: X-values from use of the FIR technique (B and D1) and by way of application of The Virial Theorem (S and D_2). The values relate to the Inner Galaxy.

et al., 1989) taken using a reasonable gas density distribution, to give "k" in $M = kR\sigma_v^2$, and the optically thin ^{13}CO velocity dispersion for σ_v gives the result denoted D2(VT) in Figure 3. Also included is an allowance for helium. The X-value relates to GMC masses of several times $10^5 M_\odot$. Also shown in Figure 3 is the estimate from Solomon et al. (1987) and Scoville et al. (1987) denoted S(VT), which has the shortcomings listed in Table 1.

4 The Mass of Molecular Gas in the Galaxy and General Conclusions

Although we would not claim that our value of $X \simeq 1$ derived for the Inner Galaxy is "proved", it does appear to allow a consistent analysis of a variety of astronomical data. So far, at least, there appear to be no features which do not fit.

If the apparent inverse correlation of X with metallicity which we have claimed elsewhere (Bhat *et al.*, 1986, Mayer, 1985) is true then there are implications for other galaxies where it will be quite inappropriate to use a fixed, universal, X-value as is commonly the case. In a recent paper some evidence from other galaxies supporting the variable X hypothesis has been advanced (MacLaren and Wolfendale 1990). Further work is needed in this area.

Another consequence of the metallicity idea (which may be accidental and due to the correct dependence being on temperature, which is commonly correlated with metallicity) is that the GMC in the Outer Galaxy are more massive than currently believed.

Finally, concerning the mass of molecular hydrogen in the Galaxy, application of the X-value as a function of R ($X \sim 1$ for $R \simeq 6$ kpc; $X \simeq 1.5$ locally) and adopting 10^9 $M_\odot$ for the mass of HI within the solar circle yields:

$$M(H_2)/M(HI) \simeq 60\%.$$

This ratio is to be compared with the more usually claimed values in the range 100 – 300%.

The consequences of the low value are manifold and relate to the topics referred to in Section 1, viz, star formation is more efficient than commonly believed, the role of other mechanisms in yielding the observed stellar velocity-age relation must be important, cosmic ray production in GMC must be relatively commonplace and cometary impacts are less frequent than often claimed.

Finally, considerable care must be exercised in drawing conclusions about astrophysical processes in other galaxies involving standard-values for conversions from observed CO line strengths to inferred molecular hydrogen masses.

References

Bhat, C.L., Issa, M.R., Houston, B.P., Mayer, C.J. and Wolfendale, A.W., **1985**, Nature, **314**, 511.
Bhat, C.L., Mayer, C.J. and Wolfendale, A.W., 1986, Phil. Trans. R. Soc. **Lond.**, **319**, 249.
Bloemen, J.B.G.M., Strong, A.W., Blitz, L., Cohen, R.S., Dame, T.M., Grabelsky, D.A., Hermsen, W., Lebrun, F., Mayer-Hasselwander, H.A. and Thaddeus, P., 1986, Astron. Astrophys., **154**, 25.
Bloemen, J.B.G.M., Deul, E.R., and Thaddeus, P., 1990, Astron. Astrophys., **233**, 437.
Bohlin, R.C., Savage, B.D. and Drake, J.F., 1978, Astrophys. J., **224**, 132.
Broadbent, A., MacLaren, I. and Wolfendale, A.W., 1988, *Dust in the Universe*, ed. M.E. Bailey, and D.A. Williams, Camb. Univ. Press, p. 435.
Broadbent, A., MacLaren, I. and Wolfendale, A.W., 1989, Mon. Not. Roy. Astron. Soc., **237**, 1075.
Bronfman, L., Cohen, R.S., Alvarez, H., May, J. and Thaddeus, P., **1988**, Astrophys. J., **324**, 248.
Cox, P., Krugel, E. and Mezger, P.G., 1986, Astron. Astrophys., **155**, 380.
Dame, T.M., Elmegreen, B.G., Cohen, R.S. and Thaddeus, P., **1986**, Astrophys. J., **305**, 892.
Hauser, M.G., Silverberg, R.F., Stier, M.T., Kelsall, T., Gezari, D.Y., Dwek, E., Walser, D., Mather, J.C. and Cheung, L.J., 1984, Astrophys. J., **285**, 74.
Issa, M.R., MacLaren, I. and Wolfendale, A.W. 1990, Astron. Astrophys., **236**, 237.
MacLaren, I. and Wolfendale, A.W., 1990, J. Phys. G. (in press).
MacLaren, I., Richardson, K.M. and Wolfendale, A.W., **1989**, Astrophys. J., **333**, 821.
Mayer, C.J., 1985, *Cosmical Gas Dynamics*, ed. F.D. Kahn, VNU Science Press, 21.

Richardson, K.M. and Wolfendale, A.W., 1988, Astron. Astrophys., **201**, 100.
Riley, P.A. and Wolfendale, A.W., 1984, J. Phys. G., **10**, 1149.
Sanders, D.B., Solomon, P.M. and Scoville, N.Z., 1984, Astrophys. J., **276**, 182.
Scoville, N.Z., Min Su Yun, Clemens, D.P., Sanders, D.B., and Waller, W.H., 1987, Astrophys. J. Suppl., **63**, 821.
Solomon, P.M., Rivolo, A.R., Barrett, J. and Yahil, A., 1987, Astrophys. J., **319**, 730.
Strong, A.W., Bloemen, J.B.G.M., Dame, T.M., Grenier, I.A., Hermsen, W., Lebrun, F., Nyman, L.A., Pollock, A.M.T and Thaddeus, P., 1987, Proc. 20th Int. Conf. on Cosmic Rays, Moscow, 1, 125.

High latitude molecular clouds

LEO BLITZ

ASTRONOMY PROGRAM, UNIVERSITY OF MARYLAND, COLLEGE PARK, MARYLAND 20742, U.S.A.

Abstract The properties of the molecular clouds found at latitudes $>25°$ from the galactic plane are reviewed. Direct distance measurements have now been made to twelve of the clouds confirming that they are objects closely confined to the galactic plane. On the other hand, three clouds have been identified with highly anomalous negative velocities which indicate that the clouds may form a separate class of objects; these are discussed in some detail. It is shown that although the clouds are not generally gravitationally bound, the clumps *within* the clouds appear to be bound by the pressure of the ambient interstellar gas. The clouds themselves appear to have ages of $\sim 2 \times 10^6$ yr. The paper also discusses the relationship of the high latitude molecular clouds to the classical diffuse molecular clouds and discusses the role of the magnetic fields in producing the CO line widths observed in the clouds. A tentative picture of the formation of the clouds is presented in the concluding section.

1 Introduction

Dark nebulae at high galactic latitudes were known in the early part of this century from the catalogues of Lundmark and Melotte (1926), and E. E. Barnard (1927). These objects were, of course, later found to consist primarily of molecular hydrogen (Wilson *et al.*, 1974), and were the first known high-latitude molecular clouds. Subsequently, Lynds (1962) produced the catalogue of dark nebulae that bears her name from the Palomar Observatory Sky Survey (POSS); it is the standard reference for dark interstellar clouds. Her catalogue added to the number of clouds found at high galactic latitudes, all of which were later found to be molecular, but she remarked that due to the paucity of background stars, the catalogue was likely to be incomplete far from the plane. These objects were the focus of numerous studies of their optical and radio properties, and of their interaction with the interstellar radiation field (Matilla, 1970, 1979, de Vries, 1986). In 1984, Blitz, Magnani and Mundy greatly added to the number of molecular clouds at high galactic latitudes by looking for radiation from the CO $J = 1 \to 0$ transition in regions that appeared to exhibit low level absorption on the POSS plates. These clouds have been found to have a number of surprising properties, a number of which are reviewed in this paper.

Some Giant Molecular Clouds (GMCs) have been known for many years to be at large angles from the midplane of the Galaxy, the most prominent of which are the L1641 cloud that is the site of the Orion Nebula, and the ρ Ophiuchus molecular cloud that contains the extensively studied star forming region. Both clouds extend to angles as large as 20° from the plane. At larger angles from the Galactic plane, the molecular clouds appear to be much smaller than the GMCs, and consequently Magnani, Blitz and Mundy (1985, hereafter MBM) sought to find an operational definition that would differentiate the small, relatively transparent clouds that dominated their study from the high latitude extensions of the GMCs in the solar vicinity. They found that by limiting their catalogue to those clouds with $|b| > 25°$, they eliminated nearly all of the molecular gas that could easily be associated with the well known GMCs. Their definition will be adopted here: a high-latitude molecular cloud (HLC) is one for which all of its emission is found at latitudes greater than 25°. This definition makes it possible to study the properties of the ensemble without significant contamination from the gas associated with the dark star-forming clouds. According to this definition, the molecular cloud mapped by Heithausen and Thaddeus (1990) in a region connected to the Cepheus flare (Lebrun et al., 1986) would not be a high latitude cloud because much of its emission is at $b < 25°$, and because it is physically associated with material close to the plane. Even though this cloud extends to a latitude more than 30° from the plane, it belongs more properly to the clouds typified by the Taurus Molecular Clouds.

The work done by many authors in the last five years demonstrates that the HLCs are not members of a homogeneous class of clouds, but are an ensemble that varies considerably in its properties. For example, some of the clouds catalogued by MBM are dark clouds catalogued by Lynds (1962), others have opacities so small that they are not seen on the POSS prints at all. A class of very small molecular clouds was found by Knapp and Bowers (1988) to be associated with Betelgeuse. Although not properly HLCs because they are found relatively close to the galactic plane, such clouds must surely exist at high galactic latitudes, but have so far escaped detection because of their extremely small size. These clouds are only a few arcminutes in diameter. Lada and Blitz (1988) have found the HLCs to be divided into CO rich and CO poor clouds (but see the criticism of this designation by van Dishoeck, 1990a), and there appear to be a few clouds with anomalously high radial velocities that may have been formed in the halo of the Galaxy. Nevertheless, it is instructive to treat all of these clouds as belonging to a single group.

In Section 2, recent results on the mapping of the clouds is reviewed. Section 3 discusses the distances to the clouds, and Section 4 their mean properties. Section 5 discusses the structure and boundedness of the clouds. Section 6 discusses the relationship to atomic hydrogen, and the infrared "cirrus". Section 7 discusses a small class of HLCs that might originate in the galactic halo. Section 8 discusses the relationship of the HLCs to the diffuse molecular clouds and Section 9 the role of magnetic fields. Section 10 presents a somewhat speculative picture of the formation of HLCs which attempts to unify the observations presented in the preceeding sections. There are a number of subjects relating to the HLCs that are not covered in this review. Among these are the chemistry of the clouds; this has been reviewed by van Dishoeck and Black (1988), and by van Dishoeck (1990a, b). There will only be passing reference to the infrared and detailed dust properties of the clouds, and no discussion of the phenomenon of broad line wings without internal energy sources.

2 Recent results on the mapping of HLCs

The first maps of the HLCs by MBM were limited to objects visible from the northern hemisphere. Shortly after these maps were published, Keto and Myers (1986) extended the survey of HLCs to the southern hemisphere and there is now a reasonably good all sky survey of these objects. In both studies the clouds are identified from perusal of the POSS prints, and both surveys are therefore insensitive to clouds that are smaller than about 15' because it is difficult to pick out smaller regions of obscuration visually. In order to determine the overall properties of the ensemble, MBM were forced to undersample the clouds by a large fraction; the clouds were mapped with a 2′.3 beam, but were sampled only every 10' or 20'. Because the MBM survey was carried out with an equatorially mounted telescope, a small region near the North Celestial Pole went unobserved. Subsequently, de Vries, Heithausen and Thaddeus (1987) found an HLC in that unobserved hole. Both the Keto and Myers and the de Vries, Heithausen and Thaddeus studies were carried out with large beam instruments (beamwidth = 8'), and their maps were sampled at full beamwidth spacing. Thus, a fairly large number of maps in CO are available at either coarse sampling or low resolution. The best maps have now been made with the Bell Labs antenna. One large cloud, MBM 12, has been mapped with a resolution of 1′.5 resolution in the generally optically thin species ^{13}CO (Pound, Bania, and Wilson, 1990). Their map is reproduced in Figure 1. Pound has made maps at the same resolution of seven more HLCs in CO which will be submitted for publication shortly. Gir and Blitz have also recently completed the mapping of several HLCs in the CO $J = 2 \rightarrow 1$ transition that are sampled every beamwidth at 2'. Thus, within the next year, there will be many clouds for which there are good maps at relatively high resolution that can be used to investigate the detailed structure of HLCs.

In order to obtain an unbiased sample of high latitude molecular gas, Désert, Bazell and Boulanger (1988) have used the IRAS survey to look for clouds that show infrared excesses in excess of what is expected from the correlations of HI and 100 μm emission observed at high galactic latitude. Although they do not show maps in their paper, they have catalogued about 500 potential molecular clouds. The implications of their study are discussed in more detail below.

3 Distances to the HLCs

The distances to the ensemble of HLCs were initially determined in two ways (MBM). The first was to *assume* that they are Galactic plane objects, and assign them a scale height comparable to the other molecular clouds in the plane (70 pc – see below). By then assuming that they are all at one mean distance from the Sun, the distribution in b gives the mean distance to the ensemble. The second method assumes that the HLCs are in equilibrium with the z-component of the gravitational potential of the disk stars. If the measured radial velocity dispersion of the disk stars is equal to the velocity dispersion in z, then using the

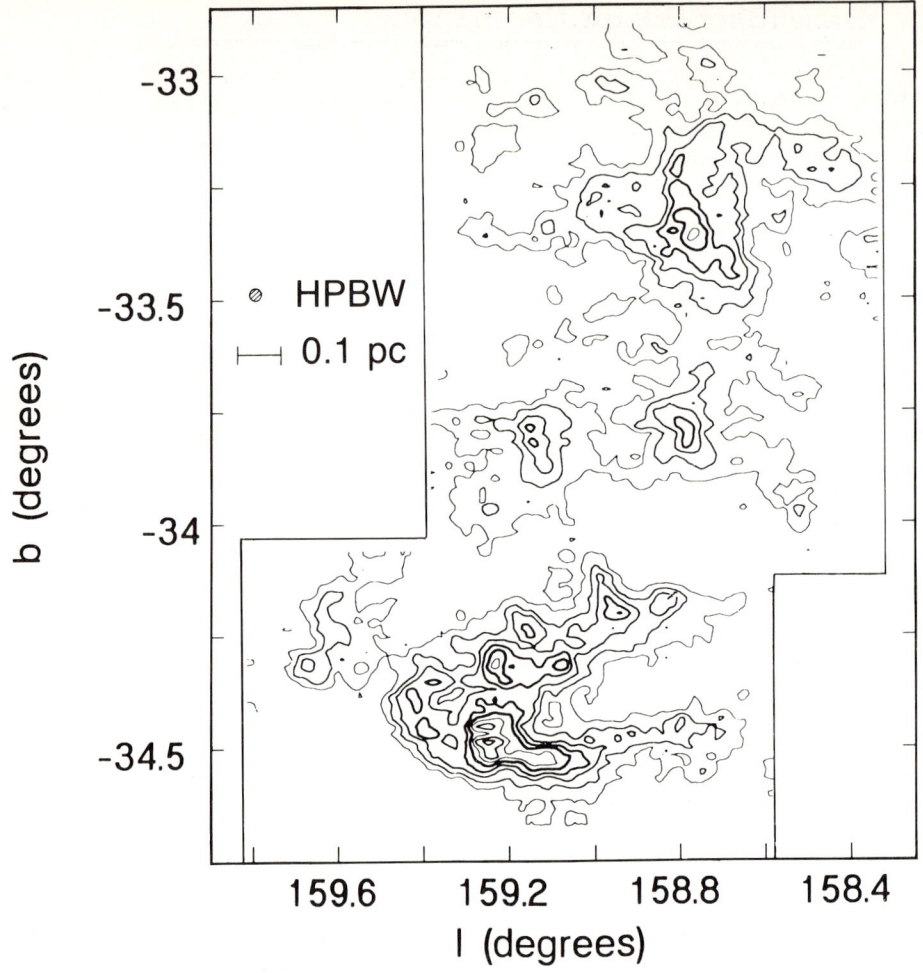

Figure 1: Map of the ^{13}CO emission from MBM 12 by Pound, Bania and Wilson (1990) from data obtained with the 7m Bell Laboratories telescope. Because ^{13}CO is optically thin at nearly all locations in the map, the map is effectively a map of H_2 column density. The data are sampled every beamwidth within the outlined region. At the 65 pc distance of this cloud the resolution is 6500 AU.

Oort limit of the density of matter in the local midplane of the Galaxy (Bahcall, 1984) gives the scale height of the HLCs. The angular distribution then gives the mean distance from the Sun. The second method gives a value similar to the one assumed in the first, which provides confidence that the mean distance to the HLC ensemble has been reasonably well determined in the initial studies.

Subsequently, a number of direct distance determinations were made to the HLCs. The first was a measurement of the distances by the classical method of Wolf diagrams (Magnani and deVries, 1986). These found the distances to three clouds and upper limits to five more. The distances determined to the eight clouds were consistent with the mean distance of 100 pc. Next, in a series of three papers, Hobbs *et al.* (1986, 1988) and Welty *et al.* (1989) determined

Table 1 *Distances to HLCs*

MBM	Distance (pc)	z (pc)	(ref.)
7	125 ± 50	-75	1
12	$60 - 70$	-35	2
16	100 ± 50	-60	1
16	$65 - 95$	~ -50	3
18	≤ 175	≥ -100	1
20	≤ 125	≥ -75	1
26	175 ± 50	95	1
32	≤ 275	≤ 180	1
40	≤ 140	≤ 100	4
41–44	(> 800)	(> 500)	5
53	$110 - 155$	~ -70	4
54	$145 - 260$	~ -120	4
55	≤ 175	≤ -110	1
55	30 to 265	-20 to -175	4
55A	≤ 275	≥ -180	1

References:
1 Magnani and deVries (1986)
2 Hobbs, Blitz, and Magnani (1986)
3 Hobbs *et al.* (1988)
4 Welty *et al.* (1989)
5 Mebold *et al.* (1985)

the distances to 6 HLCs using the presence or absence of strong Na, D and Ca H+K absorption toward stars along the line of sight to the HLCs. The stars all have known spectroscopic parallaxes and the distance determinations identified the nearest known molecular cloud (MBM 12). Table 1 gives the distances to the clouds for which measurements have been obtained to date.

To determine how well confined the clouds are to the galactic plane, it is necessary to determine σ_h, the scale height of the molecular gas not associated with the HLCs. This quantity is determined primarily by three methods:

1. The angular scale height of CO is measured at the tangent points of inner Galaxy surveys. The Galactic rotation curve is then used to convert the measurements to a linear scale height.
2. The velocity dispersion of the CO in the solar vicinity and the midplane density of matter determine the local scale height of CO if the gas is in equilibrium with the gravitational potential of the disk.
3. A direct measurement of the scale height can be made from HII region/CO complexes with known distances.

The first method can be extrapolated to the solar vicinity for comparison with the second. The third method is the only method that can be used in the outer Galaxy. The distances can, in principle, be either spectrophotometric or

Table 2. *Molecular* (CO) *Scale Height at* R_0

Reference	σ_h (pc)
Clemens et al. (1988)	68
Dame et al. (1987)	64
Knapp (1987)	48
Fich and Blitz (1984)	88
Wouterloot et al. (1990)	~85

kinematic, but the latter introduces large uncertainties near the Sun and in regions of known gas streaming such as the Perseus Arm.

Table 2 gives a number of independently determined values of the local CO scale height extrapolated from inner Galaxy CO surveys.

Taking a straight average of the five determinations gives a gaussian scale height of 71 ± 16 pc. The local scale height of the molecular gas layer may then be taken to be about 70 pc with an uncertainty of about 25%. With the possible exception of MBM 41–44, none of the measured distances imply z-heights that are more than $3\sigma_h$ from the plane. The distance to MBM 41–44 is very uncertain and is discussed separately below. Furthermore, the direct distance measurements imply that the scale height of the HLCs is the same as that determined for the overall population of molecular gas; the measured velocity dispersion of the clouds implies that they are in equilibrium with the gravitational potential of the stars in the disk. Therefore, the molecular clouds in the vicinity of the Sun are found to be galactic plane objects.

4 Mean properties

Although the HLCs are a rather heterogeneous population, it is instructive to list some of their mean properties. Table 3 provides such a listing and is taken from the data presented in MBM.

The filling fraction is the surface filling fraction on the sky determined from the blind survey of Magnani, Lada and Blitz (1985). Both $N(H_2)$ and $n(H_2)$ are averages over the cloud within the lowest CO contour in the MBM survey. These quantities assume a CO/H_2 conversion ratio taken from Magnani, Blitz and Wouterloot (1988). Observations of de Vries, Heithausen and Thaddeus (1987) and Heithausen and Thaddeus (1989) imply a lower value for two clouds; if their value is generally applicable, the densities given in the table should be correspondingly lower. Note however that de Vries and van Dishoeck (1988) support a value closer to the ones used to derive the quantities used in the table. In any event, the H_2CO maps of Magnani, Blitz and Wouterloot (1988), Turner et al. (1989) and the ^{13}CO maps of Pound, Bania and Wilson (1990) show that even though the mean densities toward clouds may be quite low, the HLCs contain dense knots where $n(H_2)$ is likely to be at least two orders of magnitude greater

Table 3. *Mean Properties of HLCs*

Distance	100 pc
Diameter	1.7 pc
Mass	40 M$_\odot$
A$_v$	0.6 mag
$N(H_2)$	$\sim 1 \times 10^{21}$ cm^{-2}
$n(H_2)$	140 cm^{-3}
filling fraction	0.5%
$\sigma(H_2)$	0.2 M$_\odot$ pc^{-2}
σ_v(internal)	1.0 km s^{-1}
σ_v(linewidth)	0.3 km s^{-1}
σ_v(cloud-cloud)	5.6 ± 1.2 km s^{-1}
v(lsr)	0.13 ± 0.11 km s^{-1}

than the mean. The quantity $\sigma(H_2)$ is the surface density of H$_2$ projected onto the galactic plane.

The velocity dispersions in Table 3 deserve some explanation. The quantity σ_v(linewidth) is the mean of a large number of individual linewidths in a number of the HLCs. It is a measure of the motions of the molecules within an identifiable clump within an HLC. Many clouds exhibit multiple lines and line centres that vary from position to position within a cloud. The quantity σ_v(internal) is determined by taking the dispersion of the line centres with respect to the mean velocity of a given cloud. It is a measure of the degree to which a cloud is self gravitating. Values for individual clouds range from 0.1 to 3.1 km s^{-1}. The quantity σ_v(cloud-cloud) is determined from the velocity differences between all of the clouds in the MBM sample. It is a measure of the degree to which the clouds respond to the gravitational potential of the galactic disk. The mean velocity of the ensemble is indistinguishable from 0 km s^{-1} and shows that there is no overall expansion or contraction of the ensemble as might occur if all of the clouds were formed in a large expanding shell of gas around the Sun, or were formed by gas raining in from the galactic halo. Evidently, the HLCs are part of the steady state gas distribution in the plane of the Milky Way.

The HLCs as a whole are small, low mass clouds, with very low values of the mean visual extinction (Magnani and deVries, 1986). It is primarily for this reason that it took such a long time to identify them in large numbers. Although the smallest clouds that MBM were able to identify are perhaps 10′ in diameter, the largest clouds have dimensions almost 20° across (MBM 53–55). Because the clouds do not seem to span a range in distance exceeding a factor of about 5 from the smallest to the largest, the corresponding range in mass is evidently quite large. As mentioned above, the very large cloud mapped by Heithausen and Thaddeus (1990) is not considered an HLC, because it appears to be contiguous with CO emission associated with a GMC of more than 10^4 M$_\odot$ at $b < 20°$.

5 Boundedness of the HLCs

As mentioned above, a large number of maps of the HLCs have already appeared in the literature at low spatial resolution and/or low spatial sampling (MBM, Keto and Myers, 1986, de Vries, Heithausen and Thaddeus, 1987, Heithausen and Thaddeus, 1990, Mebold et al., 1985). Regardless of the assumptions used to calculate the column densities and masses, all of the studies indicate that the clouds are far from being gravitationally bound. For example, MBM show that half of the clouds that they mapped have masses two orders of magnitude less than what is needed for them to be self-gravitating. The MBM analysis uses the quantity σ_v(internal) as an estimator of the internal kinetic energy of a given cloud. Because this quantity is a measure of the random velocities of the *clumps* within an HLC, it is not possible that the clouds are held together by the external pressure of the general interstellar medium. That is, independently moving clumps can only be confined by a pressure *gradient* which would equalize itself on an acoustic timescale, which is about 10^5 yr for the HLCs. The tension of a large scale magnetic field which threads the clumps can confine them in two dimensions, but not along the direction of the magnetic field, so the clumps will separate from each other at speeds comparable to the one dimensional velocity dispersion, σ_v(internal). Elmegreen (1988) has argued that the clumps can be confined by ram pressure, but the detailed density distribution of the well mapped clouds are not, in general, consistent with the large ram pressure that would be needed to confine the clouds (see, for example, the ^{13}CO maps of Pound, Bania and Wilson, 1990). This may not be true, however, for the anomalous velocity clouds discussed below. Furthermore, for some of the clouds, σ_v(internal) is nearly equal to σ_v(cloud-cloud), which indicates that for these clouds, ram pressure must be relatively unimportant.

It is for these reasons that MBM concluded that the HLCs must have ages comparable to their kinematic crossing time of $\sim 2 \times 10^6$ yr, and are thus very young. But if the clouds are not confined by either gravity, external pressure or magnetic fields, how then can they have formed in the first place? The answer comes from looking at the environment of the clouds: their relation to the atomic hydrogen in which most of the clouds are embedded.

Let us first examine, however, the pressure of the individual clumps of the HLCs. The quantity σ_v(linewidth) has a mean value of $0.3\,\mathrm{km\,s^{-1}}$ as measured from individual CO line profiles. In an investigation of the properties of the individual clumps within a cloud, Magnani (1986) estimated that the typical density within an individual clump is ~ 1000 cm^{-3}. The internal pressure within a clump is then given by

$$P = nm(\sigma_v)^2$$

where m is the mass of a hydrogen molecule and σ_v is the one dimensional velocity dispersion of an individual clump, which is the same as σ_v(linewidth). For a clump, then, the internal pressure is about 2×10^4 K cm^{-3}. There are several strong arguments that the pressure of the interstellar medium is determined by the turbulent pressure of the gas, and has a value of $1\text{–}2 \times 10^4 \, \text{K cm}^{-3}$ (Maloney, 1988). Furthermore, recent reevaluations of the hydrostatic pressure of the disk of the Milky Way at the solar circle has this same value (Bloemen, 1987). Thus if these values are correct, then the *clumps* within an HLC can be confined by the general pressure of the interstellar medium. It appears that if the clumps form by means of a thermal instablity, perhaps driven by the molecular cooling induced by a phase transition when the molecules can shield themselves from the dissociating radiation of the interstellar medium, the clumps maintain themselves in pressure balance with the material out of which they formed. This idea is examined further below.

6 The environment of the HLCs

The IRAS satellite showed that the local interstellar medium is suffused with extensive filamentary emission which is particularly bright at a wavelength of $100 \, \mu\text{m}$, and has become known as "infrared cirrus" (Low et al., 1984). Prior to the IRAS launch, the maps of Colomb, Pöppel and Heiles (1980) showed that the cold HI in the local interstellar medium is collected primarily in loops and filaments with a relatively small filling fraction. Numerous studies (*e.g.* Boulanger, Baud and van Albada, 1984) showed that there is a good correlation between the infrared cirrus and the atomic hydrogen gas, and it became clear that much of the cirrus is the dust associated with the atomic hydrogen filaments identified by Heiles. Weiland *et al.* (1986) showed that there is also a good correlation between the HLCs and the infrared cirrus. The HLCs, they argued, are the cores of the infrared cirrus, because the cirrus is in general much more extended than the molecular emission. Désert, Bazell and Boulanger (1988) were able to use these good correlations to identify molecular clouds from the IRAS survey as regions that showed infrared emission in excess of what is expected from the HI–IR correlation. Their identifications were confirmed by Blitz, Bazell, and Désert (1989) with CO measurements of the infrared peaks of the clouds showing infrared excesses.

In order to examine the relationship of the HI and CO in the HLCs in more detail, Gir, Blitz and Magnani (1991) observed the regions around a number of HLCs in the 21-cm line, and found that without exception, extended HI emission is found to be associated with the HLCs. Maps of the emission from two clouds are shown in Figures 2 and 3. Qualitatively, the maps show that although there is a general correlation of the HI with the molecular line emission at the same velocity, there is an offset between the peaks of the CO and the 21-cm emission. A similar result is obtained from maps of two individual clouds by de Vries, Heithausen and Thaddeus (1987), and Verschuur (1990). How are we to understand this result?

It was shown above that the clumps in a cloud appear to be in pressure equilibrium with the interstellar medium. Imagine, then, that a filament of HI has sufficient column density to shield itself from the dissociating radiation of the interstellar radiation field. In that case, some portion of the cloud will undergo a

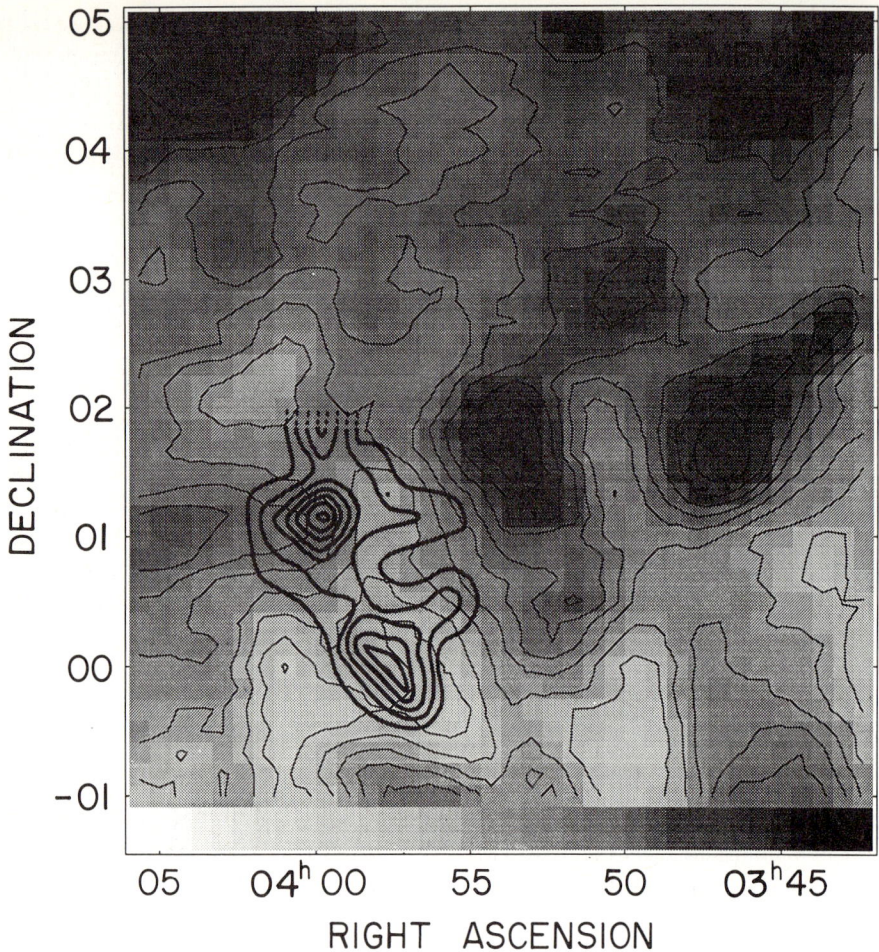

Figure 2: Map of the CO emission from MBM 18 superimposed on the HI emission in the vicinity of the HLC. The CO cloud is shown as the heavy contours and the HI is shown in gray scale. The HI is the velocity integrated emission over a velocity range of $\pm 2\sigma_v$(internal) of the CO cloud. The separation between the HI peak and the CO peak is clearly evident in the figure. The map is from Gir et al. (1991).

phase transition, and the bulk of the gas will become molecular. The molecular gas will be able to cool efficiently through radiation of the CO lines, and will attain a density determined by pressure equilibrium. The line of sight that originally had the highest HI column density will no longer have the highest column density; most of the atoms will have gone into the molecular phase. We would then expect that the highest HI column density will appear in positions *adjacent* to the CO cloud and these would represent the directions for which the column density is just below the critical value for self-shielding.

One may test this hypothesis quantitatively in at least two ways. (1) The mean distance projected along the line of sight between the CO peaks and the HI peaks should be equal to the mean radius of the CO clouds. The results of Gir

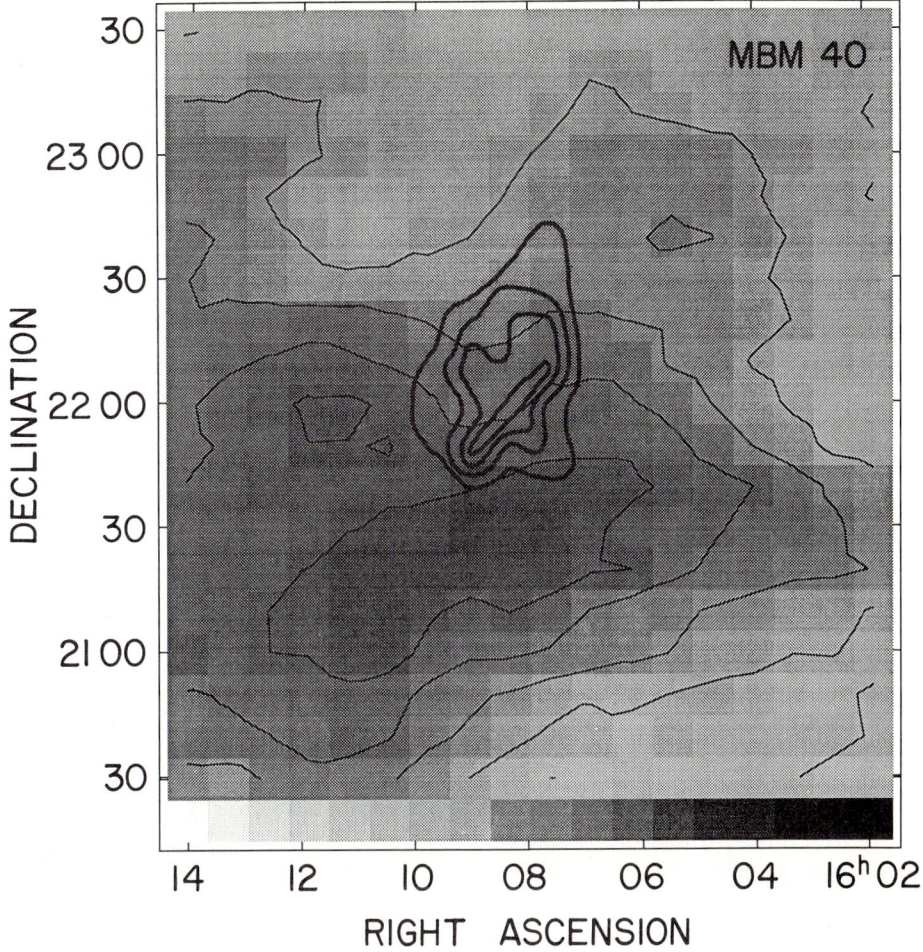

Figure 3: Same as Figure 2 for MBM 40.

et al. confirm this expectation. (2) The column density of the H_2 associated with the CO cloud should not exceed the value of the maximum HI column density which is compressed by an amount consistent with pressure balance. That is, if the HLCs form through a process of condensation from an ambient HI cloud in pressure equilibrium, there is a relation between the column density in the molecular cloud and the uncondensed HI cloud given by

$$\frac{N_1}{N_2} = (\frac{T_2}{T_1})^{2/3}.$$

where the subscripts 1 and 2 refer to the atomic and molecular phases of the cloud respectively. The 2/3 power occurs because gas compressed along the line of sight does not change the column density, but for a given beam, molecules are brought into the line of sight from the two directions perpendicular to the line of sight. For typical observed temperatures in the atomic phase of 100 K, and 10 K in the molecular phase,

$$\frac{N_2}{N_1} = 4.6.$$

For most of the clouds observed by Gir et al. this ratio ranges from 2 to 6, but a few pathological clouds have significantly higher values in the sense that there is an overabundance of molecular gas.

7 Molecular clouds in the halo?

The cloud complex with the highest ratio of $N(H_2)/N(HI)$ in the Gir et al. study encompasses the clouds MBM 41–44, where the mean ratio is about 33. This complex does not therefore appear to have formed *in situ* by condensing from the atomic phase. This complex is peculiar in that it has a velocity that is 4 times the rms value of σ(cloud-cloud). It is the prototype of a group of three HLCs with highly anomalous radial velocities that may be interacting with coronal gas in the disk, or alternatively may be in the halo. These are G90+38, also known as the Draco Cloud (Georigk et al., 1983) and MBM41–44 (MBM), with a radial velocity of -23 km s^{-1}, G211+63 with a velocity of -39 km s^{-1} (Désert, Bazell and Blitz 1990), and G135+55 with a velocity of -45 km s^{-1} (Heiles, Reach and Koo, 1988). These clouds have velocities that range from 4–8 σ(cloud-cloud), and are clearly different from the remainder of the HLC population. If the velocities are representative of motions in equilibrium with the stellar gravitational potential, these clouds would be in the halo of the Galaxy. On the other hand, that all of the velocities are negative suggests that the velocities may represent some other kind of motion, perhaps related to the motions of the classical high-velocity clouds. Whatever their origin, these clouds appear to be a group that is distinct from the other catalogued HLCs. All three clouds have been the subject of previous studies, and I will summarize the observational data for each of them.

7.1 G90+38; The Draco Cloud; MBM 41–44

Georigk et al. (1983) first called attention to this object because they noticed a high latitude HI feature corresponding to the position of a faint high latitude nebulosity. Furthermore, star counts indicated that there was more extinction than could be accounted for by the dust associated with HI; CO was then sought and detected. Subsequently, Mebold et al. (1985), and MBM independently observed the CO associated with the object. MBM called attention to the kinematic anomaly of the CO clouds, and Mebold et al. have attempted to establish a distance to the cloud with the specific aim of trying to ascertain whether the Draco cloud is a molecular cloud in the halo. Using UBV photometry and estimates of the extinction to the cloud, Georigk and Mebold (1986) argued that the distance to the cloud is $> 800\,\text{pc}$ implying that z is $> 500\,\text{pc}$. If correct, it would place this molecular cloud unequivocally in the galactic halo.

The existing evidence is, however, weak. On the basis of two lines of sight for the CO as well as the HI observations, Mebold *et al.* argue that the extinction (A_v) to the region is $\gtrsim 2\,\text{mag}$. The star counts imply much lower extinctions and Mebold *et al.* concluded that the cloud is more distant than the stars used for the counts. However, the HI data of Georigk *et al.* (1983) indicate that the mean extinction due to HI is only $0.3\,\text{mag}$. Data given in MBM suggest that the *mean* extinction due to H_2 is no greater than $0.7\,\text{mag}$. The mean extinction over the region in which the star counts were made is probably no greater than 0.5 because of the limited extent of the CO emission. Since the distance of $> 800\,\text{pc}$ was derived by comparing the star count data to $A_v = 2\,\text{mag}$, it is probably true that the true distance is less than that derived by Mebold *et al.*.

Subsequently, Georigk and Mebold (1986) did photometry to 56 stars projected on the Draco atomic hydrogen cloud, found that the reddening to the stars is inconsistent with an extinction of $2\,\text{mag}$ and concluded that the stars must be objects in the foreground of the Draco cloud. However, inspection of their colour-colour plot and the reddening line indicates that the locations of the stars may not be inconsistent with a mean extinction of $0.5\,\text{mag}$, and thus many of the stars, whose distances are determined from spectrophotometry and argued to be in the foreground, may indeed be background objects. In any event, the question of the distance can be decided by interstellar absorption line studies of stars projected along the line of sight to the cloud, especially in the direction of the molecular emission. These have yet to be done.

Regardless of whether this cloud is in the halo or not, the morphology resembles that of the other two anomalous velocity clouds found to date. That is, the clouds have a cometary morphology suggesting that the region of highest density is at the leading portion of the feature. Odenwald and Rickard (1987), and Odenwald (1988) have catalogued high-latitude cometary objects gleaned from the IRAS data base in order to find other objects that resemble G90+38. They find that the morphology of their objects can be explained by assuming that dense clouds are moving through the ambient gas with varying Reynolds numbers. For G90+38 Odenwald argues that the data suggest that the dense portion of the cloud is moving subsonically through an ambient interstellar medium of low Reynolds number, which suggests that the cloud is indeed moving through coronal gas. Other objects appear to be moving through ambient gas of higher viscosity producing turbulent wakes suggestive of Mach cones. In the case of G90+38 the CO has been extensively mapped, the strongest lines are near the ends of elephant trunks (Mebold *et al.*, 1985, Rohlfs *et al.*, 1989), and the ends of the trunks point in the general direction of the galactic plane. Odenwald's work suggests that much can be learned about the ambient gas and the dynamics of the disk-halo interface if one can obtain good estimates of the densities in the ambient gas of the general class of cometary structures seen at high galactic latitudes.

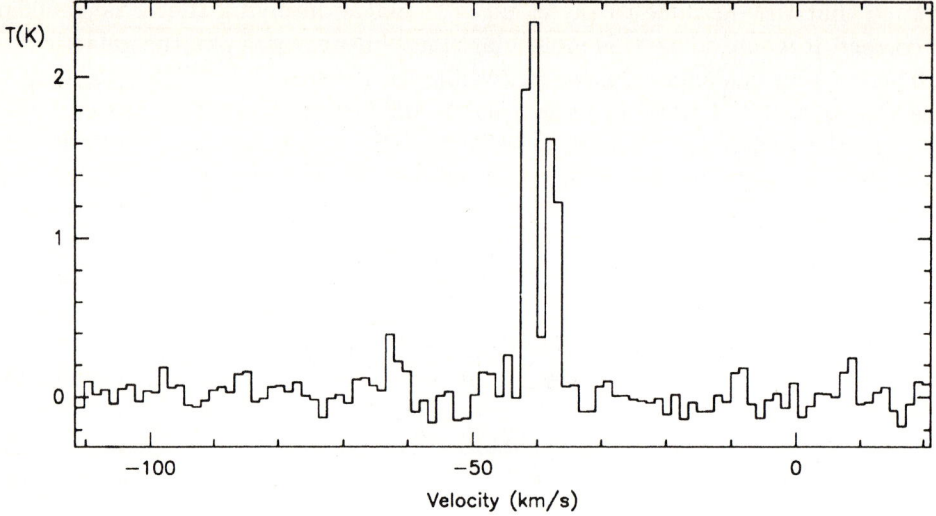

Figure 4: The CO J = 2 →1 spectrum at the secondary peak of the IR emission seen in Figure 5; the position of the spectrum is the centre of the map. Note the two velocity components.

7.2   G135+55

This object has the highest radial velocity, $-45\,\mathrm{km\,s^{-1}}$, of any molecular cloud detected at high galactic latitude and was discovered by Heiles, Reach and Koo (1988). The Heiles, Reach and Koo study sought to identify the properties of a sample of 26 isolated clouds at high galactic latitude from the IRAS survey. In their kinematic analysis they concluded that at least two of the clouds in their sample have undergone shocks, and that the shock has modified the grain size distribution. Specifically, they suggest that the very small grains responsible for 12 μm emission are destroyed by shocks with velocities of $\sim 10\,\mathrm{km\,s^{-1}}$ and formed in shocks with somewhat higher velocities. Fast shocks, $\gtrsim 40\,\mathrm{km\,s^{-1}}$, are argued to preferentially destroy large grains, thereby elevating the 60 μm/100 μm flux ratio in clouds. Thus one would expect that the anomalous velocity clouds as a group would exhibit 60 μm/100 μm flux ratios with values significantly greater than those found in the general interstellar medium. The G135+55 cloud was not otherwise explicitly analyzed in their study.

7.3   G211+63

This cloud has been detected by Désert, Bazell and Blitz (1990), but was independently observed by Heiles, Reach and Koo (personal communication). The cloud was found by Désert *et al.* in a CO survey of interstellar clouds that exhibit infrared excesses (Désert, Bazell and Boulanger, 1988). This cloud is the only one of about 30 detections in their study that has a radial velocity more than $2\sigma_v$(cloud-cloud).

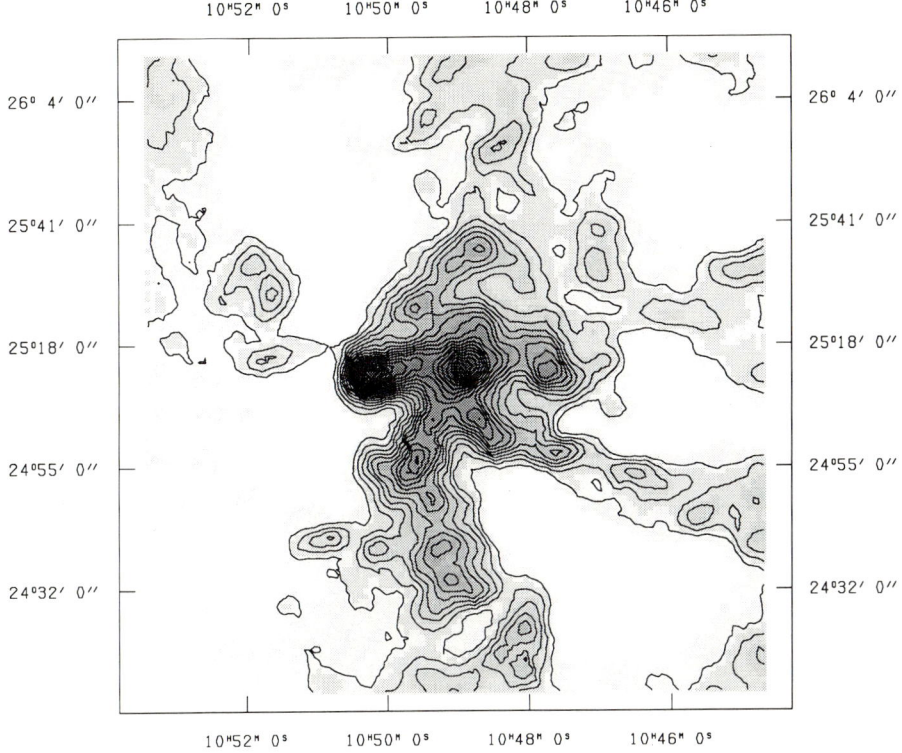

Figure 5: Map of the 100 μm emission from G211+63. Note the cometary morphology of the emission with a strong peak in the east and more diffuse emission to the west. Figures 4 and 5 are from Désert *et al.* (1990).

Désert, Bazell and Blitz obtained CO (1 → 0) and CO (2 → 1) spectra at the position of peak infrared emission. The 2 → 1 line observation was made with high frequency resolution and shows a striking double peaked structure. The CO spectrum is shown in Figure 4. However, no CO mapping of the source was made. Instead, the authors made carefully constructed maps at 12 μm, 60 μm, and 100 μm, and they compared the colour ratios of the resulting maps. The 100 μm map is shown in Figure 5. They conclude that like the cometary globules catalogued by Odenwald (1988), the interaction with the ambient gas is observed as plumes trailing from the back side of the two main dense clumps from which the most intense radiation is detected. The colours suggest that there has been grain processing as the result of shocks that have destroyed the small grains at the leading edge of the object. The brighter clump of dust shows an elevated 60 μm/100 μm flux ratio as suggested by Heiles, Reach and Koo (1988) for fast shocks. The mass of the dense portion of the cloud is estimated to be $1.7\,M_\odot$ and the mass of the associated HI (Verschuur, 1971) is estimated to be $8.7\,M_\odot$ at an assumed distance of 100 pc. The diameter of the cloud at this distance is 1.2 pc.

Based on their velocities, their morphologies, and their infrared properties, the anomalous velocity HLCs seem to be good candidates to be molecular clouds at the interface between the Galactic disk and the halo. Their interaction with the ambient gas can provide useful information about the shock processing of dust and the production of molecules in shocks. In any event, these clouds appear to be a group that is distinct from the remainder of the HLCs and deserve special attention in their own right.

8 HLCs and diffuse molecular clouds

One of the puzzling results about the IRAS survey of Désert, Bazell and Boulanger (1988 – hereafter DBB) to detect HLCs is that the clouds they identified have a surface filling fraction on the sky of 4%. Compare this value to the value obtained by Magnani, Lada and Blitz (1985) listed in Table 2 of only 0.5%. How could this discrepancy be so big?

To try to arrive at an answer, Blitz, Bazell and Désert (1989) observed about 300 of the DBB clouds at their infrared peaks in CO to determine whether the blind survey of Magnani, Lada and Blitz was somehow biased and undercounted the molecular clouds at high galactic latitude. They found that only one cloud in eight of the DBB sample was in fact detectable in CO, and that the blind survey correctly gave the surface filling fraction of CO at latitudes above 25°. What then, are the remaining clouds identified by DBB? Blitz, Bazell and Désert noticed that there is a cutoff in infrared surface brightness below which CO is never detected and above which CO is almost invariably detected. This cutoff value is 4 MJy ster^{-1} which corresponds to $A_v = 0.25$ mag. They postulated that below this cutoff, the CO has insufficient column density to be self-shielding, and $N(CO)/N(H_2)$ has a value close to the value of 10^{-6} found in diffuse molecular clouds (*e.g.* Federman *et al.*, 1980). This is to be compared to the value of $\sim 5 \times 10^{-5}$ for the HLCs detected in CO (Magnani, Wouterloot and Blitz, 1988). The DBB clouds that were not detected in CO were then compared to the diffuse molecular clouds identified in the *Copernicus* observations (Bohlin, Savage and Drake, 1978), and were found to have both numbers and column densities similar to the clouds identified in the UV.

Thus, there appears to be reasonable evidence that the DBB clouds below the CO cutoff are the classical diffuse clouds found in absorption against bright early-type stars. Nevertheless, there are no definitive observations of stars behind the DBB clouds that show molecular interstellar lines. The suggestion must therefore be considered tentative until a "smoking gun" is found.

9 Line widths and the role of magnetic fields

As discussed above in Section 5, the mean internal pressure of the clumps in the HLCs is 2×10^4 K cm^{-3}, which results from a mean σ_v(linewidth) of 0.3 km s^{-1}. From the analysis of Weiland *et al.* (1986), the dust temperature of the HLCs is found to be about 20 K, and the maximum line strength of the CO never exceeds 10 K. Thus, the kinetic tempertures in the HLCs are unlikely to exceed $\sim$ 20 K, and may be even lower. However, even at a kinetic temperature of 20 K, the thermal width (σ_v(linewidth)) of a CO line should only be 0.05 km s^{-1}, a factor of 6 less than what is observed. On the other hand, the thermal line width for H$_2$ is 0.3 km s^{-1}, just the value observed! That is, although the CO lines are supersonic with respect to the kinetic temperature of the gas, the lines are not broader than the thermal line widths of the unobservable H$_2$. This leads one to the suggestion that the various molecular species are not in thermal equilibrium, but that they move under some other perturbing influence. Circumstantial evidence suggests that this influence is the magnetic field.

Let us postulate that magnetic fields thread both the clump and the interclump gas. If the clump and interclump gas are in thermal equilibrium, then the magnetic field pressure in the interclump gas should be approximately the same as that in the molecular gas. If the interclump gas can be identified with the remnant atomic gas, then it should be possible to measure the magnetic fields by means of the Zeeman effect in the 21-cm line. For a pressure of 2×10^4 K cm^{-3} (see Section 5), the expected magnetic field is about 9 μG.

The HLCs have been shown to be preferentially located in loops and filaments of the HI (Blitz, 1987, Gir *et al.*, 1991). Heiles (1989) has measured the Zeeman effect and thus has made magnetic field measurements at a large number of positions in some of these loops. In a few cases, the loops are the sites of HLCs, but the HLCs are not, in general, within the beam of the Heiles measurements. Nevertheless, these measurements should give a good indication of the approximate magnetic field strength in the vicinity of the clumps. Heiles found that the magnetic fields are significantly larger in the loops than outside them, and the mean of the measured fields is 6.4 μG, close to the value expected if the clump pressure is in equilibrium with the external magnetic field pressure. Heiles measures only one component of the magnetic field of course, and the true mean magnetic field strength will be higher than the observed value.

Pressure equilibrium with the magnetic field implies that the turbulent velocity is the same as the Alfvén speed; that is, one expects that the line widths are equal to the Alfvén speed, which is 0.3 km s^{-1} in one dimension. The point here is that, although the CO line widths are supersonic, they are not super-Alfvénic. Indeed, they are equal to the Alfvén speed in the dense gas. The large observed (relative to thermal) CO line widths are good evidence that the motions within the molecular clouds are dominated by magnetic fields.

10 The formation of HLCs

The observations presented above, when taken together, suggest a fairly simple (though still fairly tentative) picture for the formation of HLCs. First, a disturbance propagates through the general interstellar medium in the form of some sort of outflow or ejection event. This disturbance compresses the atomic hydrogen into loops and filaments that are seen in both the HI and the IR. The HI is manifested by the loops and filaments in the Colomb *et al.* (1980) maps, and the IR is manifested by the cirrus. Eventually, a large enough column density of the atomic gas is collected for the H_2 to become self-shielding, forming numerous diffuse molecular clouds in the filaments. When there is sufficient column density for the CO to become self-shielding, the CO (and the associated coolants such as C^+), allow the cloud to cool and become denser. It is at this stage that the cloud becomes observable in the millimeter lines of CO and other molecular tracers. The instability that allows the cooling to take place presumably takes place on many spatial scales. The clumpy structure would then be a result of the formation process and would therefore be primordial. The magnetic field in these clumps is still likely to be connected to the diffuse atomic gas out of which they formed. If the magnetic field is in pressure equilibrium with the gas, the internal motions within a clump will be reflected in the motions of the detected molecules which are in turn controlled by the Alfvén waves which permeate both the atomic and molecular gas. The clumps therefore remain pressure bound within the loop or filament of atomic gas out of which it formed. The motions of the ensemble of clumps that make up a particular HLC on the other hand reflects the turbulent motions that exist within the primordial HI gas; this would have no relationship to and would normally exceed the velocities that would be expected if an HLC were to be gravitationally bound. Thus, the majority of HLCs are born with values of σ_v(internal) larger than permitted if they were in virial equilibrium. In some cases however, the clumps may collide to form gravitationally bound entities. The ultimate fate of the HLCs in this picture would depend on the ultimate fate of the pressure within the surrounding HI gas. If that pressure were to decrease, an HLC could reexpand and once again become atomic, especially if the shielding layer were to be destroyed.

Since the processes that form the HLCs presumably occur everywhere within the disk of the Milky Way and spiral galaxies in general, the HLCs are presumably part of the steady state partition of the phases of the interstellar medium, and their relative abundance is determined by local conditions. However, even if this picture is correct or even approximately so, it does not explain the presence of the anomalous velocity HLCs which are likely to have a different origin.

Acknowledgements

Funding for this work is partially provided by the USNSF grant AST-8918912, and the contribution of the State of Maryland to the Laboratory for Millimeter-wave Astronomy.

References

Bahcall, J.N., 1984, Astrophys. J., **276**, 169.
Barnard, E.E., 1927, *Carnegie Institute of Washington Publication No. 247, Part 1.*
Blitz, L., 1987, in *The Evolution of Galaxies*, ed. J. Palous, (Publications of the Czechoslovak Academy of Sciences: Prague), p. 201.
Blitz, L., Maganani, L. and Mundy, L., 1984, Astrophys. J., **282**, L9.

Blitz, L., Bazell, D. and Désert, F.X., 1988, Astrophys. J., **352**, L13.
Bloemen, J.B.G.M., 1987, Astrophys. J., **322**, 694.
Bohlin, R.C., Savage, B.D. and Drake, J.F., 1978, Astrophys. J., **224**, 132.
Boulanger, F., Baud, B. and van Albada, T., 1985, Astron. Astrophys., **144**, 9.
Clemens, D.P., Sanders, D.B. and Scoville, N.Z., 1988, Astrophys. J., **327**, 139.
Colomb, F.R., Pöppel, W.G.L. and Heiles, C., 1980, Astron. Astrophys. Suppl. **40**, 47.
Dame, T.M., Ungerechts, H., Cohen, R.S., de Geus, E.J., Grenier, I.A., May, J., Murphy, D.C., Nyman, L.-Å. and Thaddeus, P., 1987, Astrophys. J., **322**, 706.
Désert, F.X., Bazell, D. and Boulanger, 1988, Astrophys. J., **334**, 815.
Désert, F.X., Bazell, D. and Blitz, L., 1990, Astrophys. J., **355**, L51.
van Dishoeck, E.F., 1990a, in *The Evolution of the Interstellar Medium*, ed. L. Blitz, (PASP, San Francisco, in press).
van Dishoeck, E.F., 1990b, in *Molecular Astrophysics*, ed T.H. Hartquist, (Cambridge University Press), p. 55.
van Dishoeck, E.F. and Black, J.H., 1988, Astrophys. J., **334**, 771.
de Vries, C.P, 1986, Ph.D. Dissertation, University of Leiden.
de Vries, C.P. and van Dishoeck, E., 1988, Astron. Astrophys., **203**, L23.
de Vries, H.W., Heithausen, A. and Thaddeus, P., 1987, Astrophys. J., **319**, 723.
Elmegreen, B.G., 1988, Astrophys. J., **326**, 616.
Federman, S.R., Glassgold, A.E., Jenkins, E.B. and Shaya, E.J., 1980, Astrophys. J., **242**, 545.
Fich, M. and Blitz, L., 1984, Astrophys. J., **279**, 125.
Georigk, W. and Mebold, U., 1986, Astron. Astrophys., **162**, 279.
Georigk, W., Mebold, U., Reif, K., Kalberla, P.M.W. and Velden, L., 1983, Astron. Astrophys. **120**, 63.
Gir, B-Y., Blitz, L. and Magnani, L., 1991, Astrophys. J., (submitted).
Heiles, C., 1989, Astrophys. J., **336**, 808.
Heiles, C., Reach, W.T. and Koo, B.-C., 1988, Astrophys. J., **322**, 313.
Heithausen, A. and Thaddeus, P., 1990, Astrophys. J., **353**, L49.
Hobbs, L.M., Blitz, L. and Magnani, L., 1986, Astrophys. J., **306**, L109.
Hobbs, L.M., Blitz, L., Penprase, B.E., Magnani, L. and Welty, D.E., 1988, Astrophys. J., **327**, 356.
Keto, E.R. and Myers, P.C., 1986, Astrophys. J., **304**, 466.
Knapp, G.R., 1987, P.A.S.P., **99**, 1134.
Knapp, G.R. and Bowers, P.F., 1988, Astrophys. J., **330**, 684.
Lada, E.A. and Blitz, L., 1988, Astrophys. J., **326**, L69.
Lebrun, F., 1986, Astrophys. J., **306**, 16.
Low, F.J., Beintema, D.-A., Gautier, T.N., Gillett, F.C., Beichman, C.A., Neugebauer, G., Aumann, H.H., Boggess, N., Emerson, J.P., Habing, H.J., Hauser, M.G., Houck, J.R., Rowan-Robinson, M., Soifer, B.T., Walker, R.G. and Wesselius, P.R., (1984), Astrophys. J., **278**, L19.
Lundmark, K. and Melotte, P.K., 1926, *Upps. Medd.*, No.12.
Lynds, B.T., 1962, Astrophys. J. Suppl., **7**, 1.
Magnani, L., 1986, Ph.D. Dissertation, University of Maryland.
Magnani, L., Blitz, L. and Mundy, L., 1985, Astrophys. J., **295**, 402 (MBM).
Magnani, L., Lada, E.A. and Blitz, L., 1985, Astrophys. J., **301**, 395.
Magnani, L. and de Vries, C.P., 1986, Astron. Astrophys., **168**, 271.
Magnani, L., Blitz, L. and Wouterloot, J.G.A., 1988, Astrophys. J., **326**, 909.
Maloney, P., 1988, Astrophys. J., **334**, 761.
Mattila, K., 1970, Astron. Astrophys., **9**, 53.
Mattila, K., 1979, Astron. Astrophys., **78**, 253.
Mebold, U., Cernicharo, J., Velden, L., Reif, K., Crezelius, C. and Georigk, W., 1985, Astron. Astrophys., **151**, 427.
Odenwald, S.F., 1988, Astrophys. J., **325**, 320.
Odenwald, S.F. and Rickard, L.J, 1987, Astrophys. J., **318**, 703.
Pound, M., Bania, T.M. and Wilson, R.W., 1990, Astrophys. J., **351**, 165.

Rohlfs, R., Herbstmeier, U., Mebold, U. and Winnberg, A., 1989, Astron. Astrophys., **211**, 402.
Turner, B.E, Rickard, L. J. and Xu, L.-P., 1989, Astrophys. J., **344**, 292.
Verschuur, G.L., 1971, Astron. J., **76**, 317.
Verschuur, G.L., 1990, Astrophys. J., **361**, 497.
Welty, D.E., Hobbs, L.M., Blitz, L. and Penprase, B.E., 1989, Astrophys. J., **346**, 232.
Weiland, J., Blitz, L., Dwek, E., Hauser, M.G., Magnani, L. and Rickard, L.J., 1986, Astrophys. J., **306**, L101.
Wilson, W.J., Schwartz, P.R., Epstein, E.E., Johnson, W.A., Etcheverry, R.D., Mori, T.T., Berry, G.G. and Dyson, H.B., 1974, Astrophys. J., **191**, 357.
Wouterloot, J.G.A., Brand, J., Burton, W.B. and Kwee, K.K., 1990, Astron. Astrophys., **230**, 21.

Galactic cirrus clouds in an HI loop

HORST MEYERDIERKS and VOLKMAR GROSSMANN

RADIOASTRONOMISCHES INSTITUT, AUF DEM HÜGEL 71, D-5300 BONN 1, F.R.G.

1 Introduction

The North Celestial Pole Loop (NCP Loop) is an HI loop 20° in diameter located near the north equatorial pole. It is also a prominent feature in the far infrared cirrus detected with the IRAS satellite. From the HI velocity field of the loop Meyerdierks et al. (1990b) derive an expansion velocity of approximately $20\,\mathrm{km\,s^{-1}}$. Whereas the individual clouds belong to one common loop structure, they differ in size, opacity, extent of molecular emission etc. The physics and chemistry may be influenced by a non-dissociative shock. We report here on observations of dust, HI, ^{12}CO, ^{13}CO, OH, and H_2CO.

2 Dust, atoms, and molecules

It is well known that gas and dust are associated in the interstellar medium and that their mass ratio is roughly constant. This is born out also in the morphological similarity of the far infrared and HI emissions at high galactic latitudes. When looked at in detail, however, the HI cannot account for all of the 100 μm intensity. In fact, if one includes CO as a tracer of molecular hydrogen in the comparison, the problem is solved. A striking example is given by de Vries et al. (1987) who depict two dust filaments in the NCP Loop, one associated with purely atomic gas and the other having a significant though minor fraction of molecules in the gas phase.

For low-extinction cirrus clouds, the scaling factor $X = N(H_2)/W(^{12}CO)$ from CO line integrals to H_2 column densities can be derived from the comparison of infrared, HI, and CO (Heithausen and Mebold, 1989). Surprisingly, the X factors turn out to be rather low at $0.2 - 0.5\ 10^{20}\ \mathrm{cm^{-2}\ K^{-1}\ km^{-1}\ s}$, several times less than derived on a galactic scale from comparing gamma rays with HI and CO (cf. Bloemen, Wolfendale, this volume).

High-sensitivity ^{12}CO observations (de Vries et al., 1987, Heithausen and Thaddeus, 1990) show that CO emission is more extended than previously thought and that it is observed also where extinction is low ($< 0.3^m$). The ratio of molecular to atomic hydrogen mass column density varies significantly from 0.4 to 3. This has to be taken into account when abundances of other molecules are determined.

3   Cloud parameters

The CO emission regions in the NCP Loop have sizes between $0.1°$ and $0.8°$ and CO excitation temperatures of 5 to 8 K. The densities derived from column densities are between 200 and 500 cm^{-3}. The extinctions and 100 μm intensities are 0.3^m to 2^m and up to 13 MJy ster^{-1} respectively. The column densities of HI and H$_2$ both have similar ranges of $3 - 10 \times 10^{20}$ cm^{-2} and $2 - 15 \times 10^{20}$ cm^{-2} respectively. The masses range from 0.3 to 30 M$_\odot$ (at 100 pc). The virial masses are a factor of 10 to 200 larger than the observed ones. The clumps and clouds are not gravitationally bound at all observed size scales, their life times (size divided by velocity dispersion) are on the order of 10^5 to 10^6 yr. If the Polaris Flare (Heithausen and Thaddeus, 1990) is taken as one structure, its size and mass are $8°$ and 1000 M$_\odot$ (at 100 pc).

4   Carbon monoxide and hydroxyl

Meyerdierks et al. (1990a) apply the IR-HI-^{12}CO comparison to a diffuse cloud in order to determine the column density of molecular hydrogen and then to derive the abundance of ^{13}CO, which turns out to be very high at 10^{-5}. For two other clouds observed in ^{13}CO the H$_2$ column density cannot be derived by the above method, because the extinction is too high or the cloudlet too small compared to the HI and IRAS resolutions. The CO abundance derived by Meyerdierks et al. is incompatible with steady-state chemistry for a diffuse cloud (van Dishoeck and Black, 1988) yet can be accounted for by assuming a shock going through the diffuse cloud (Mitchell and Watt, 1985).

In a translucent cloud, Großmann et al. (1990) observe extended OH emission. In the cloud centre they detect a main line anomaly, i.e. the levels involved in the 1667 MHz and 1665 MHz transitions are not populated according to LTE. This could result from far-infrared pumping due to a velocity gradient, which might be another signature of the NCP Loop shock. This is vindicated by the OH abundance of 2.5×10^{-6} being even somewhat higher than in the diffuse cloud ζ Oph. Magnani et al. (1988) observed OH in a different more diffuse part of the NCP Loop. Their spectra indicate smaller OH column densities due to coarser resolution.

5   Formaldehyde

The 1_{10}-1_{11} transition of H$_2$CO was detected in a variety of clouds in the NCP Loop (Heithausen et al., 1987, Mebold et al., 1987, Magnani et al., 1988, Meyerdierks et al., 1990a, Großmann et al., 1990). Often the lines are too broad to resolve the hyperfine structure, or they are too weak to determine a significant optical depth. Yet where strong narrow lines are observed, the excitation temperature is always between 2 K and 2.5 K, which is larger than the usually assumed 1.7 K (Cohen *et al.*, 1983) and consequently leads to H_2CO abundances of up to 1.6×10^{-6}, similar to or higher than in dark clouds.

Some positions were also observed in the 2_{11}-2_{12} transition by Turner *et al.* (1989). These authors apply an LVG analysis to reproduce the observed brightness temperatures of the 2_{11}-2_{12} and 1_{10}-1_{11} transitions. Their result is a very low H_2CO abundance. We note, however, that this result depends strongly on the assumed gas temperature entered into the LVG analysis. The solution for $T_{kin} = 20$ K given by Turner *et al.* cannot reproduce an excitation temperature for the 1_{10}-1_{11} transition as high as 2.2 K.

In our view, gas temperatures of ≈ 10 K and densities of at most a few thousand are more realistic: the observed parameters accord better in the LVG analysis, the ^{12}CO excitation temperatures are between 5 K and 8 K, and the densities as estimated from CO column densities are below 1000 cm^{-3}. In consequence the H_2CO abundances should be high.

The very detection of formaldehyde in diffuse clouds is surprising, because the photo destruction rate for this molecule corresponds to only about 100 years if shielded by 1^m extinction. Indeed formaldehyde is not observed in diffuse clouds. Again, shock chemistry can resolve this contradiction. In the "unshielded" model ($A_V = 1^m$) of Mitchell and Deveau (1983), a high formaldehyde abundance persists for 10^4 yr behind the shock.

References

Cohen, R.J., Matthews, N., Few, R.W. and Booth, R.S., 1983, Monthly Notices Roy. Astron. Soc., **203**, 1123.
de Vries, H.W., Heithausen, A. and Thaddeus, P., 1987, Astrophys. J., **319**, 723.
Großmann, V., Heithausen, A., Meyerdierks, H. and Mebold, U., 1990, Astron. Astrophys. (in press).
Heithausen, A. and Mebold, U., 1989, Astron. Astrophys., **214**, 347.
Heithausen, A., Mebold, U. and de Vries, H.W., 1987, Astron. Astrophys., **179**, 263.
Heithausen, A. and Thaddeus, P., 1990, Astrophys. J., **353**, L49.
Magnani, L., Blitz, L. and Wouterloot, J.G.A., 1988, Astrophys. J., **326**, 909.
Mebold, U., Heithausen, A. and Reif, K., 1987, Astron. Astrophys., **180**, 213.
Meyerdierks, H., Brouillet, N. and Mebold, U., 1990a, Astron. Astrophys., **230**, 172.
Meyerdierks, H., Heithausen, A. and Reif, K., 1990b, Astron. Astrophys., (submitted).
Mitchell, G.F. and Deveau, T.J., 1983, Astrophys. J., **266**, 646.
Mitchell, G.F. and Watt, G.D., 1985, Astron. Astrophys., **151**, 121.
Turner, B.E., Rickard, L.J. and Xu, Lan-ping, 1989, Astrophys. J., **344**, 292.
van Dishoeck, E.F. and Black, J.H., 1988, Astrophys. J., **334**, 771.

The high-latitude cloud towards HD210121

ROLAND GREDEL[1], EWINE F. VAN DISHOECK[2], COR P. DE VRIES[2] and JOHN H. BLACK[3]

Introduction

The high-latitude cloud towards the star HD210121 was observed in ^{12}CO $(1 \rightarrow 0)$ with the Swedish-ESO Submillimetre Telescope (SEST) with a spatial resolution of 45″. Low-level, extended emission with $T_A^* = 1-2\,\mathrm{K}$ was detected in a region of 1 square degree. Embedded are individual clumps with significantly higher antenna temperatures, $T_A^* = 4-5\,\mathrm{K}$. The ^{12}CO $(1 \rightarrow 0)$ line profile towards HD210121 shows wings with an extension of a few $\mathrm{km\,s^{-1}}$ from the line core and changes significantly at scales of 0.1 pc. Limited maps in ^{13}CO $(1 \rightarrow 0)$ were obtained towards selected positions. The clumps seen in ^{13}CO are smaller in extent and have steeper gradients towards their boundaries. ^{12}CO $(3 \rightarrow 2)$ observations towards HD210121 were performed with the Caltech Submillimeter Observatory (CSO). From the ^{12}CO $(1 \rightarrow 0)/^{12}$CO $(3 \rightarrow 2)$ intensity ratio, a density of $1000\,\mathrm{cm^{-3}} < n < 8000\,\mathrm{cm^{-3}}$ was inferred. Optical absorption lines of C_2 in the spectrum of HD210121 were detected with the ESO Coude Auxiliary Telescope. From the analysis of the C_2 data, a gas kinetic temperature of 15 K and a density of $200\,\mathrm{cm^{-3}}$ was determined.

Observations and discussion

Molecular emission in the $J = 1 \rightarrow 0$ line of ^{12}CO was detected in a region of approximately 1 square degree around the high-latitude star HD210121 ($\alpha_{1950} = 22^h\,05^m\,36.1^s$, $\delta_{1950} = -3°\,46'\,35.5''$) with SEST. Figure 1 shows a contour map of the peak antenna temperature T_A^*. The region is characterised by widespread emission with $T_A^* = 1-2\,\mathrm{K}$. Embedded into the low-level emission are a number of individual clumps with significantly higher antenna temperatures of $T_A^* = 4-5\,\mathrm{K}$. The clumps have spatial dimensions of a few times 0.1 pc at an adopted distance of 200 pc of the cloud. The line profile of ^{12}CO $(1 \rightarrow 0)$ towards HD210121 exhibits wings with an extension of a few $\mathrm{km\,s^{-1}}$. The line profile changes significantly at spatial scales of 1′. Line wings and changes in the ^{12}CO profiles at small spatial scales are well modelled by Keto and Lattanzio (1989). The authors proposed that collisions among high-latitude clouds are responsible for the observed features.

Selected regions were observed in ^{13}CO $(1 \rightarrow 0)$. The clumps seen in ^{13}CO are

[1] I. PHYSIKALISCHES INSTITUT, ZÜLPICHER STR. 77, D-5000 KÖLN 41, F.R.G.
[2] STERREWACHT LEIDEN, P.O. BOX 9513, 2300 RA LEIDEN, THE NETHERLANDS.
[3] STEWARD OBSERVATORY, UNIVERSITY OF ARIZONA, TUCSON, ARIZONA 85721, U.S.A.

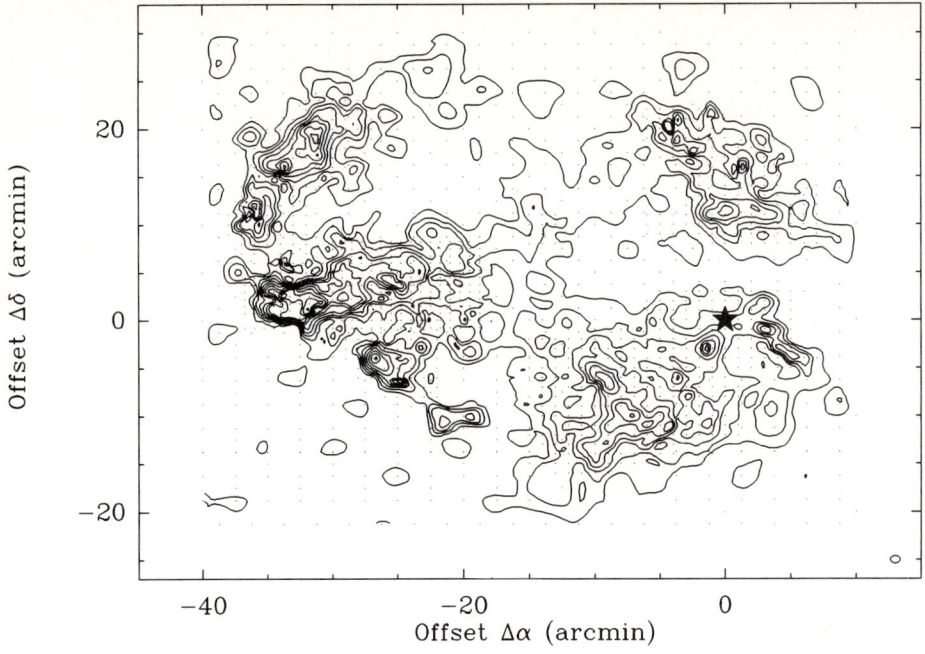

Figure 1. The peak antenna temperature T_A^* of ^{12}CO $(1 \to 0)$ measured towards HD210121. The lowest contour is $T_A^* = 1$ K, each additional contour is incremented by 0.5 K. Points indicate the position of individual ^{12}CO observations. The position of HD210121 is indicated by a star. Map coordinates are given in offsets in arcmin from the star.

smaller and have steeper gradients towards their boundaries. Because of the clumpy structure of the HD210121 cloud, far UV radiation of the interstellar radiation field may penetrate deep into the molecular material. Selective photodestruction is held responsible for the observed differences.

^{12}CO $(3 \to 2)$ observations were performed towards selected positions with the CSO. CO excitation calculations were performed, where the radiative transfer was treated in terms of mean escape probabilities. From the ^{12}CO $(3 \to 2)/^{12}$CO $(1 \to 0)$ ratio, a mean density of $1000 \, \text{cm}^{-3} < n(\text{H}_2) < 8000 \, \text{cm}^{-3}$ was inferred.

Optical absorption lines of C_2 around $8765 \, \text{Å}$ were detected in the spectrum of HD210121 using the ESO Coude Auxiliary Telescope (CAT) with the Coude Echelle Spectrograph (CES). The C_2 observations indicate a gas kinetic temperature of 15 K and a density of $200 \, \text{cm}^{-3}$. The differences in the density derived from the CO and C_2 measurements may indicate a different distribution of the respective molecules in the cloud.

Reference

Keto, E.R. and Lattanzio, J.C., 1989, Astrophys. J., **346**, 184.

Structure and Energy Balance of Molecular Clouds

REINHARD GENZEL

MAX-PLANCK INSTITUT FÜR EXTRATERRESTRISCHE PHYSIK,

D-8046 GARCHING, FRG

Abstract Maps of millimeter molecular line emission clearly show that molecular clouds have a complex, clumpy spatial structure. The mm data and maps of the submm/far-infrared atomic fine structure lines of C^+ and C^0 indicate a density contrast of more than an order of magnitude between clumps and the material in between them. Large scale statistical correlations between velocity width, size and mean density suggest that the molecular gas is self-gravitating and that typical column densities do not change much as a function of scale size. Sections 1 and 2 of this review discuss some of the relevant recent observations of the structure and energetics of molecular clouds, as well as possible physical models, such as the effects of radiation and mass outflows from embedded stars. The emphasis of the discussion is on clouds in the vicinity of regions of active OB star formation.

The topic of section 3 is the question of how the smallest, high density condensations in cloud cores can be investigated. Recent high resolution maps of submillimeter emission show very compact concentrations of cool dust in several OB star forming regions. Are these possible candidates for high mass protostars? What is the role of molecular spectroscopy in studying such objects?

Section 4 is on the recent discovery of unusually warm molecular gas near the surfaces of molecular clouds and its possible explanations.

1   The spatial structure of molecular clouds

1.1   Molecular line maps

It was realized as early as the mid-seventies that molecular clouds have substantial structure on scales smaller than the arcmin beam sizes of typical single dish telescopes (e.g. Barrett *et al.*, 1977). Overwhelming evidence for the clumpiness of molecular clouds on all accessible scale sizes (0.003 to 30 pc) has recently come from large scale maps of clouds with moderate size and large single dish millimeter telescopes, as well as from the VLA and millimeter interferometers (Wilson and Walmsley, 1989). Much of the recent progress is based on mapping of sufficiently large areas that contain many resolution elements and on selecting optically thin (or not too thick) lines that give much more intensity contrast than the very optically thick ^{12}CO transitions.

A good illustration of the complex and highly inhomogeneous distribution of the gas in molecular clouds is contained in Figure 1 (left inset) which gives a ^{13}CO $1 \to 0$ velocity channel map (6.5 to 7.5 km s^{-1} LSR) of the entire Orion A molecular cloud at a resolution of 90″ (0.19 pc at 450 pc) from the work of Bally *et al.* (1987). Bally *et al.* emphasize the presence of dense clumps and filaments as well as those of bubbles and cavities. They conclude that the massive

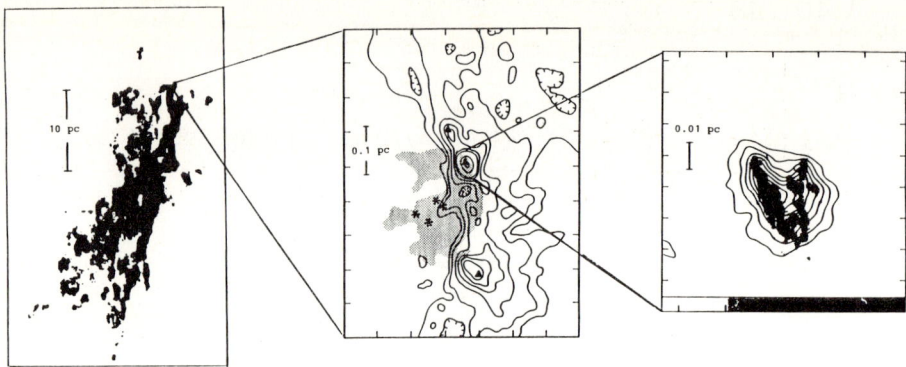

Figure 1: Molecular line maps of Orion A over four orders of magnitude in spatial scale. Left: ^{13}CO $1 \rightarrow 0$ map between LSR 6.5 and 7.5 km s^{-1} (Bally et al., 1987, Bell Labs telescope, 90″ beam). Middle: CS $2 \rightarrow 1$ map between -0.2 and 18 km s^{-1} LSR (Mundy et al., 1987, OVRO mm interferometer, 7.5″ beam). Right: NH$_3$(3, 2) map (grey shading) between 7.4 and 8.6 km s^{-1} (Migenes et al., 1989, VLA, 1.2″ beam). Contours represent a velocity averaged map, smoothed to 2″.

stars of the Orion OB associations as well as embedded lower mass stars have a strong dynamical effect on the cloud structure. Bally (1989) points out a possible correlation between a kinematic "twist" motion in the southern part of the cloud and a helical structure of the magnetic field wrapping around the cloud that may be inferred from HI Zeeman and optical polarization measurements.

Is Orion typical? The answer is probably yes, as the spatial structures of other clouds with different environments appear qualitatively rather similar. As a demonstration of this conclusion, Figures 2 and 3 show maps of isotopic CO rotational lines of the dark and "cold" cloud(s) in Taurus (distance 140 pc) and in the interface of the giant "warm" cloud M17 (distance 2.2 kpc), taken at about the same spatial resolution (0.13 pc for M17, Stutzki and Güsten 1990, 0.16 pc for Taurus, Cernicharo and Guélin, 1987).

Is the clumpiness a strong function of distance from the nearest OB stars? The answer is probably no, a conclusion that is already plausible from inspection of the Bally et al. maps. As an additional confirmation consider the C^{18}O $2 \rightarrow 1$ channel maps of a "random" piece of the M17 molecular cloud shown in Figure 4 (from Genzel et al., 1990). This region, referred as M17SW C on the ^{12}CO map of Elmegreen and Lada (1976), is located at a distance of about 25′ (15 pc) from the M17 OB cluster and at least 5′ from any compact HII region or luminous far-infrared source. Size of the maps and spatial resolution are essentially identical to the C^{18}O map of the cloud/HII region interface shown in Figure 3. The appearances of the C^{18}O maps in these two regions are quite similar; clumps of size $\leq 15″$ to 40″ are embedded in a ridge-like structure. In all other respects the two regions are different. M17SW C is significantly colder ($T_{kin} \approx$ 20 K instead of 60 K in the interface), of lower H$_2$ volume density ($\leq 10^4$ cm^{-3} instead of a few 10^5 cm^{-3}) and column density (a few 10^{22} cm^{-2} instead of several 10^{23} cm^{-2}). The mean energy density of the far-ultraviolet radiation field (a measure of the effect of O and B stars) in M17SW C is probably between 10 and 100 times that of the

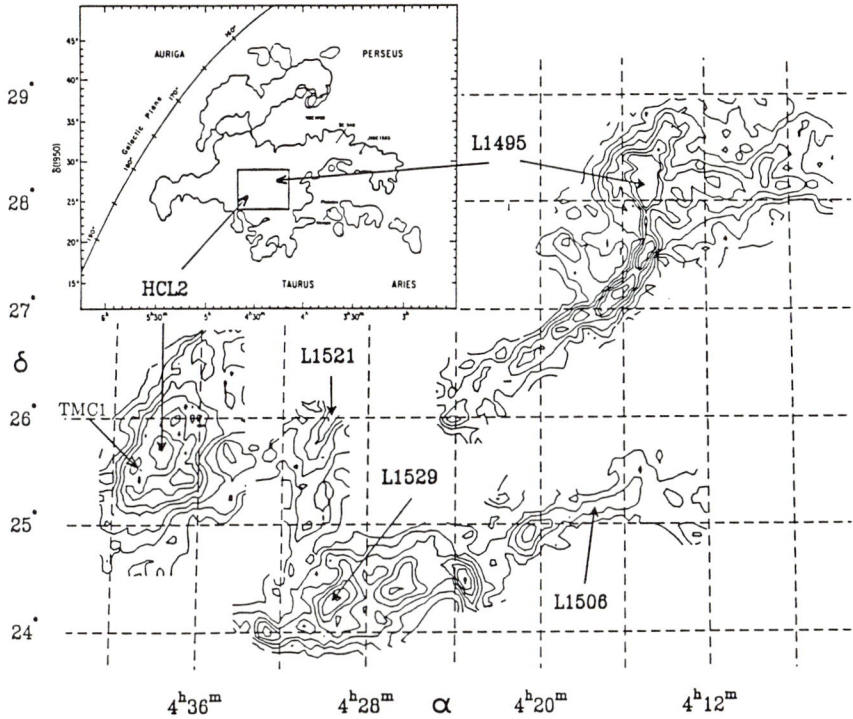

Figure 2: ^{13}CO $1 \to 0$ integrated velocity map of the central Taurus region (Cernicharo and Guélin, 1987, 2.5m Bordeaux, 3.6' beam). The upper left inset shows the region mapped on the outlines of the ^{12}CO map of the Taurus-Perseus complex (Ungerechts and Thaddeus 1987).

solar neighborhood ($\chi_0 \approx 2\times 10^{-3}$ ergs/s/cm^2/sr). For comparison, the radiation field intensity at the interface is a few 10^4 times χ_0. These findings strongly suggest a *local* mechanism for the creation and maintainance of clumpiness.

What is the scale size of clumping? The maps in Figures 1 to 4 show structure on all scales from the map size (typically 10 pc) to the beam size (10^{-1} pc). No preferred scale emerges in these studies, while Perault et al. (1985) had found preferred clumping on the 1 pc scale from ^{13}CO maps in several other dark clouds. The trend of scale-independent structures continues to significantly smaller scales (see Perault and Falgarone, 1989). Figure 1 is a composite of molecular line maps of the Orion A region covering about four orders of magnitude in spatial scale. In addition to a channel map from the Bally et al. (1987) ^{13}CO survey Figure 1 gives a CS $2 \to 1$ interferometer map of the central "ridge" of the Orion A molecular cloud (Mundy et al., 1988) and a NH$_3$(3,2) interferometer channel map of the BN-KL "hot core" (Migenes et al., 1989). The CS data have a resolution of 7.5" (0.016 pc), or less than one tenth of the Bally et al. (1987) data. The CS map shown in the middle inset of Figure 1 corresponds to only 3 linear resolution elements of the ^{13}CO map on the left and clearly has significant substructure down to the scale of the interferometer beam. The Orion-KL hot core region (the brightest CS emission spot in Figure 1) again shows considerable stucture

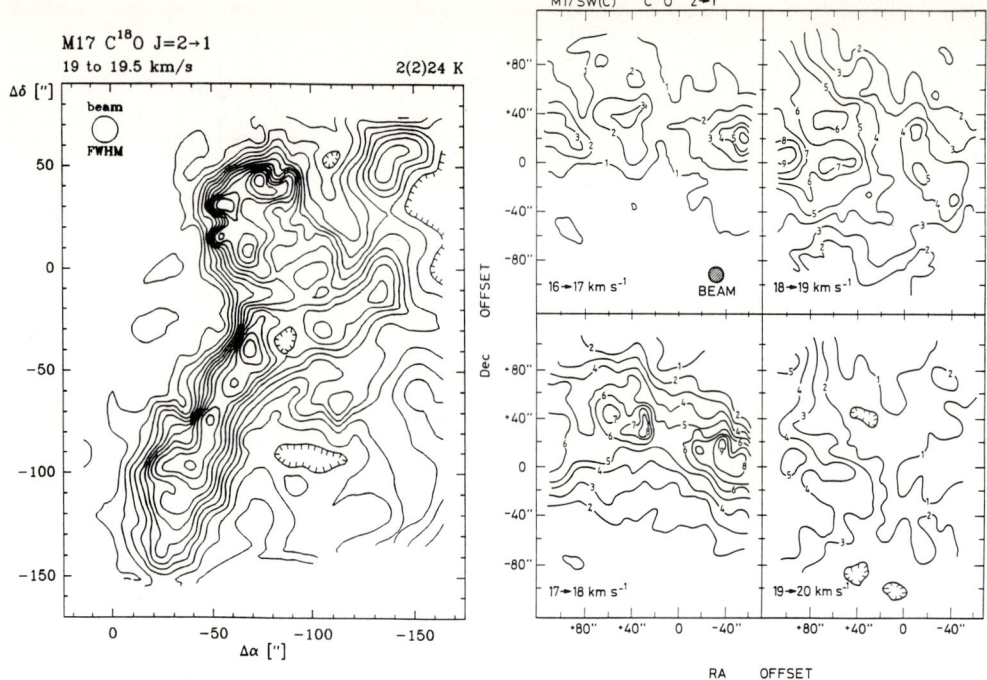

Figure 3: $C^{18}O$ $2 \to 1$ channel map between 19 and 19.5 km s^{-1} LSR of the M17 interface region (Stutzki and Güsten, 1990, IRAM 30m, 13″ beam). The base position is at R.A.= 18^h 17^m 34.5^s, Dec.=-16° 13′ 24″ (1950).

Figure 4: $C^{18}O$ $2 \to 1$ channel maps of the M17SW C region. Map center is at R.A.=18^h 16^m 13.5^s, Dec.=-16° 32′ 00″ (1950) (Genzel et al., 1990, IRAM 30m, 13″)

to the resolution limit of the VLA images of about one arcsec (2.3 × 10^{-3} pc, Genzel et al., 1982, Pauls et al., 1983, Migenes et al., 1989).

Do the clumps represent stable, physical entities or are they temporary fluctuations in an ever shifting dynamical balance? In favor of the first interpretation speaks the fact that the clumps are not just column density fluctuations in projection on the sky but appear to be well defined also in the velocity domain. Figure 5 shows a position-velocity diagram for the $C^{18}O$ emission in the M17 interface region from Stutzki and Güsten (1990). The individual clumps visible on the column density map in Figure 2 can be easily separated from other velocity components in Figure 5 and thus represent high-contrast condensations in three dimensional phase space (see also Blitz and Stark, 1986). Furthermore, in most cases an increase in local line intensity is also accompanied by an increase in local line width suggesting that gravity plays a role and that the clumps are actual mass concentrations. It is plausible from the various observations that the line widths follow virial equilibrium, to within the measurement accuracies (about a factor $\approx$ 2 in mass). There are, however, significant differences in the spatial distributions of different molecules suggesting that spatial variations in chemical

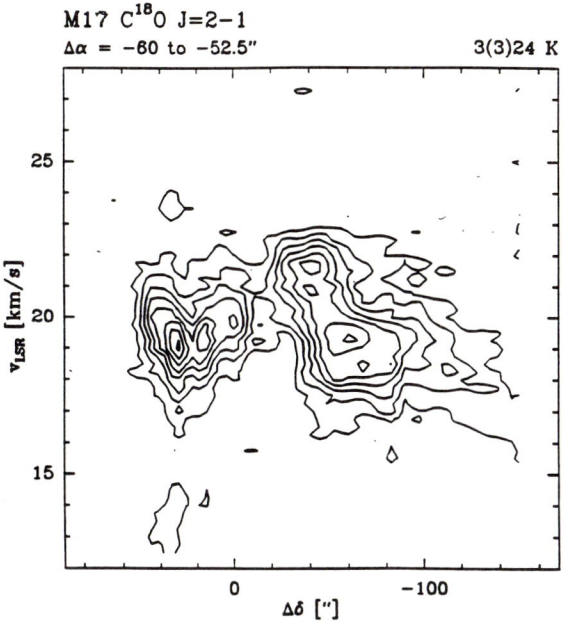

Figure 5: Position velocity diagram of the $C^{18}O$ $2 \to 1$ emission in the M17 interface. Same base position and resolution as in Figure 3.

abundances can play a role (Plambeck, 1988, Swade, 1989). For this reason, it is probably preferable to investigate molecular cloud structure in species whose abundances are not expected to vary much with environment, such as CO and its isotopes, or in submillimeter dust emission. Finally, molecules may freeze out on dust grains and deplete the gas phase in cold and very dense condensations (Mezger et al., 1987, 1988, see section 3).

Scalo (1990) has challenged the interpretation that intensity maxima on molecular line maps can be categorized and interpreted as stable physical entities, such as "cores", clumps or fragments. He points out that limited spatial resolution and dynamic range, selection bias and the natural tendency of astronomers to put things in categorized boxes and tables will conspire and lead to false, quasi-static evolutionary models. He shows that the spatial structure of the Taurus cloud(s) as derived from a large scale IRAS 100 μm map has some features of a random, fractal structure. He also points out that uncertainty and spread of the data in the velocity width vs. size relationship are too large to deduce virial equilibrium. Falgarone and Phillips (1990) find that the velocity field in molecular clouds with a wide range of physical conditions follows turbulence with intermittency. Compressible turbulence with moderately large Mach numbers can perhaps account for the observed filamentary structures.

1.2 What is in and between Clumps ?

Important criteria for assessing the reality of individual emission peaks as

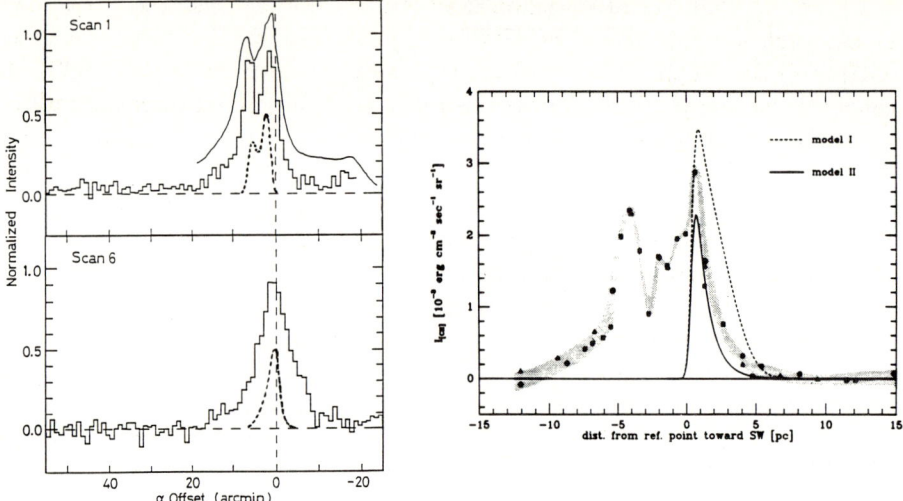

Figure 6: Cross cuts of 158 μm $^2P_{3/2} \to {}^2P_{1/2}$ [CII] line emission in M17. Left, top: NE-SW cut. The histogram represents the balloon data of Matsuhara et al. (1989, 3.4' beam), the continuous line the KAO data of Stutzki et al. (1988, 55" beam smoothed to 3.4') and the dashed line is a schematic representation of the distribution of the radio continuum emission. The base position is the 100 μm continuum peak. Left, bottom: SE-NW crosscut. Right: [CII] crosscut and two models of UV radiation of the M17 OB cluster penetrating a clumpy molecular cloud (Stutzki et al., 1988).

stable physical objects are the spatial correlation between clumps and recently formed stars and the density contrast between clumps and the medium in between them. T-Tauri stars or embedded IRAS point sources in Taurus are often located in or close to NH_3 cores (Myers 1986, Emerson 1987). The densest part of the ρ Oph molecular cloud contains a cluster of embedded stars (Wilking and Lada, 1983). Lada (1990) finds clusters of embedded stars associated with four of the five major CS $2 \to 1$ emission knots in the Orion B molecular cloud. It is likely that IRc2, a very young and luminous star in the Orion-KL star forming region, has formed in the Orion-KL hot core (right inset of Figure 1), the most prominent gas concentration in the Orion A ridge (Masson and Mundy 1988).

The clump to interclump density contrast in the warm and dense clouds that form O and B stars appears to be remarkably large. Blitz and Stark (1986) infer that ratio to be 10 or larger from ^{13}CO and ^{12}CO $1 \to 0$ observations of the Rosette molecular cloud. Stutzki and Güsten (1990) derive a density contrast of at least 20 from their $C^{18}O$ $2 \to 1$ maps of the M17 interface region. The contrast may be at least 5 in the M17SW C region shown in Figure 4. There is evidence that the clump/interclump contrast is significantly smaller in dark clouds, (Perault et al., 1985, Falgarone and Puget, 1988, Myers, this volume, p. 133).

Clump to interclump contrasts of up to a factor of 100 are inferred from the large observed spatial extent of the C^+/C^0 regions around OB stars (Stutzki et al., 1988). Figure 6 shows crosscuts of the 158 μm $^2P_{3/2} \to {}^2P_{1/2}$ line emission of [CII] through the M17 HII region (located at the base position, from Matsuhara et al., 1989 and Stutzki et al., 1988). Figure 7 shows two dimensional maps of [CII] emission in the W3 star forming region (from Howe et al., 1990). In both cases the far-infrared line emission is much more extended than the radio continuum emission and appears to extend several parsecs from the central luminosity sources into the surrounding molecular clouds. [CII] emission is almost certainly excited by far-UV (λ = 912 to 1200 Å) radiation from O and B stars. Detectable [CII] emission requires UV fluxes greater than a few 100 times χ_0, the UV flux in the solar neighborhood (Tielens and Hollenbach, 1985, Sternberg and Dalgarno, 1989). The large extent of the [CII] emission region is then interpreted as a large penetration depth of the UV radiation. The average column density of neutral gas, as measured from isotopic CO emission, between the central OB stars in the HII regions and the point where the [CII] emission has dropped significantly corresponds to $A_v \approx$ 50 to 300 in W3 and M17 (Stutzki et al., 1988, Howe et al., 1990). Without obscuration, the diluted UV flux from the central OB stars corresponds to a few $10^3\chi_0$ at a distance of 2 pc, not much larger than the UV flux necessary to account for the observed [CII] flux at that radius. Hence the average visual extinction toward the central OB stars must be less than $A_v \approx$ a few. This is in contradiction with the above mentioned measured column density unless most of the column density is concentrated in dense clumps with a clump to interclump contrast between 10 and 100. Numerical calculations confirm this argument and completely exclude homogeneous cloud models (see Figures 6 and 7, Stutzki et al., 1988, Boisse, 1990, Howe et al., 1990). Clumpy cloud models also explain the similarly extended submillimeter [CI] line emission first discovered by Phillips and Huggins (1981, see also Genzel et al., 1988). Presently there appear to be only two ways around this argument. If the clouds contain a distributed population of B stars, enhancing the far-UV radiation throughout, clouds could be more homogeneous. Second, it is conceivable that the UV dust absorptivity per hydrogen nucleus is lower in interclump gas than it is in the dense clumps. Léger (1990) and Desert et al. (1990) have argued that a significant fraction of the far-UV extinction could be due to very small dust grains or large molecules, such as poly-cyclic aromatic hydrocarbons (PAH's). These could have a much lower abundance in the interclump medium, due to photodestruction by UV radiation. It is very unlikely, however, that any one of these two effects can fully account for the observations.

Direct evidence for an interclump medium may come from the detection of faint ^{12}CO non-Gaussian emission components in the wings of line profiles (Blitz and Stark, 1986, Falgarone and Phillips, 1990). As an example Figure 8 shows ^{12}CO and ^{13}CO profiles toward a "clump-free" zone in the Rosette molecular cloud from the work by Blitz and Stark (1986).

In summary then, keeping Scalo's (1990) comments in mind as an important warning and caveat it appears quite convincing that many of the molecular column density peaks are self-gravitating physical entities with a large density contrast relative to the interclump medium. This finding puts strong constraints on

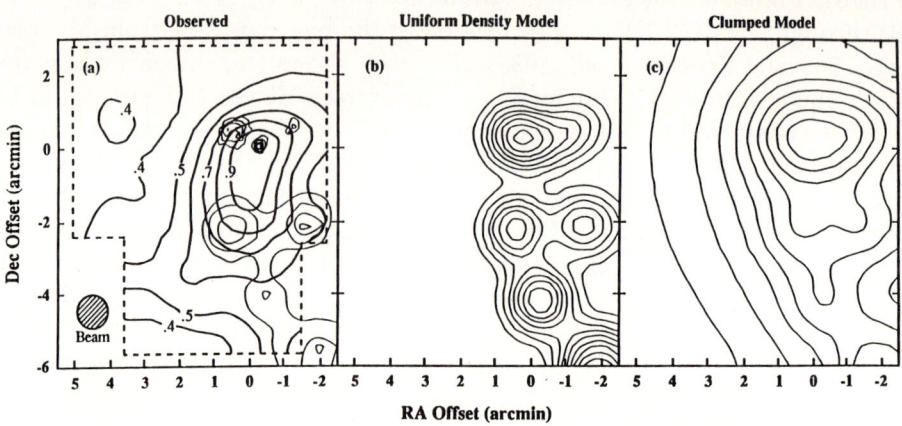

Figure 7: Two dimensional map of the [CII] emission in W3 (left, heavy contours, KAO, 55″) superposed on the radio continuum emission (light contours). Models of a homogeneous (middle) and clumpy cloud (right) (from Howe *et al.*, 1990).

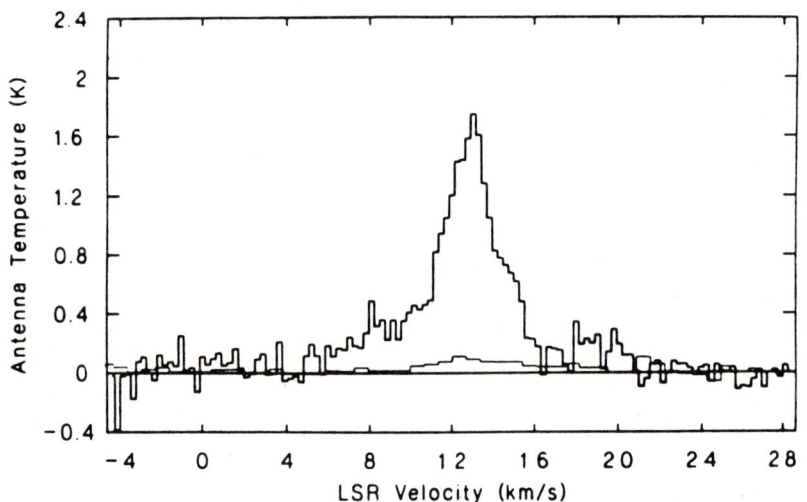

Figure 8: Non Gaussian wing emission in a ^{12}CO $1 \to 0$ profile toward a clump-free region in the Rosette molecular cloud (heavy histogram, Bell Labs antenna 90″). A ^{13}CO $1 \to 0$ profile is shown for comparison (from Blitz and Stark 1986).

molecular cloud models. Next we have to adress the clump's mass spectrum and stability.

1.3 Mass Spectrum of Clumps

Stutzki and Güsten (1990) fitted Gaussian clumps to the data cube of $C^{18}O$ spectra in the M17 interface. They decompose the emission into 179 clumps of size $\leq 10''$ to $60''$, FWHM velocity width 0.5 to 3 km s^{-1} and mass 10 to 10^3

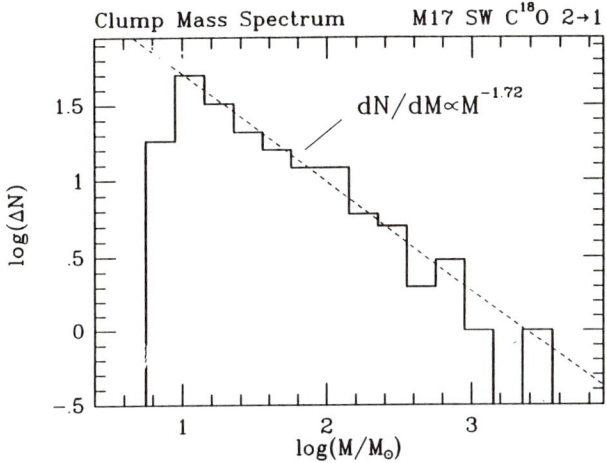

Figure 9: Clump mass spectrum derived from the $C^{18}O$ data in the M17 interface (from Stutzki and Güsten 1990).

$M_\odot$. The clumps have a molecular hydrogen density between 10^5 and 10^6 cm^{-3} and fill about 30% of the volume. They fit virial equilibrium ($\Delta v^2 \approx GM(R)/R$) quite well, as is the case for other regions and scale sizes as well (e.g. Solomon et al., 1987). Figure 9 shows the derived clump mass distribution which can be fit by a single power law with index α of about 1.7 ($dN/dM \propto M^{-\alpha}$). Note that the Jeans mass for the typical parameters of the emitting gas in M17 is about 10 $M_\odot$. Similar power law spectra with exponents between 1.1 and 1.6 have been found for the Rosette cloud by Blitz (1987), for ρ Oph by Loren (1989) and for the general mass spectrum of giant clouds by Casoli et al. (1984) and Sanders et al. (1985).

A power law spectrum with $\alpha \approx 1.5$ is the plausible result of an equilibrium of coagulation and fragmentation (Spitzer, 1982). It may also be consistent with a Salpeter type stellar mass spectrum ($dN(M)/dM \propto M^{-2.35}$) as the eventual outcome of that fragmentation if the fraction of the clump's mass that ends up in the star decreases with increasing mass (Zinnecker, 1989).

2 Energy balance of molecular clouds

2.1 Stability of Clumps

One of the fundamental problems of molecular cloud research has been the question of how the clouds and clumps are stabilized against immediate gravitational collapse. Assuming that they are indeed self-gravitating on all scales, they should collapse on a few times the free fall time scale $t_{ff} \approx (3G\rho)^{-1/2}$ or less than a few 10^6 years (note that there definitely exist clouds, such as the high latitude clouds discussed by Blitz (this volume, p. 49) which are not gravitationally bound although the majority of the molecular mass is probably in entities close to gravitational equilibrium). The result would be a star formation rate much larger than observed in the Galaxy. One common explanation is that magnetic fields are approximately in balance with gravity (e.g. Myers and Goodman, 1988) and that gravitational collapse is prevented by magnetic pressure until the field has slowly diffused out by ambipolar diffusion (Mestel and Spitzer 1956, Shu et al., 1987). Star formation will then proceed (with low efficiency) on a time scale (Zweibel, 1987, Shu et al., 1987)

$$t_{amb} \approx 2 \times 10^7 \left(\frac{[x_e]}{10^{-7}}\right) yr \approx 10 t_{ff} \qquad (1)$$

$[x_e]$ is the fractional electron abundance. Another mechanism is the maintenance of clouds and clumps by the observed "turbulent" motions. As these motions are supersonic in almost all cases, dissipation of kinetic energy by shocks in cloud-cloud or clump-clump collisions takes place on a time scale proportional to the free-free collapse or dynamical time scales ($t_{ff} \approx t_{dyn} = R/v$). Scalo and Pumphrey (1982) estimate the dissipation time scale to be about 5×10^6 years. Supernovae, Galactic tidal shear and stellar winds may all contribute at some level to feeding of the "turbulent cascade" (Falgarone and Puget, 1988, Wilson and Walmsley, 1989). Promising local sources of kinetic energy that could create the observed turbulence are outflows from young stars (Norman and Silk, 1980). Fukui et al. (1986) and Margulis and Lada (1986) have carried out unbiased surveys for outflows in Orion and Monoceros. In the surveyed regions of the Orion molecular cloud alone there are more than 20 flows, approximately evenly distributed with a mean size of about 1 pc and a separation of 5 pc (Fukui, 1989). Fukui (1989) estimates that all stars more massive than about 1 M$_\odot$ go through a few 10^4 year period of pre-main sequence mass loss and that mass outflows can significantly contribute to the turbulence and cloud support. However, the mass outflow time scales are highly uncertain and the estimate of the dissipation time scale by Scalo and Pumphrey sensitively depends on the assumptions in their numerical models. Clouds without outflows have line widths not very much less than clouds which do contain outflow sources (Myers, this volume, p. 133). Shu et al. (1987) have argued that the turbulence is supersonic but sub-Alfvenic. In this case magnetic fields mediate clump-clump collisions with a low rate of dissipation. The observed "turbulence" may just be the result of propagating magnetic waves (Arons and Max 1975). Shu et al. also point out that observed polarization maps of stars near clouds often indicate coherent large scale magnetic fields suggesting that magnetic and not turbulent energy dominates (see also Myers, this volume). The origin of the turbulent motions is clearly not understood yet.

The high mass clumps observed in Orion-KL (Figure 1 middle and right insets), M17 (Figure 3) and other OB star forming regions may be in a different regime as their masses are typically larger than the mass that can be supported by

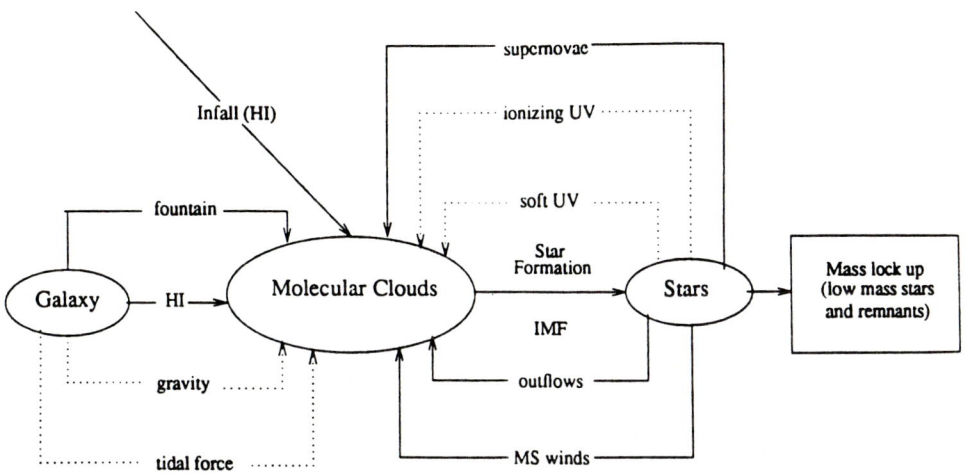

Figure 10: An illustration of the energy transport into molecular clouds (adapted from Bally 1989).

magnetic fields. This critical mass (or column density) is given by (Mouschovias and Spitzer, 1976, Shu et al., 1987)

$$M_{cr} \approx 25 \left(\frac{B}{300\ \mu G}\right)\left(\frac{R}{0.1\ pc}\right)^2 M_\odot$$

or

$$N(H_2)_{cr} \approx 4 \times 10^{22} \left(\frac{B}{300\ \mu G}\right)\ cm^{-2} \qquad (2)$$

The consequence is that most clumps in Orion-KL and M17 can, therefore, not be supported by magnetic fields and must collapse, unless they are not bound by gravitation to start with. If the latter is the case what keeps them from dispersing on a dynamic time scale (5×10^4 yr)? If they do collapse it is likely that a very high density stellar system emerges with a high rate of star formation efficiency as most of the gas in these regions is contained in the clumps. Support for this expectation comes from the work by Herbig and Terndrup (1986) and McCaughrean et al. (1990) who find that the stellar density in the stellar cluster surrounding the θ^1 OB cluster in Orion (in excess of 10^3 stars per pc^3) is equivalent to the mean gas density of the surrounding molecular cloud (n(H$_2$) $\approx$ 10^5 cm^{-3}).

It is fairly clear that an explanation of the energy balance of molecular clouds and of the time evolution of clumps must involve a number of factors. In addition to mass outflows and magnetic fields discussed so far, there are the effects of external pressure and UV radiation (see next paragraph and section 4), of supernovae and tidal forces. The energy pathways into molecular clouds are schematically indicated in Figure 10 (adapted from Bally, 1989).

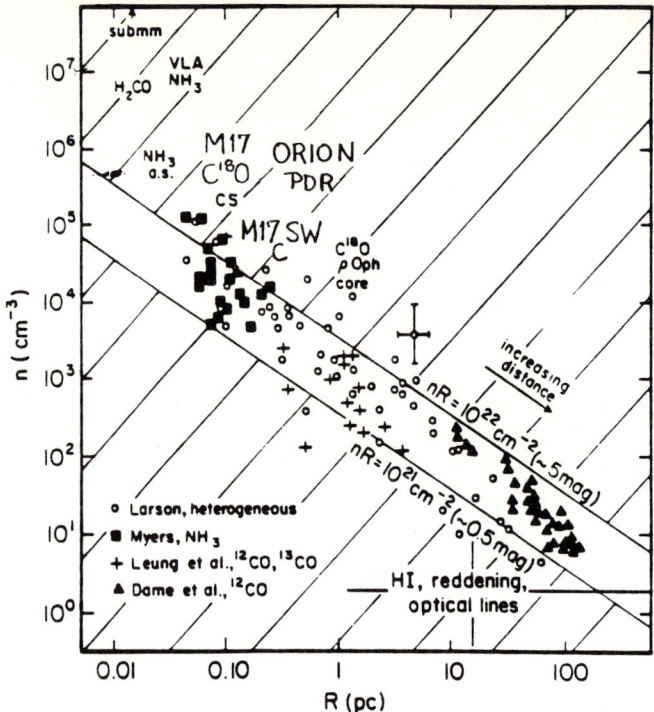

Figure 11: Relationship between molecular hydrogen volume density and size for selected cloud structures (from Scalo, 1990).

2.2 Larson's Correlations

Larson (1981) was the first to point out that average molecular hydrogen volume density $\langle n \rangle$ and velocity width Δv are correlated with scale size R via simple power laws: $\langle n \rangle \propto R^{-\gamma}$ and $\Delta v \propto R^{\delta}$. Most recent evaluations give $\gamma \approx 1$ (Figure 11, adapted from Scalo, 1990) and $\delta \approx 0.5$. These two relationships are consistent with just two intrinsic conditions; virial equilibrium and an approximately constant column density, independent of scale size. Figure 11 indicates $N(H_2)$ of a few 10^{21} cm^{-2} as the average proportionality constant. There are currently four reasonable competing explanations for the approximately constant column densities.

i) most clouds have column densities near the critical value for support by magnetic fields (eq.2, Shu *et al.*, 1987). The higher column densities of clumps near OB stars (upper left corner in Figure 11) are then a result of super-critical collapse and/or much higher fields at higher densities.

ii) molecular clouds are in equilibrium with external pressure (Maloney, 1988, Elmegreen, 1989). A simple estimate of hydrostatic equilibrium gives

$$\rho \frac{d}{dR}\left(\frac{GM(R)}{R}\right) = \frac{dP}{dR}$$

or, at the surface

$$P_{ext} \approx \frac{GM^2}{R^4} \propto N^2 \qquad (3)$$

Maloney (1988) and Elmegreen (1989) show that the observed correlations can be met by virialized clouds and their atomic envelopes that are in virial equilibrium with the general pressure of the ISM, $P_{ext} \approx$ 4000 cm^{-3} K. The higher column densities of the gas in OB star forming regions (up to 2 orders of magnitude) then correspond to correspondingly higher pressures. Since the gas pressure of the atomic gas is directly coupled to the external radiation field, the higher column densities and volume densities of these clumps should then scale with the square root of the density of the UV radiation field $\chi^{1/2}$. This is in reasonable agreement with the observations and may explain why the densest and most massive clumps are typically seen in the immediate vicinity of luminous O stars. An attractive scenario would be that the high degree of clumpiness and the concept of an external pressure regulated cloud structure are coupled through the presence of a hot (8000 K), interclump phase with densities

$$n_{intercl} \approx \frac{T_{clump}}{T_{intercl}} n_{clump} \approx 0.01 \text{ to } 0.03 \, n_{clump} \qquad (4)$$

It is at present unclear what the heating mechanism of the hot, interclump phase would be as photoelectric heating of bulk grains does not work at temperatures above a few 10^3 K (Draine, 1978). For such a two phase medium to work over a wide range of pressures, a heating mechanism is required that approximately scales with the intensity of the radiation field (Sternberg, 1990).

iii) the third explanation by McKee (1989) rests on the fact that the time scale for ambipolar diffusion is proportional to the fractional electron abundance $[x_e]$ (eq.1). Low mass star formation (sub-critical in the sense of eq.2) is only possible in regions of low electron abundance, that is, regions where UV radiation cannot photoionize carbon ($A_v >$ a few, or $N(H_2) >$ a few 10^{21} cm^{-2}). In the scenario of "photoionization-regulated" star formation proposed by McKee an initially diffuse cloud contracts quickly to $A_v \approx 4$. At that point its central parts are shielded from external UV radiation, the ambipolar diffusion time drops rapidly to the value determined by cosmic rays alone and star formation commences. Outflows from newly-formed low mass stars then stabilize the cloud and prevent further collapse. On average most clouds are then expected to have column densities near the "onset" value of a few 10^{21} cm^{-2}.

iv) the fifth explanation excludes any physical basis of the correlation (Kegel, 1989, Scalo, 1990). The apparently constant column density is a direct consequence of observational bias and sensitivity limits (intensity limit = column density limit).

3 How to find protostars ?

Finding protostellar condensations after gravitational collapse has formed a stellar core but before nuclear reactions have set in has always been a major goal of millimeter and infrared astronomy. Attempts based on observations of the mid-infrared dust emission have largely been unsuccessful since the dust condensations turned out to already contain hot, central stars (Wynn-Williams, 1982). Observations in the millimeter and submillimeter range until recently either did not have the necessary spatial resolution, or were hampered by optical depth and chemical effects. Recently Mezger and colleagues (Mezger et al., 1987, 1988, 1990) have found very compact (size $< 10''$) condensations of 1 mm dust emission from high spatial resolution observations with the IRAM 30m telescope toward a number of OB star forming regions (Orion, NGC 2024, S255, S106, see also Jaffe et al., 1984). Figure 12 shows the map of NGC 2024 (Mezger et al., 1988). There are about half a dozen compact dust emission knots along the north-south dust band across the face of the NGC 2024 HII region. Mezger et al. deduce very high column densities ($N(H_2) \approx$ a few 10^{24} cm^{-2} beam averaged) and low dust temperatures (16 K) from the 1 mm fluxes and a decomposition of the overall millimeter, submillimeter and far-infrared spectrum into 3 different components (Table 1). A key assumption is that the submm/mm absorption efficiency of dust scales like λ^{-2}. Mezger et al. infer masses for the individual condensations of 10 to 60 $M_\odot$ and interpret them as massive isothermal protostars without luminous central stellar cores. Mezger et al. also conclude that the knots were not recognized before in molecular line observations because at the low temperatures molecules are frozen out on dust grains and are significantly depleted in the gas phase.

If correct, the observation by Mezger et al. clearly is a key result in understanding protostellar evolution and shows the way to future searches for protostellar candidates from observations of small scale structure in molecular cloud cores.

A detailed comparison with molecular data is necessary in assessing the conclusions of Mezger et al. Moore et al. (1989) have presented a 15″ resolution JCMT map of CS $7 \rightarrow 6$ emission of the NGC 2024 ridge. Figure 13 is a 14″ resolution IRAM map of integrated $C^{17}O$ $2 \rightarrow 1$ emission from Graf et al. (1990b). The molecular maps are in reasonable agreement with each other. They clearly show condensations in the ridge close to but not identical with the submillimeter peaks. It appears that molecular line data of optically thin or "thinnish" species, if taken at about the same resolution as the dust observations, reproduce the gross features found by Mezger et al. (1988) but leave some significant differences. In particular, there is only one major peak in between FIR 5 and 6 and the density contrast between clumps and the extended emission is less in the molecular maps than in the 1 mm dust emission. While it is possible that some of the CS $7 \rightarrow 6$ emission is optically thick (Moore et al., 1989), the optical depth of the $C^{17}O$ $2 \rightarrow 1$ line is low (Table 1). The ratio of the main to satellite hyperfine emission in the $C^{17}O$ emission toward FIR 5 and the ratio of $C^{17}O$ to $C^{18}O$ line intensities across the source are fully consistent with optically thin emission (Graf et al., 1990b).

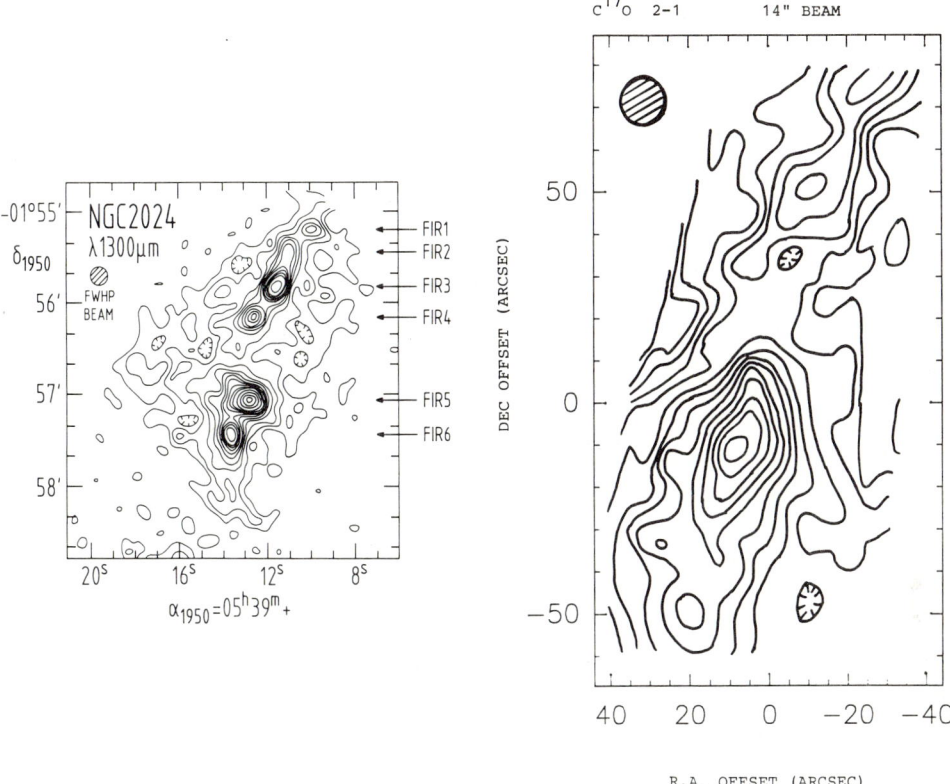

Figure 12: 1.3 mm continuum map of NGC 2024 (Mezger et al., 1988, IRAM 30m, 13″ beam).

Figure 13: Velocity integrated $C^{17}O$ $2 \to 1$ map of the same region as shown in Figure 12 (Graf et al., 1990b, IRAM 30m, 13″). The base position is FIR 5 at R.A.= 05^h 39^m 12.8^s, Dec. = $-01°$ $57'$ $04''$ (1950). Contour steps are 1.5 K main beam brightness temperature times km s^{-1}, peak is at 23.5 K km s^{-1}.

The temperature of the gas emitting in CS and the CO isotopes is high and inconsistent with the 16 K derived for the dust emission. From an analysis of various CS lines Moore et al. (1989) and Schulz et al. (1990) derive a lower limit to the gas temperature of about 28 K and a most probable value near 45 K (see also Evans et al., 1987). Graf et al. (1990a, see also Stutzki et al., this volume, p. 17) find a gas temperature near 100 K from observations of the ^{13}CO $6 \to 5$ and $3 \to 2$ lines. Graf et al. (1990b) find no evidence for a cold ($T_{ex} < 20$ K) component in available molecular line data, including the optically thin $C^{17}O$ measurements. Column densities toward FIR 5/6 derived from the optically thin $C^{17}O$ data for any assumed temperature greater than 20 K are within a factor of 2 of the column densities derived from the ^{13}CO measurements that measure the warm gas. This is shown in Table 1. On the other hand, Moore et al. (1989) have shown that a single component model of temperature 47 K can be fitted to

Table 1. *Column Densities Toward* NGC 2024 *FIR 5/6*[a)].

Measurement	$N(H_2)$ [cm^{-2}]	Comments
1 mm Dust Emission (Mezger et al., 1988)	$2...4 \times 10^{24}$	Multicomponent fit $T_d = 16$ K, $Q_d \propto \lambda^{-2}$
Dust Spectrum (Moore et al., 1989)	$3...6 \times 10^{23}$	Single component fit $T_d = 47$ K, $Q_d \propto \lambda^{-1.6}$
^{13}CO $6 \to 5/3 \to 2$[b)] (Graf et al., 1990a)	$1...2 \times 10^{23}$	$T_{gas} = 95$ K [^{13}CO]/[H$_2$] = 1.2×10^{-6}
C^{17}O/C^{18}O $2 \to 1$[c)] (Graf et al., 1990b)	$2.5..4 \times 10^{23}$	$T_{gas} \geq 20$ K, $n(H_2) = 10^7$ cm^{-3} [C^{17}O]/[H$_2$]=4.4×10^{-8}

a) 10″ to 15″ beam
b) $\tau(^{13}$CO $2 \to 1) \approx 1$ for $N(H_2) \approx 2 \times 10^{23}$ cm^{-2} at $\Delta v = 2.5$ km s^{-1}
c) $\tau($C^{17}O$) \approx 1$ for $N(H_2) \approx 6 \times 10^{24}$ cm^{-2}

all dust measurements if the assumption of the λ^{-2} wavelength dependence of the dust emissivity is relaxed (Table 1). The resulting column densities of the one component model are then quite comparable to the column densities derived from the isotopic CO data (Table 1). While a single component fit to the entire dust emission spectrum is probably not physically reasonable (Mezger et al., 1988), the argument of Moore et al. is nevertheless a "proof of feasibility" of a solution without a large amount of cold dust. There is other evidence suggesting a weaker than λ^{-2} dependence of submm/mm dust emissivity. Wright and Vogel (1985) and Woody et al. (1989) derive $Q_d \propto \lambda^{-1.2...1.7}$ from 3 mm to 350 μm continuum observations in several star forming regions. Beckwith et al. (1990) come to similar conclusions from 1.3 and 2.7 mm observations of the dust disks around a sample of pre-main sequence stars. Tielens and Allamandola (1987) comment that layered, amorphous dust particles with a large surface to volume area are theoretically expected to have a λ^{-1} emissivity dependence in the far-infrared.

Several of the FIR peaks of Mezger et al. (1988) may be associated with stellar activity. Richer et al. (1990) find a compact outflow associated with FIR 6. Moore and Chandler (1989) report a 2 μm stellar source at the position of FIR 4.

The argument on the nature of the dust clumps in NGC 2024 and elsewhere is certainly not closed. Molecules are expected to freeze out on dust grains at low temperatures and the resulting near-infrared absorptions by grain mantles have been seen toward a number of embedded compact near-infrared sources (e.g. Roche, 1989). It is suspicious, however, that no trace of any cold molecular material has been seen in the vicinity of the NGC 2024 condensations, or that no compact dust peaks have yet been reported at some distance from luminous OB stars which are are likely to substantially heat up the material in their vicinity.

4 Warm gas at the surfaces of molecular clouds

There is now a substantial amount of evidence for the presence of > 100 K atomic and molecular gas at the surfaces of molecular clouds. The warm atomic gas traced in the bright fine structure lines of [OI], [CII], [SiII] and [CI] has been recently reviewed by Genzel et al. (1989) and will not be discussed here in detail. The far-infrared atomic line emission almost certainly comes from dense photodissociation regions (PDR's) excited by the far-UV radiation impinging on the cloud's surface (Tielens and Hollenbach, 1985, Sternberg and Dalgarno, 1989). Harris et al. (1987) found intense submillimeter CO emission (CO $7 \to 6$ brightness temperatures near 100 K) of moderately narrow velocity width ($\Delta v \approx$ 5 to 10 km s^{-1}) toward S106 and the M17 interface region. Figure 14 shows the CO line fluxes as a function of rotational quantum number J toward the Orion Trapezium and Bar regions (Stacey et al., 1990). As in M17 and S106, the CO fluxes rise all the way to the far- infrared transition with $J \geq 14$ (see also Schmid-Burgk et al., 1989, Boreiko et al., 1989) indicating temperatures in excess of 200 K and densities of at least 10^6 cm^{-3}. The derived gas temperatures are high enough above the temperatures of the dust in the same regions that heating of the gas by gas-dust collisions can be excluded with some certainty (Harris et al., 1987, Graf et al., 1990a). Spatial distributions of the warm molecular gas are similar to that of the warm atomic material (Stutzki et al., 1988, Jaffe et al., 1990, Schmid-Burgk et al., 1989, Stacey et al., 1990). It is thus highly likely that the warm molecular gas is a surface phenomenon that requires an external heating mechanism.

The warm molecular gas at the cloud surfaces is not just relevant when observing star forming regions like Orion A. Dutrey et al. (this volume, p. 13), Castets et al. (1989) and Gierens (1990), for example, show that an externally heated warm cloud surface is a key assumption in a quantitative interpretation of the ^{12}CO and ^{13}CO emission from the entire Orion molecular cloud. The warm molecular gas found in starburst galaxies may also be a direct consequence of the enhanced heating rate at the cloud surfaces (e.g. Genzel 1990).

How much material is associated with the warm gas component? Estimates based on the brightness of the mid-J ^{12}CO lines give a lower limit of $\approx 10^{18}$ cm^{-2} ($\tau(^{12}\text{CO } 7 \to 6) \approx 1$) to the CO column density in gas at temperatures of at least 100 K. This lower limit corresponds to between 5 and 20% of the mass in these regions. The ^{12}CO mid-J lines are likely optically thick. Recently Graf et al. (1990a, see also Stutzki et al., this volume, p. 17) detected bright ^{13}CO $6 \to 5$ emission with the JCMT toward the Orion Bar and NGC 2024. Together with ^{13}CO $3 \to 2$ data obtained in the same positions, they were able to derive better column densities, as well as an independent estimate of the kinetic temperature of the isotopic CO emission. The results are shown in Figure 15. Toward the HII region/molecular cloud interfaces in the Orion Bar and NGC 2024 Graf et al. estimate CO column densities between 2.5 and 10×10^{18} cm^{-2} (N(H$_2$) $\approx$ 3 to 13×10^{22} cm^{-2}) for gas at temperature ≥ 70 K. This is about a factor of 3 to 5 more than obtained from the ^{12}CO data and corresponds to at least 30% of the total molecular column densities in these regions.

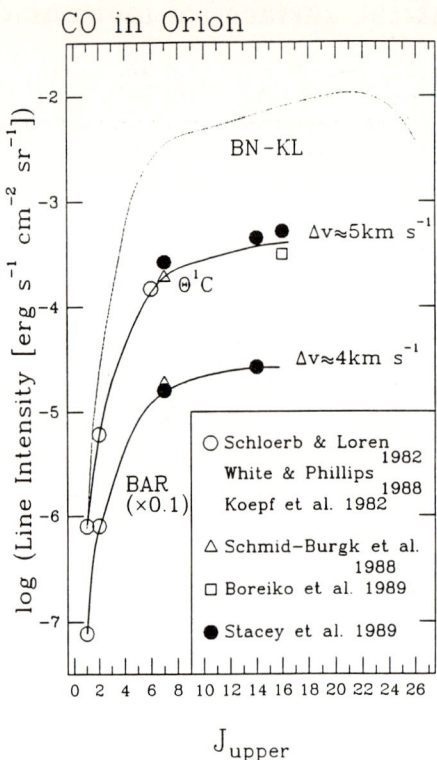

Figure 14: CO line fluxes as a function of rotational quantum number, toward Θ^1 C (middle) and the bright Bar (bottom). For comparison, the distribution of CO emission in the shocked region near BN-KL is given at the top (from Stacey et al., 1990).

The temperatures and column densities of the warm, quiescent molecular gas are very difficult to explain with any reasonable heating mechanism. The high temperatures almost certainly exclude gas-dust collisions, as mentioned above. Slow shocks consistent with the line width fail to explain the line intensities by several orders of magnitude. The spatial correlation with the atomic emission suggests an interpretation in terms of photodissociation regions. The "standard" models of Tielens and Hollenbach (1985) and Sternberg and Dalgarno (1989) fail to explain the submm and far-infrared ^{12}CO emission in Orion by an order of magnitude and the ^{13}CO emission by more. The basic reason is that in the standard models ($n(H_2) \leq 10^5$ cm^{-3}) CO is mostly dissociated in the zones that have temperatures > 100 K. A similar and possibly related problem is the failure of current models to account for the CO abundance in diffuse clouds (van Dishoeck and Black, 1986, 1989).

Burton et al. (1990) have proposed that a very dense ($n(H_2) \approx 10^6$ to 10^7 cm^{-3}) photodissociation region accounts for the measurements. In this case the combination of accelerated CO formation rate, enhanced heating rate and H_2/CO self-shielding can keep a significant column of molecular gas at temperatures ≥ 100 K. As a demonstration, Figure 16 shows a PDR model calculation for a case of $n(H_2) = 10^7$ cm^{-3} from the work by Sternberg (1990). While the ^{12}CO

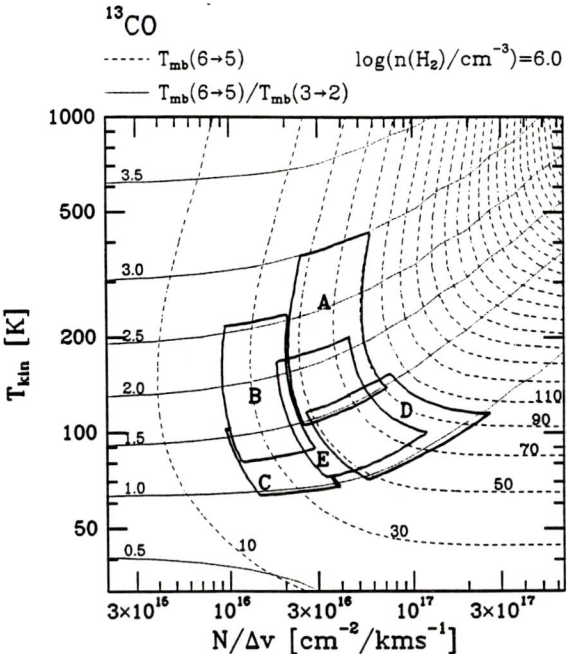

Figure 15: Kinetic temperatures and ^{13}CO column densities derived from the ^{13}CO $6 \to 5$ and $3 \to 2$ observations in NGC 2024 (regions E and A) and the Orion Bar (C,B) by Graf et al. (1990a).

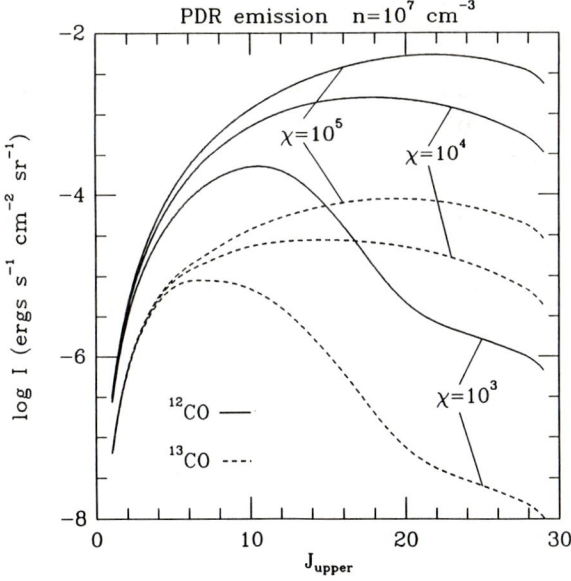

Figure 16: CO emission in a high density photodissociation region model ($n(H_2) \approx 10^7$ cm^{-3}). χ is the far-UV radiation field intensity in units of 2.5×10^{-3} ergs/cm^2/s/sr (from Sternberg, 1990).

data of Figure 14 are plausibly fitted by such models, the models still do not account for the large column densities inferred from the ^{13}CO $6 \to 5$ measurements $\{I(^{13}\text{CO } 6 \to 5) \approx 30\% \text{ to } 100\% \text{ of } I(^{12}\text{CO } 6 \to 5)\}$. If the observed ^{13}CO $6 \to 5$ emission is formed in PDR's, the effective far-UV dust absorption cross section per hydrogen nucleus must be 5 to 10 times smaller than presently assumed ($\approx 10^{-21}$ cm^{-2}) in order to heat the gas to an H$_2$ column density of a few times 10^{22} cm^{-2} (Sternberg 1990). An interesting speculation is whether the clumpy (fractal) structure of molecular clouds increases the surface to volume ratio so much that the UV penetration is significantly enhanced (cf. Boisse, 1990).

Acknowledgements I thank my colleagues A.Eckart, U.Graf, R.Güsten, A.Harris, J.Howe, D.Jaffe, G.Stacey, A.Sternberg and J.Stutzki whose work I have presented in several places of this review.

References

Arons, J. and Max, C.E. 1975, Astrophys. J., **196**, L77.
Bally, J., Langer, W.D., Stark, A.A. and Wilson, R.W. 1987, Astrophys. J., **312**, L45.
Bally, J. 1989, in *Low Mass Star Formation and Pre-Main Sequence Objects*, ed. B.Reipurth, ESO Conf.Proc., **33**, 1.
Barrett, A.H., Ho, P.T.P. and Myers, P.C. 1977, Astrophys. J., **211**, L39.
Beckwith, S.V.W., Sargent, A.I., Chini, R. and Güsten, R., 1990, (preprint).
Blitz, L. 1987, in *Physical Processes in Interstellar Clouds*, ed. G. Morfill and M. Scholer, Reidel(Dordrecht), p. 35.
Blitz, L. and Stark, A.A. 1986, Astrophys. J., **300**, L89.
Boisse, P. 1990, Astron. Astrophys., **228**, 483.
Boreiko, R., Betz, A. and Zmuidzinas, J. 1989, Astrophys. J., **337**, 332.
Burton, M., Hollenbach, D. and Tielens, A.G.G.M. 1990, Astrophys. J., (in press).
Casoli, F., Combes, F. and Gerin, M. 1984, Astr. Astrophys. Suppl., **58**, 327.
Castets, A., Duvert, G., Bally, J., Wilson, R.W. and Langer, W.D. 1989, in *The Physics and Chemistry of Interstellar Molecular Clouds*, ed. G. Winnewisser and J.T. Armstrong, Springer(Berlin), p. 133.
Cernicharo, J. and Guélin, M. 1987, Astr. Astrophys., **176**, 299.
Desert, F.-X., Boulanger, F. and Puget, J.-L. 1990, Astr. Astrophys., **237**, 215.
Draine, B.T. 1978, Astrophys. J. (Suppl.), **36**, 595.
Elmegreen, B.G. and Lada, C.J. 1976, Astron. J., **81**, 1089.
Elmegreen, B.G. 1989, Astrophys. J., **338**, 178.
Emerson, J. 1987, in *Star Forming Regions*, ed. M.Peimbert and J.Jugaku, Reidel(Dordrecht), p. 19.
Evans, N.J., Mundy, L.G. Davis, J.H. and Vanden Bout, P. 1987, Astrophys. J., **312**, 344.
Falgarone, E. and Puget, J.L. 1988, in *Galactic and Extragalactic Star Formation*, ed. R. Pudritz and M. Fich, Kluwer(Dordrecht), p. 195.
Falgarone, E. and Phillips, T.G. 1990, Astrophys. J., **359**, 344.
Fukui, Y., Sugitani, K., Takaba, H., Iwata, T., Mizuno, A., Ogawa, H. and Kawabata, K. 1986, Astrophys. J., **311**, L85.
Fukui, Y. 1989, in *Low Mass Star Formation and Pre-Main Sequence Objects*, E.S.O. Conf. Proc., ed. B. Reipurth, p. 95.
Genzel, R., Downes, D., Ho, P.T.P. and Bieging, J. 1982, Astrophys. J., **259**, L103.

Genzel, R., Harris, A.I., Jaffe, D.T. and Stutzki, J. 1988, Astrophys. J., **332**, 1049.
Genzel, R., Harris, A.I. and Stutzki, J. 1989, in *Infrared Spectroscopy in Astronomy*, ed. B.H. Kaldeich, ESA SP-290, p. 115.
Genzel, R. 1990, in *Space Chemistry*, ed. J.M. Greenberg, (in press).
Genzel, R., Eckart, A., Graf, U.U., Harris, A.I., Jackson, J., Stutzki, J. and Wild, W. 1990, (in prep).
Gierens, T. 1990, Ph.D. Thesis, Univ. Cologne.
Graf, U.U. et al., 1990b, (in prep).
Graf, U.U., Genzel, R., Harris, A.I., Hills, R.E., Russell, A.P.G. and Stutzki, J. 1990a, Astrophys. J., **358**, L49.
Harris, A.I., Stutzki, J., Genzel, R., Lugten, J.B., Stacey, G.J. and Jaffe, D.T. 1987, Astrophys. J., **322**, L49.
Herbig, G.H. and Terndrup, D.M. 1986, Astrophys. J., **307**, 609.
Howe, J.E., Jaffe, D.T., Genzel, R. and Stacey, G.J. 1990, Astrophys. J. (submitted).
Jaffe, D.T., Genzel, R., Harris, A.I., Howe, J., Stacey, G.J. and Stutzki, J. 1990, Astrophys. J., **353**, 193.
Jaffe, D.T., Davidson, J.A., Dragovan, M. and Hildebrand, R.H. 1984, Astrophys. J., **284**, 637.
Kegel, W.H. 1989, Astr. Astrophys., **225**, 517.
Lada, E. 1990, Ph.D. Thesis, Univ. of Texas.
Larson, R.B. 1981, Mon. Not. Roy. Astron. Soc., **194**, 809.
Léger, A. 1990, in *Space Chemistry*, ed. J.M. Greenberg, (in press).
Loren, R.B. 1989, Astrophys. J., **338**, 902.
Maloney, P.1988, Astrophys. J., **334**, 761.
Margulis, M and Lada, C.J. 1986, Astrophys. J., **309**, L87.
Masson, C.R. and Mundy, L.G. 1988, Astrophys. J., **324**, 538.
Matsuhara, H. et al., 1989, Astrophys. J., **339**, L67.
McCaughrean, M.J. et al., 1990, (in prep).
McKee, C.F. 1989, Astrophys. J., **345**, 782.
Mestel, L. and Spitzer, L. 1956, Mon. Not. Roy. Astron. Soc., **116**, 503.
Mezger, P.G., Wink, J.E. and Zylka, R. 1990, Astr. Astrophys., **228**, 95.
Mezger, P.G., Chini, R., Kreysa, E., Wink, J.E. and Salter, C.J. 1988, Astr. Astrophys., **191**, 44.
Mezger, P.G., Chini, R., Kreysa, E. and Wink, J.E. 1987, Astr. Astrophys., **182**, 127.
Migenes, V., Johnston, K.J., Pauls, T.A. and Wilson, T.L. 1989, Astrophys. J., **347**, 294.
Moore, T.J.T., Chandler, C.J., Gear, W.K. and Mountain, C.M. 1989, Mon. Not. Roy. Astron. Soc., **237**, 1P.
Moore, T.J.T. and Chandler, C.J. 1989, Mon. Not. Roy. Astron. Soc., **241**, 19P.
Mouschovias, T. and Spitzer, L. 1976, Astrophys. J., **210**, 326.
Mundy, L.G., Cornwell, T.J., Masson, C.R., Scoville, N.Z., Baath, L.B. and Johansson, L.E.B. 1988, Astrophys. J., **325**, 382.
Myers, P. 1986, in *Star Forming Regions*, ed. M. Peimbert and J. Jugaku, Reidel(Dordrecht), p. 33.
Myers, P.C. and Goodman, A.A. 1988, Astrophys. J., **329**, 392.
Norman, C. and Silk, J. 1980, Astrophys. J., **238**, 158.
Pauls, T.A., Wilson, T.L., Bieging, J.H. and Martin, R.N. 1983, Astr. Astrophys., **124**, 23.
Perault, M., Falgarone, E. and Puget, J.L. 1985, Astr. Astrophys., **152**, 371.
Perault, M. and Falgarone, E. 1989, in *Molecular Clouds in the Milky Way and External Galaxies*, ed. R.L. Dickman, R.L. Snell and J.S. Young, (Lecture Notes in Physics, volume 315, Springer-Verlag, Berlin), p. 233.
Phillips, T.G. and Huggins, P. 1981, Astrophys. J., **251**, 533.
Richer, J., Hills, R. and Padman, R. 1990, (in prep).
Roche, P.F. 1989, in *Infrared Spectroscopy in Astronomy*, ed. B.H. Kaldeich, ESA SP-290, p. 79.
Sanders, B.B., Scoville, N.Z. and Solomon, P.M. 1985, Astrophys. J., **289**, 373.
Scalo, J. 1990, in *Physical Processes in Fragmentation and Star Formation*, ed. R. Capuzzo-Dolcetta, C. Chiosi, and A. DiFazio, (Kluwer, Dordrecht, in press).
Scalo, J. and Pumphrey, W.A.1982, Astrophys. J., **258**, L29.
Schmid-Burgk, J. et al., 1989, Astr. Astrophys., **215**, 150.

Schulz, A. et al., 1990, (in prep).
Shu, F.H., Adams, F.C. and Lizano, S. 1987, Ann. Rev. Astr. Astrophys., **25**, 23.
Solomon, P.M., Rivolo, A.R., Barrett, J. and Yahil, A. 1987, Astrophys. J., **319**, 730.
Spitzer, L. 1982, *Searching Between Stars*, (Yale Univ.Press), p. 148.
Stacey, G.J. et al., 1990, (in prep).
Sternberg, A. and Dalgarno, A. 1989, Astrophys. J., **338**, 197.
Sternberg, A. 1990, (in prep).
Stutzki, J. and Güsten, R., 1990, Astrophys. J., **356**, 513.
Stutzki, J., Stacey, G.J., Genzel, R., Harris, A.I., Jaffe, D.T. and Lugten, J.B. 1988, Astrophys. J., **332**, 379.
Swade, D. 1989, Astrophys. J. Suppl., **71**, 219.
Tielens, A.G.G.M. and Allamandola, L.J. 1987, in *Interstellar Processes*, ed. D.Hollenbach and H.A.Thronson, (Kluwer, Dordrecht), p. 397.
Tielens, A.G.G.M. and Hollenbach, D. 1985, Astrophys. J., **291**, 722.
Ungerechts, H. and Thaddeus, P. 1987, Astrophys. J. Suppl., **63**, 645.
van Dishoeck, E. and Black, J. 1988, Astrophys. J., **334**, 711.
van Dishoeck, E. and Black, J. 1987, in *Physical Processes in Interstellar Clouds*, ed. G.Morfill and M.Scholer, (Reidel, Dordrecht) p. 241.
Wilking, B. and Lada, C.J. 1983, Astrophys. J., **274**, 698.
Wilson, T.L. and Walmsley, C.M. 1989, Astr. Astrophys. Rev., **1**, 141.
Woody, D.P., Scott, S.L., Scoville, N.Z., Mundy, L.G., Sargent, A.I., Padin, S., Tinney, C.G. and Wilson, C.D., 1989, Astrophys. J., **337**, L41.
Wright, M.C.H. and Vogel, S.N. 1985, Astrophys. J., **297**, L11.
Wynn-Williams, C.G. 1982, Ann. Rev. Astr. Astrophys., **20**, 587.
Zinnecker, H. 1989,, in *Evolutionary Phenomena in Galaxies*, ed. J.Beckman, (Cambridge Univ.Press) p. 115.
Zweibel, E.G. 1987, in *Interstellar Processes*, ed. D.Hollenbach and H.A.Thronson, (Kluwer, Dordrecht), p. 195.

Millimetre-wave observations of interstellar molecules

M. GUÉLIN[1], C. RIST[1,2] and J. CERNICHARO[1,3]

Abstract Observations at millimetre wavelengths and their impact on astrochemistry are a rapidly evolving subject well documented in the literature. Referring the reader elsewhere for comprehensive reviews (e.g. Hjalmarson, 1989, Guélin and Cernicharo, 1989, Omont, 1989) we focus here on interstellar water, hydrogen sulphide and refractory molecules, three topics which have much progressed recently (and are likely to do so in the future).

1 Interstellar and circumstellar water

Since the pioneering works with the Kuiper Airborne Observatory (Waters *et al.*, 1980, Phillips *et al.*, 1978), water vapour has remained a poorly studied interstellar molecule, except in untypical regions observed via its 22 GHz ($6_{16} - 5_{23}$) transition (the 6_{16} level lies 643 K above the ground level). The early KAO detections, restricted to one single transition ($3_{13} - 2_{20}$) of $H_2^{16}O$ and $H_2^{18}O$ in one source, Orion A (the DR21 detection seems now confirmed by Jacq *et al.*, 1988), were plagued by low angular resolution (7′), insufficient bandwidth, poor signal-to-noise ratio and unreliable calibration. They yielded very rough estimates of the water vapour abundance ($10^{-6} - 10^{-5}$ of that of H_2) consistent with theoretical predictions for dense clouds.

Progress in instrumentation, in particular the installation of large telescopes on high altitude sites, and access to space have changed (and will change even more) this picture.

Detection of the 1 mm and 2 mm water lines lying not too high in the energy ladder is now possible from the ground with resolutions of 6–15″. The $3_{13} - 2_{20}$ line of the rare $H_2^{18}O$ isotope (203.4 GHz) has now been observed in half a dozen sources, using the NRAO 12m and the IRAM 30m telescopes (Jacq *et al.*, 1988, 1990). Although the energy of the 3_{13} level ($E/k = 200$ K) is too large to allow a reliable determination of the H_2O column density, these observations open an interesting path for investigating interstellar water, as they benefit from a good resolution, are well calibrated and are easily repeated. Jacq and co-workers have been able to observe several lines of HDO in the same sources and to estimate the degree of excitation of water (although the different symmetries of HDO and H_2O and the probable enhancement of deuterium in the cold gas make these estimations disputable). They derive a fractional abundance $[H_2O]/[H_2] = 10^{-5}$ and $[HDO]/[H_2O] = 3 - 6 \times 10^{-4}$ (*i.e.* a ratio 10 – 100 times larger than the interstellar D/H ratio). The above ratios assume that $[H_2^{16}O]/H_2^{18}O]$ is equal to the terrestrial $^{16}O/^{18}O$ ratio, 490.

[1] IRAM, 300 RUE DE LA PISCINE, 38406 ST. MARTIN D'HÈRES, FRANCE.
[2] OBSERVATOIRE DE GRENOBLE, 38402 ST. MARTIN D'HÈRES, FRANCE.
[3] CENTRO ASTRONOMICO DE YEBES, OAN, AP 148, 19080 GUADALAJARA, SPAIN.

Detection from the ground of 2 mm and 1 mm lines of the main $H_2{}^{16}O$ isotope has been reported by Menten and Melnick (1989), Menten *et al.* (1990a, b) and Cernicharo *et al.* (1990). Compared with the KAO observations, these observations, although weather dependent, are easier to repeat and, above all, have a much higher angular resolution. Menten and co-workers have observed at the CSO (altitude 4000 m) the $10_{29} - 9_{36}$ transition of the ground vibrational state of H_2O (321 GHz) toward several star forming regions and late-type stars, as well as two rotational transitions of the ν_2 vibrational state toward the supergiant star VY CMa. Because the 10_{29} level lies 1860 K above the ground rotational level and the ν_2 levels lie even higher, the observed lines are most likely masers and yield little information on the water abundance.

Cernicharo *et al.* (1990) have re-observed from the ground with the IRAM telescope the more fundamental $3_{13} - 2_{20}$ transition (183 GHz), previously detected by Waters *et al.* (1980). Contrary to the cases of the weak and high excitation 22 GHz and 321 GHz lines, the 2_{20} level is low enough to be much populated in the terrestrial atmosphere and the 183 GHz line is usually optically very thick. The conjunction of a high altitude site (2900 m) and of dry winter conditions, however, lowered the amount of precipitable water vapour above the telescope to less than 1 mm. The zenith atmospheric transmission at the center of the water line was close to 30%, allowing the mapping of interstellar water emission.

A map of the Orion A (KL) nebula shows that the broad spectrum, originally observed by Waters *et al.*, is a blend of many narrow spatial and spectral features with high brightness temperatures (up to 2000 K, averaged over the 15″ telescope beam) which are almost certainly masers. Similar 183 GHz masers are also detected in a dozen other interstellar (W3(OH), W49, CepA, HH 7 – 11) or circumstellar (RX Boo,...) sources known to host 22 GHz masers. Their spectral components do not always coincide with the 22 GHz maser components.

The 183 GHz emission observed in W3 and W49N, two sources with high velocity molecular outflows, is particularly interesting since it exhibits a broad and smooth velocity component underneath the narrow maser features. If thermal, this component could arise from the compact disks thought to surround the outflow sources. Alternatively, it could be the blurred image of a myriad of unresolved masers.

Very recently, Menten *et al.* (1990b) have reported the detection in 4 star-forming regions of another water line (also, presumably, masing): the $5_{15} - 4_{22}$ transition, at 325 GHz.

The new 321 GHz and 183 GHz masers can be explained by collisional pumping. Excitation calculations of ortho-water by Neufeld and Melnick (1990), using the collisional rate coefficients of Palma *et al.* (1988), reproduce the population inversions needed for the 321 GHz and 22 GHz masers over a wide range of physical conditions (*e.g.* shocked gas). Similar calculations for para-water by Cernicharo and Gonzalez show that the 183 GHz maser can also be collisionally pumped.

Thermal emission of water should logically be searched for in the fundamental rotational transitions: the $1_{10} - 1_{01}$ transition of the ortho state (557 GHz for the main isotope) and the $1_{11} - 0_{00}$ transition of the para state (1114 GHz). Clearly, these can only be accessed from space. The more favourable $1_{10} - 1_{01}$ line of the rare isotope $H_2^{18}O$, at 547 GHz can be observed from an aircraft. This line, of course, is much weaker than the 557 GHz line, but has the advantage of being optically thin in all interstellar clouds. An attempt to detect it in 9 astronomical sources (7 star-forming regions and 2 evolved stars) has been made with the KAO (Wannier et al., 1989, 1991), with no clear positive result (hints of lines have been seen in NGC 2264 and NGC 7538). The upper limits on the $[H_2O]/[CO]$ ratio are < 0.03 in most sources (3 – 10 times lower than in Orion-KL). The atmospheric opacity above the KAO (altitude 13000 m) was only 0.2 in the wings of the telluric $H_2^{18}O$ line; this shows that observations from an extremely dry ground site, such as the Antarctic Plateau, could be possible 10 GHz away from the fundamental water line (557 GHz).

Balloon (e.g. PRONAOS-CNES: diameter = 2 m, altitude $\sim$ 40 km) and satellite (e.g. SWAS-NASA: diameter = 0.55 m) borne telescopes will be able in the next 3 – 4 years to map the 380 GHz ($4_{24} - 3_{21}$), 557 GHz and 547 GHz lines, yielding, at last, a large body of reliable data on the distribution and abundance of interstellar water.

2 Hydrogen sulphide

The fundamental $1_{10} - 1_{01}$ transition of ortho H_2S (168 GHz) lies only 15 GHz away from the 183 GHz water line. Atmospheric absorption and the location of this transition at the end of the band where most 2 mm receivers operate, prevented it from being much observed until now. Except for its original detection toward a few star-forming regions (Thaddeus et al., 1972) and for a first crude map of Orion KL and a (negative) search toward dark clouds (Guélin and Thaddeus, 1977), both made with dedicated receivers, few studies have been reported.

Ukita and Morris (1983) were able to detect the 168 GHz line in the Egg Nebula, NGC 2688 and, recently, Minh et al. (1989, 1990), using the FCRAO 14m telescope, have observed this line towards a few positions in L134N, TMC1 and Orion A (and the corresponding $H_2^{34}S$ line in Orion A). They derive $[H_2S]/[H_2]$ = 1 – few $\times 10^{-9}$ in these dark clouds and the quiescent regions of Orion.

The IRAM 30m telescope is an interesting tool for investigating the H_2S emission. Its site altitude ensures (most of the time) a good atmospheric transparency at 168 GHz and its surface accuracy a high aperture efficiency. Full advantage can thus be taken of the good angular resolution (16" at 168 GHz). A survey of the 168 GHz line emission in circumstellar envelopes (Omont et al., 1991) and star-forming regions (Rist et al., 1991) has been carried out with this instrument. The results on the circumstellar envelopes, which also include $2_{20} - 2_{11}$ (217 GHz) line observations, have been briefly discussed by Omont (1989). The 168 GHz and 217 GHz H_2S lines were securely or tentatively detected in all 15 surveyed O-rich envelopes, as well as in two C-rich envelopes (including IRC+10216). Both line shapes and the 168 GHz/217 GHz line intensity ratio (assuming an ortho/para abundance ratio of 3) suggest that the H_2S abundance is enhanced in the warm

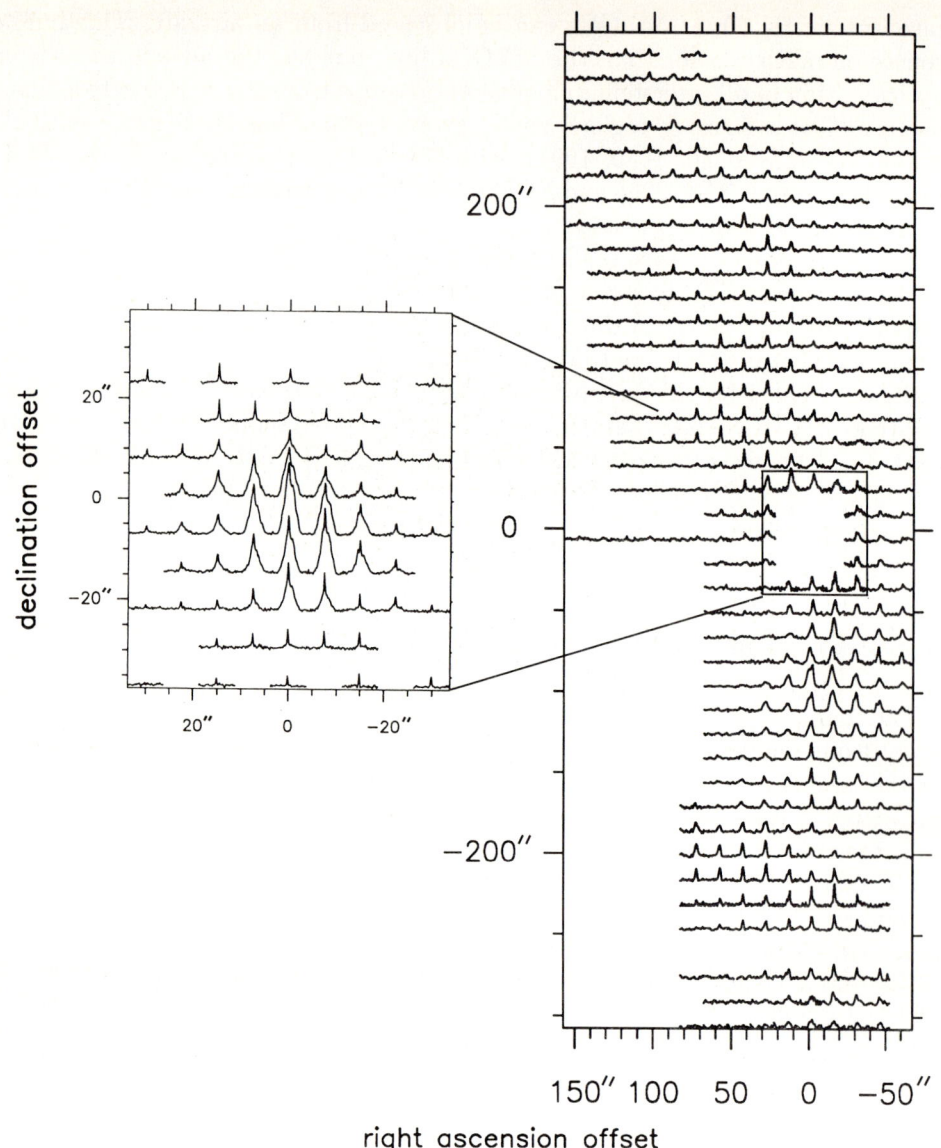

Figure 1: The profiles of the $1_{10} - 1_{01}$ line of ortho H_2S (168 GHz) in the Orion-MC1 region, plotted against the offset from IRc 2 (offsets are in arc seconds) **right:** the Orion ridge area and S6 (box scale: $\Delta v = 18\,\mathrm{km\,s^{-1}}$, $T_A^* = 7\,\mathrm{K}$); **left:** the central KL region (box scale: $\Delta v = 100\,\mathrm{km\,s^{-1}}$, $T_A^* = 22\,\mathrm{K}$). The line profiles towards IRc 2 are 5 times stronger and 10 times broader than towards the ridge. The observations, made with the IRAM 30m telescope, have an angular resolution of 16″.

internal layers where dust is formed. The first results of the dense cloud survey are presented below.

The 168 GHz line emission has been detected towards 20 sources and mapped in 7 of them. The 8 GHz spacing between the receiver upper and lower sidebands

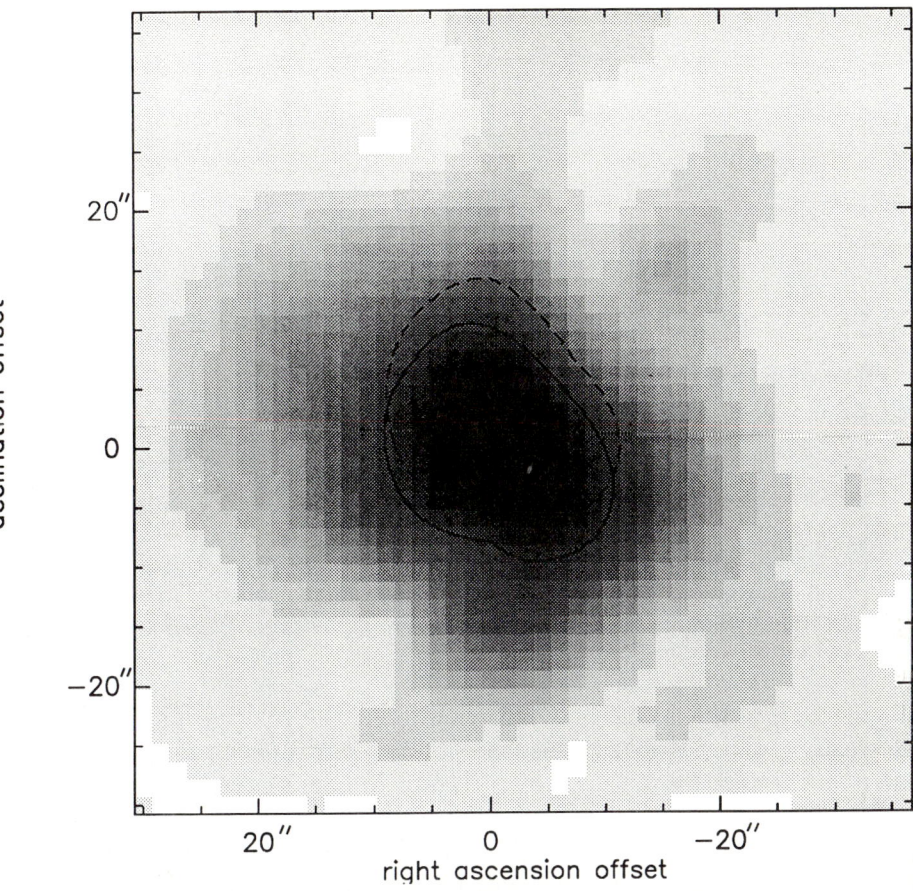

Figure 2: The bipolar outflow associated with Orion IRc 2, observed in the $1_{10} - 1_{01}$ line of H_2S. **Gray levels:** the velocity-integrated line intensity, $\int T_A^* dv$ around IRc 2 (position 0,0). (The Line profiles are presented in Figures 1a and 3). **Full line:** the red high velocity gas ($28 < v < 46\,\mathrm{km\,s^{-1}}$) half-power contour; **dashed:** the blue high velocity gas ($-25 < v < -10\,\mathrm{km\,s^{-1}}$) half-power contour. Offsets are in arc seconds, the angular resolution is 16″.

allowed simultaneous mapping with this line and the $10_{010} - 9_{19}$ (161 GHz) line of SO_2. The largest map pertains to the Orion ridge region near the KL nebula (Figure 1). Clearly, the H_2S emission there is bi-modal. The lines are broad and very intense around IRc 2 and the BN object (the *plateau* component, first observed by Thaddeus *et al.*, 1972). They are narrow and relatively weak over most of ridge. Intermediate width profiles are observed toward S6. Judging from the $H_2{}^{32}S/H_2{}^{34}S$ line intensity ratio, the 168 GHz line is optically thin or nearly optically thin in the broad plateau component and in the ridge. Adopting for H_2S the collisional excitation rates calculated for H_2O by Palma *et al.* (1988) and the physical conditions derived from the NH_3 (2,2) and (1,1) lines (Batrla *et al.*, 1983), Rist *et al.* find $[H_2S]/[H_2] = 10^{-9}$ in the ridge and toward S6, and a few $\times 10^{-7}$ in the *plateau*. In all cases $[H_2S] \ll [CS]$, *i.e.* H_2S is not an important reservoir of sulphur.

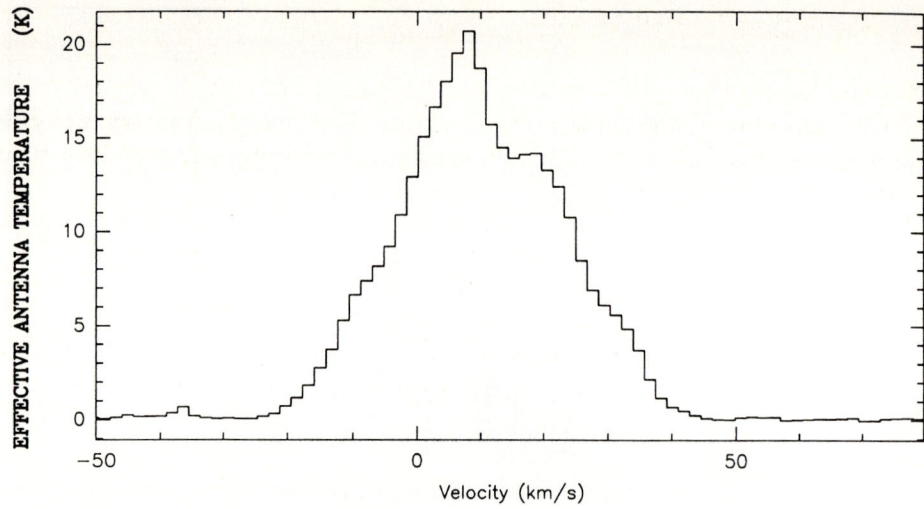

Figure 3: The profile of the $1_{10} - 1_{01}$ H$_2$S line towads IRc 2, observed with the IRAM 30m telescope. The "plateau" ($\Delta v = 40\,\mathrm{km\,s^{-1}}$) and "hot core" ($v = 5\,\mathrm{km\,s^{-1}}$, $\Delta v = 10\,\mathrm{km\,s^{-1}}$) components completely dominate the narrow ($\Delta v = 4\,\mathrm{km\,s^{-1}}$) component associated with the "ridge".

Figure 2 (gray levels) shows the distribution of the velocity-integrated 168 GHz line emission around IRc 2. This emission is seen to peak right on the infrared source (to better than 2″). There, (Figure 3), the plateau dominates the line profile completely (it contains 95% of the line integrated intensity) and extends from $-25\,\mathrm{km\,s^{-1}}$ to $+46\,\mathrm{km\,s^{-1}}$. The high velocity gas is part of a bipolar outflow, just resolved by the 16″ HPW beam: the full and dashed contours in Figure 2, which correspond respectively to the half-intensity contours of the red ($v > 28\,\mathrm{km\,s^{-1}}$) and blue ($v < -10\,\mathrm{km\,s^{-1}}$) lobes, are shifted by 4″ along the north-south direction. IRc 2 lies between the two lobe centers. It has been suspected for a long time (*e.g.* Genzel *et al.*, 1981) that IRc 2 is at the origin of the H$_2$O maser outflow in OMC 1 and of the high velocity gas corresponding to the *plateau* component observed in several molecular lines. Figure 3 supports this view and shows that the outflow is bipolar and, probably, not much inclined along the line of sight. As noted above, the H$_2$S abundance is $\sim$ 100 times higher in the outflow than in the quiescent (yet relatively warm) ridge, suggesting that $30\,\mathrm{km\,s^{-1}}$ shocks can efficiently form H$_2$S.

Besides OMC 1, the 168 GHz H$_2$S line has been observed toward a score of star-forming regions and mapped in W3(OH), NGC 2024, NGC 2264, W 75S and NGC 7538. Although less conspicuous than in OMC 1, broad "plateau" lines and extended "spike" components are observed in several cases.

3 Refractory molecules

Most molecules detected in cool ($<$ 200 K) interstellar or cirumstellar gas, obviously, have low condensation temperatures. Yet, many refractory molecules remain in the cold gas phase. These molecules offer a unique way to investigate the interactions between the interstellar dust and the gas.

The largest crop of refractory molecules comes from the carbon-rich circumstellar envelope IRC+10216. The gas temperature in this envelope is 10^3 K at a few times 10^{-1} arcseconds from the central star, and decreases to $\sim$ 70 K or less in the outer layers (*e.g.* at 10 – 30 arcseconds from the star). Most molecules formed in the outer atmosphere of the star condense onto dust grains in the inner envelope, a few species with low condensation temperature (*e.g.* CO) remaining in the gas phase and forming the basis of a rich ion-molecule chemistry (*e.g.* Omont, 1989).

Most silicon-bearing species are refractory and condense early in this process. SiO, SiS, SiC_2 and SiH_4 have been known already for some time and, at least for the first three of them (the non polar molecule SiH_4 being observed only in absorption against the central infrared source), appear to have an abundance which decreases with the distance from the central star. The SiS $J = 6 \rightarrow 5$ line interferometric map of Bieging and Rieu (1990) shows a centrally-peaked source of radius 6'' (2×10^{16} cm), the SiS abundance decreasing by a factor of 10 between 1'' and 10''. IRAM 30m telescope maps (Kahane *et al.*, 1991) show that SiC_2 is also centrally-peaked with a radius of 15''; the SiO distribution is even more compact.

SiC is expected to be one of the first molecules to condense onto dust grains (as soon as $T < 1300$ K) and, indeed, the 11 μm silicon carbide grain signature is observed around many carbon stars, among which is IRC+10216. Not much of this molecule is predicted to remain in the gas phase in the outer parts of circumstellar envelopes (McCabe, 1982). Yet the free SiC radical is observed to be relatively abundant, $[SiC]/[H_2] \sim 10^{-8}$ in the outer layers of IRC+10216 (Cernicharo *et al.*, 1989). It is probable that, rather than being a residue of the stellar atmosphere, the observed SiC is formed *in situ* by the photodissociation of SiC_2. Another silicon-bearing molecule recently discovered in IRC+10216 is l-SiC_4 (Ohishi *et al.*, 1990), which could be the first member of a new family of linear circumstellar molecules (l-C_3 and l-C_5 are observed in this source). Since the size of the SiC_4 source is unknown we cannot discriminate between a formation by neutral-neutral and ion-molecule reactions in the outer envelope (similar to those of C_5 and C_5H) or by thermochemical equilibrium reactions in the stellar atmosphere. A seventh silicon-bearing molecule, $HSiC_2$, may also be present in IRC+10216 (Guélin *et al.*, 1986, Largo, 1989).

The rotational lines of four metal halides NaCl, AlCl, KCl and AlF are detected in IRC+10216 (Cernicharo and Guélin, 1987). Crude mapping of the AlF and AlCl millimetre wave emission, with the IRAM 30m telescope and the IRAM interferometer, show that these refractory species remain abundant up to $\sim 6''$ (2×10^{16} cm) from the central star.

Finally, a phosphorus molecule, CP, has been observed in IRC+10216 (Guélin *et al.*, 1990). CP, contrary to the isostructural CN radical, which very probably results from the photodissociation of the abundant molecule HCN, seems confined to the inner envelope ($\leq 6'' - 7''$), where it is probably more abundant than HCP (which is not detected).

A survey of phosphorus molecules (PN, HPO, HCP, PH$_3$) in the dense interstellar clouds has been carried out by Turner *et al.* (1990). Only PN was detected, the limits set (on HPO in particular) being quite low. PN was found in 6 (out of 12) warm star-forming regions, but in none of the 9 cold dark clouds or 2 circumstellar envelopes surveyed. The fractional abundance of PN in the warm clouds ($\leq 10^{-10}$) and the upper limits on HPO, PH$_3$, HCP, PO, CP,.. suggest that either phosphorus is depleted by a factor of $\simeq 1000$ in the "warm" dense gas (and up to 10^5 in the cold dense gas), or that the atomic P abundance is much larger than predicted by steady state models (*e.g.* Millar *et al.*, 1987).

It has been overlooked for some time that the refractory molecule SiO, known to be abundant in circumstellar envelopes and in a few outflows near hot cloud cores (*e.g.* Orion-IRc 2), was also present in relatively cold regions: its 86 GHz (J = $2 \to 1$) line is indeed observed in absorption in the direction of the main galactic centre continuum sources. A recent survey of the 86 GHz ($v = 0$, J = $2 \to 1$) line emission toward 9 warm and cold clouds (Ziurys *et al.*, 1989) gives a better idea of the distribution of SiO in the gas phase. The 86 GHz line is detected in all clouds with $T_k > 30$ K (with, usually, [SiO]/[H$_2$] $\simeq 10^{-10}$), but not toward the cold dark clouds [SiO]/[H$_2$] $\leq 10^{-11}$), which suggests (see Ziurys *et al.*, 1989) that interstellar SiO is formed in high temperature regions and traces shocked gas (for the formation of SiO in shocks, see Neufeld and Dalgarno, 1989). Alternatively, as argued by Langer and Glassgold (1990), SiO, which is difficult to form in the warm gas by ion-molecule reactions, could be formed by the reactions of atomic Si with O$_2$ or OH. These reactions should be relatively fast and have the required temperature dependence, owing to the large splitting of the Si atom fine structure.

Further high resolution studies (in particular interferometric studies) of refractory molecules in the dense clouds will be important to better understand the condensation/evaporation process, as well as to trace the regions where the gas has recently been shocked.

Acknowledgement

We thank Dr. L. Pagani for communicating his results prior to publication.

References

Batrla, W., Wilson, T.L., Bastien, P. and Ruf, K., 1983, Astron. Astrophys., **128**, 279.
Bieging, J.H. and Nguyen-Q-Rieu, 1989, Astrophys. J., **343**, L25.
Cernicharo, J., Thum, C., Hein, H., John, D., Garcia, P. and Mattiocco, F., 1990, Astron. Astrophys., **231**, L15.

Genzel, R., Reid, M.J., Moran, J.M. and Downes, D., 1981, Astrophys. J., **244**, 884.
Guelin, M. and Thaddeus, P., 1977, unpublished results.
Guélin, M., Cernicharo, J., Kahane, C. and Gomez-Gonzalez, J., 1986, Astron. Astrophys., **157**, L17.
Guélin, M. and Cernicharo, J., 1989, in *Physics and Chemistry of Interstellar Molecular Clouds*, ed. G. Winnewisser and J.T. Armstrong (Springer-Verlag), p. 337.
Guélin, M., Cernicharo, J., Paubert, G. and Turner, B.E., 1990, Astron. Astrophys., **230**, L9.
Hjalmarson, A., 1989, in *Physics and Chemistry of Interstellar Molecular Clouds*, ed. G. Winnewisser and J.T. Armstrong (Springer-Verlag), p. 73.
Jacq, T., Jewell, P.R., Henkel, C., Walmsley, C.M. and Baudry, A., 1988, Astron. Astrophys., **199**, L5.
Jacq, T., Walmsley, C.M., Henkel, C., Baudry, A., Mauersberger, R. and Jewell, P.R., 1990, Astron. Astrophys., **228**, 447.
Langer, W.D. and Glassgold, A.E., 1990, Astrophys. J., **352**, 123.
Menten, K.M. and Melnick, G.J., 1980, Astrophys. J., **341**, L91.
Menten, K.M., Melnick, G.J. and Phillips, T.G., 1990, Astrophys. J., **350**, L41.
Menten, K.M., Melnick, G.J., Phillips, T.G. and Neufeld, D.A., 1990, Astrophys. J., **363**, L27.
Millar, T.J., Bennett, A. and Herbst, E., 1987, Mon. Not. Roy. Astron. Soc., **229**, 41P.
Minh, Y.C., Irvine, W.M. and Ziurys, L., 1990, Astrophys. J., **345**, L63.
Minh, Y.C., Ziurys, L., Irvine, W.M. and McGonagle, D., 1990, Astrophys. J., **360**, 136.
Neufeld, D.A. and Dalgarno, A., 1989, Astrophys. J., **340**, 869.
Neufeld, D.A. and Melnick, G.J., 1990, Astrophys. J., **352**, L9.
Ohishi, M., Kaifu, N., Kawaguchi, K., Murakami, A., Saito, S., Yamamoto, S., Ishikawa, S., Fujita, Y., Shiratori, Y. and Irvine, W.M., 1989, Astrophys. J.,**345**, L83
Omont, A., 1989, in *Physics and Chemistry of Interstellar Molecular Clouds*, ed. G. Winnewisser and J.T. Armstrong (Springer-Verlag), p. 361.
Omont, A., Lucas, R., Guilloteau, S. and Morris, M., 1991, (in preparation).
Palma, A., Green, S., DeFrees, D.J. and McLean, A.D., 1988, Astrophys. J. Suppl., **68**, 287.
Phillips, T.G., Scoville, N.Z., Kwan, J., Huggins, P.J. and Wannier, P.G., 1978, Astrophys. J., **222**, L59.
Rist, C., Cernicharo, J. and Guélin, M., 1991, (in preparation).
Thaddeus, P., Kutner, M.L., Penzias, A.A., Wilson, R.W. and Jefferts, K.B., 1972, Astrophys. J., **176**, L73.
Turner, B.E., Tsuji, T., Bally, J., Guélin, M. and Cernicharo, J., 1990, Astrophys. J. (in press).
Ukita, N. and Morris, M., 1983, Astron. Astrophys., **121**, 15.
Wannier, P.G., Pagani, L., Encrenaz, P.J., Frerking, M.A., Gulkis, S., Kuiper, T.B.H., Lecacheux, A., Pickett, H.M. and Wilson, W.J., 1989, in *Physics and Chemistry of Interstellar Molecular Clouds*, ed. G. Winnewisser and J.T. Armstrong (Springer-Verlag), p. 222.
Wannier, P.G., Pagani, L., Kuiper, T.B.H., Frerking, M.A., Gulkis, S., Encrenaz, P.J., Pickett, H.M., Lecacheux, A. and Wilson, W.J., 1991, (in preparation).
Waters, J.W., Gustincic, J.J., Kakar, R.K., Kuiper, T.B.H., Roscoe, H.K., Swanson, P.N., Rodriguez Kuiper, E.N., Kerr, A.R. and Thaddeus, P., 1980, Astrophys. J., **235**, 57.
Ziurys, L.M., Friberg, P. and Irvine, W.M., 1989, Astrophys. J., **343**, 201.

Newly discovered CO outflows in OMC-1

T. L. WILSON[1], J. SCHMID-BURGK[1], R. MAUERSBERGER[1], R. GÜSTEN[1], A. SCHULZ[1] and L.M. ZIURYS[2]

Abstract   12″ resolution measurements of the $J = 2 \to 1$ line of CO and the $J = 5 \to 4$, $v = 0$ line of SiO are used to investigate the three features referred to as "WJ5", "Jet" and "S6". The locations are shown in Figure 1.

1   Introduction

The Orion Molecular Cloud is the nearest region of O-B star formation. This has been and remains the object of intense study by optical, infrared and radio astronomers (see Glassgold *et al.*, 1983). The Orion-KL outflow has been observed in many radio transitions (see Plambeck *et al.*, 1985). Because of the high intensity of spectral line emission from the CO outflow associated with the Orion-KL nebula, the search for nearby outflow features has been difficult. These must be either considerably weaker or compact, and searches with sensitive receivers and large radio telescopes are required. We present here a summary of results of surveys in millimeter spectral lines made with the IRAM 30-m telescope.

2   Observations and results

All observations were made with the IRAM 30-m telescope on Pico Veleta. The search for outflow sources was carried out by mapping the $J = 2 \to 1$ line of CO. The spacing between spectra in our map was 15″, and the region mapped covered 2.5′ of R.A. and 4′ in Dec. from the position of IRc2. At the line rest frequency, the FWHP telescope beam size is 12″. The quiescent CO emission in this part of the OMC-1 cloud has a radial velocity between 6 (south of Orion-KL) and 10 km s^{-1} (north of Orion-KL). The peak intensity of the quiescent CO emission is about 100 K. To locate outflows, we have searched velocity ranges more than 12 km s^{-1} from the radial velocity of the quiescent gas. In Figure 1 we show a plot of the lower contours of CO line intensity, integrated over radial velocity range from 19 to 28 km s^{-1}. The corresponding blue-shifted feature shows only the KL outflow and the outflow 100″ south of Orion-KL; channel maps for lower radial velocities show a large number of weaker, extended features. As can be seen from Figure 1, there are 4 discrete sources. Most intense is the KL outflow which has been investigated in detail in the radio and IR range.

In Figure 1 we also plot the positions of the M42 Herbig-Haro objects, as listed

---

[1] MAX-PLANCK-INSTITUT FÜR RADIOASTRONOMIE, AUF DEM HÜGEL 69, D-5300, BONN, FRG.
[2] CHEMISTRY DEPT., ARIZONA STATE UNIVERSITY, TEMPE, ARIZONA, U.S.A.

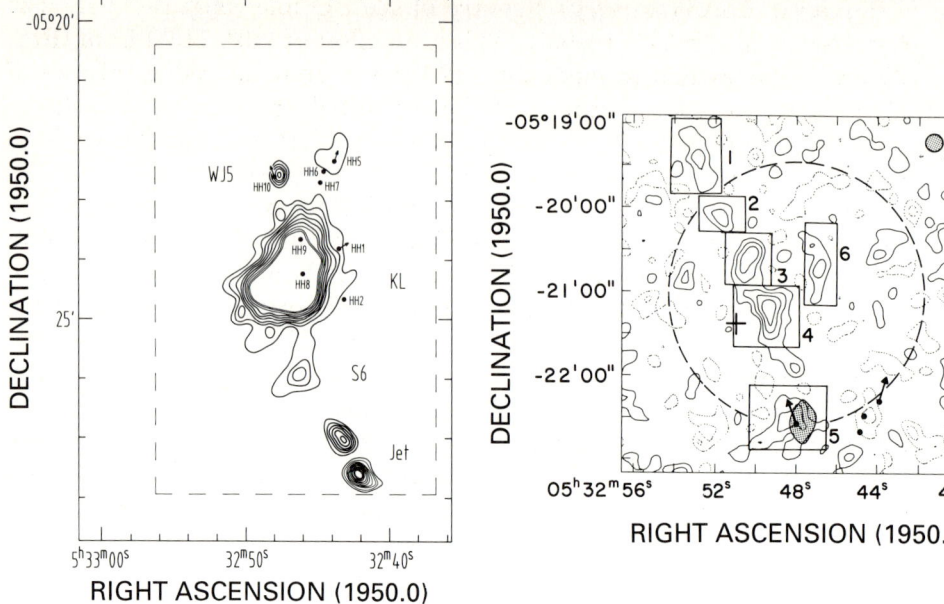

Figure 1: A contour map of the $J = 2 \to 1$ emission, integrated between 19 and 28 km s^{-1}, V_{lsr}, in the region around Orion-KL. The contour units are K km s^{-1}, where the intensity is T_R^*, the equivalent of main-beam brightness temperature. The lowest contour shown is 4 K km s^{-1}, the spacing between levels is 4 K km s^{-1}, and the highest contour shown is 40 K km s^{-1}. Data were taken with 12″ resolution. Within the dashed region, the spacing between spectra is 15″. Outside this, the spacing is 30″ between spectra. The sources discussed in this report are named. The positions of the HH objects catalogued by Axon and Taylor (1984) are shown as dots and arrows.

Figure 2: The shaded area shows the location of the northern most CO outflow found in Figure 1, on the 12.5″ resolution maps of the 2 cm line of H$_2$CO (Wilson and Johnston, 1989, WJ). The cross marks the location of the $v = 0$, $J = 2 \to 1$ line of SiO reported by Ziurys and Friberg (1987). The dots and dots connected to arrows mark the location of the M42-HH objects (Axon and Taylor, 1984)

by Axon and Taylor (1984). In Figure 2, we plot as a shaded region the CO high velocity emission on the VLA map of Wilson and Johnston (1989, hereafter WJ). We have searched for SiO emission in the $v = 0$, $J = 5 \to 4$ line toward the peaks of sources 1, 3, 4 and 5 in Figure 2. The only positive result was for WJ5. Because of limited observing time, no attempt was made to map SiO emission for sources north of Orion-KL. For the region between Orion-KL amd S6, a map was made in the $v = 0$, $J = 2 \to 1$ line of SiO. Emission was found only close to these two peaks. An additional map in the $v = 0$, $J = 5 \to 4$ line of SiO was made for the S6 region.

The WJ5 source: This is a localized source in the CO line wings, at R.A.=05^h 32^m 48.2^s, Dec.=-05° 22' 34" (1950.0). This is an offset of (10", 110") from IRc2. The CO line wings extend to more than 12 km s^{-1} from the radial velocity of the ambient material. The intensity of the redshifted line wing is three times larger than the blueshifted wing. This source is coincident with the Herbig-Haro Object M42-HH10 (Axon and Taylor, 1974) and the peak of a 25" (FWHP) source of 2 cm H$_2$CO line emission (Wilson and Johnston, 1989). Toward this position, emission from the v = 0, J = 5 → 4 line of SiO has been detected. From the ratio of the intensities of the J = 3 → 2 and 5 → 4 lines of C^{34}S, the average H$_2$ density for this region is 2×10^6 cm^{-3}. The kinetic energy in the CO outflow is more than 10^{43} ergs, about 0.01% of the energy in the high velocity CO outflow in Orion KL (see e.g. Snell *et al.*, 1984, for estimates of the outflow energy from the Orion-KL source). Weaker non-gaussian linewings are found in the CO emission from nearby extended regions; for these however, the blue line wing is more intense than the red wing. WJ5 may be the result of interaction between a high velocity flow with dense quiescent gas. Shock waves caused by this collision would liberate SiO. It is more likely, however, that this region contains a young star. The detection of $v = 0$ quasi-thermal SiO emission has been interpreted as evidence for an embedded star, since Ziurys *et al.* (1989) have argued that the presence of SiO is caused by high kinetic temperatures.

The jet-like feature: The second source is a remarkable jet-like feature, consisting of two peaks, extended NE–SW (Schmid-Burgk *et al.*, 1990). More detailed maps made in late 1989 show that these peaks are better aligned than in Figure 1. The material in this jet may have been ejected from S6, the source discussed next. The jet is 120" long, and less than 8" wide. The matter in the jet appears to be accelerated all along the jet. Lower velocity material appears to surround a higher velocity core. Halfway along the jet, there is a sharp break, beyond the break, the acceleration continues without a change in direction or strength. The detection of a ^{13}CO line indicates an H$_2$ density of a few times 10^5 cm^{-3}. The maximum peak temperature is about 45K.

The S6 source: The third source referred to as S6 (Batrla *et al.*, 1983) or OMC-S (Ziurys *et al.*, 1990). Johnston *et al.* (1983) have suggested that this region is inside the Orion A HII region. On the basis of 12" resolution SiO maps, Ziurys *et al.* (1990) have reported that OMC-S is a bipolar outflow source in the v = 0, J = 5 → 4 line of SiO. The center is located at R.A.=05^h 32^m 45.5^s, Dec.=-05° 26' 05" (1950.0). The maximum outflow velocity is about 25 km s^{-1} (FWZP). The blue lobe is located about 5" to the north, and the red-shifted lobe about 10" to the south of the center. Ziurys *et al.* (1990) estimate that the energy of this outflow is about 10^{45} ergs, about 1% of the energy in the Orion-KL outflow. From a multilevel analysis of the SiO and C^{34}S data, the density in the outflow is about 3×10^5 cm^{-3}. The direction of this outflow is nearly perpendicular to the direction of the Orion-KL outflow. This difference indicates that magnetic fields do not align the direction of all of the outflows in the Orion Molecular Cloud (see Dragovan, 1986).

3   Other outflow features

This report has concentrated on regions of small angular extent. The most remarkable feature is the Jet $>3'$ from Orion-KL. On a more extended scale, there is evidence for non-gaussian line wings extended NS over arc minute size regions near Orion-KL. These have been discovered by Hasegawa (1987) in the $J = 1 \rightarrow 0$ line of CO. Additional CO measurements in the $J = 2 \rightarrow 1$ line of CO have been made by Martin-Pintado et al. (1990), in NH_3 by Murata et al. (1990), and in the vibrationally excited line of H_2 by Lane (1990). It is possible that this extended CO is related to the source of the high velocity outflow in Orion-KL. This material may be related to the M42-HH objects. It is less likely, but still possible that a few of the discrete sources in the North of OMC-1, such as WJ5 are affected by this outflow.

References

Axon, D.J. and Taylor, K., 1984 Mon. Not. Roy. Astron. Soc., **207**, 241.
Batrla, W., Wilson, T.L., Bastien, P. and Ruf, K., 1983, Astron.Astrophys., **128**, 279.
Dragovan, M., 1986, Astrophys. J., **308**, 270.
Glassgold, A.E., Huggins, P.J. and Schucking, E.L., 1983, *Symp. on the Orion Nebula to honour H. Draper*, N.Y. Acad. of Sciences (N.Y., New York).
Hasegawa, T., 1987, in *Star Forming Regions*, Proc. of IAU Symp. 115, ed M. Peimbert and J. Jugaku, (Reidel, Dordrecht) p. 123.
Johnston, K.J., Palmer, P., Wilson, T.L. and Bieging, J.H., 1983, Astrophys.J., **271**, L89.
Lane, A. 1989 in ESO Workshop on *Low Mass Star Formation and Pre- Main Sequence Objects*, ed. B. Reipurth, p. 331.
Martin-Pintado, J., Rodriguez-Franco, A. and Bachiller, R., 1990, Astrophys. J., **357**, L49.
Murata, Y., Kawabe, R., Ishiguro, M., Morita, K.-I., Kasuga, T., Takano, T. and Hasegawa, T., 1990, Astrophys. J., **359**, 125.
Plambeck, R.L., Vogel, S.N., Wright, M.C.H., Bieging and Welch, W.J., 1985, in *The International Symposium on Millimeter and Submillimeter Wave Radio Astronomy*, ed. J. Gomez-Gonzales, (Publ. by URSI), p. 236.
Schmid-Burgk, J., Güsten, R., Mauersberger, R., Schulz, A. and Wilson, T.L., 1990, Astrophys. J., **362**, L25.
Snell, R.L., Scoville, N.Z., Sanders, D.B. and Erickson, N.R., 1984, Astrophys. J., **284**, 176.
Wilson, T.L. and Johnston, K.J., 1989, Astrophys. J., **340**, 894.
Ziurys, L.M. and Friberg, P., 1987, Astrophys. J., **314**, L49.
Ziurys, L.M., Friberg, P. and Irvine, W.M., 1989, Astrophys. J., **343**, 201.
Ziurys, L.M., Wilson, T.L. and Mauersberger, R., 1990, Astrophys.J., **356**, L25.

Detailed structure of the NGC 7538 molecular cloud

O. KAMEYA[1], N. HIRANO[1], R. KAWABE[1] and B. CAMPBELL[2]

Abstract High-spatial-resolution observations of the NGC 7538 molecular cloud have been performed with the NRO 45m telescope, JCMT, Nobeyama Millimeter Array, and VLA. We report on the detailed structure of this region based on these observational results. We have found an expanding molecular shell around the NGC 7538 HII region indicating some shock compression of the cloud due to the expansion of the HII region. A high-density region ($> 10^4$ cm^{-3}) is found at the south of the HII region. It contains some infrared sources, the IRS1 complex, IRS9, and IRS11. These infrared sources are associated with bipolar outflows, H_2O and OH masers, ultracompact HII regions, and high-density cores ($> 10^6$ cm^{-3}). We have also found four new H_2O maser sources in the high-density core suggesting star-forming activities.

Introduction. The high-density molecular cloud associated with the HII region, NGC 7538 (e.g. Israel, 1977), is a good example for investigating a formation process of OB stars. This cloud contains at least three massive star forming regions, the IRS1 complex, IRS9, and IRS11 (Werner et al., 1979). To resolve activity at distances of 2.7 kpc, high angular resolution (<30″) is required. We will describe new results from high-angular resolution observations.

Observations. The CO J = 1 → 0, ^{13}CO J = 1 → 0, HCN J = 1 → 0, HCO$^+$ J = 1 → 0 lines were observed in July 1986 using the Nobeyama Radio Observatory 45m telescope. The mapped areas were 10′x5′ for the CO, HCN, HCO$^+$ lines, and 8′x5′ for the ^{13}CO line. The data was obtained with 15″ spacing. We used 2048 channel Acousto Optical Spectrometers with a frequency resolution of 37 kHz. The CO J = 3 → 2 line was observed by using the 15m James Clerk Maxwell Telescope at Mauna Kea in Hawaii in December 1987. The beam size was 15″ and the grid spacing was 20″. The angular resolution was about the same as that of Kameya et al. (1989). The 1.3 cm H_2O line was observed with Nobeyama Millimeter Array in May 1987 in order to search for new H_2O masers in the NGC 7538 region. The sysnthesized beam was 6″x5″, and the absolute positional accuracy was 0.2″. The details of the observations are described in Kameya et al. (1990). The 6 cm continuum and 18 cm OH line were observed with Very Large Array in April 1989 in order to search for new OH masers and new ultracompact HII regions. Data were obtained in the B-configuration, which provides the sysnthesized beam of 3″x3″ in 18 cm and 1″x1″ in 6 cm. The derived rms noise of 6 cm continuum was 0.06 mJy/beam, which was at least 5 times better than those of the former observations.

An expanding region around NGC 7538. Figure 1 shows the maps of the peak antenna temperature of J = 1 → 0 of the CO, ^{13}CO, HCN and HCO$^+$ lines. There is a clear cavity at the HII region NGC 7538 itself, and there is a ridge of

---

[1] NOBEYAMA RADIO OBSERVATORY, NOBEYAMA, MINAMISAKU, NAGANO 384-13, JAPAN
[2] UNIVERSITY OF NEW MEXICO, ALBUQUERQUE, NM 87131, U.S.A.

enhanced antenna temperature at the boundary of the HII region. The thickness of the enhancement is 30" – 50" (0.4 – 0.7pc). The existence of HCO$^+$ and HCN emission suggests that this region is dense (n(H$_2$) > 10^4 cm^{-3}). The velocity structure of the ^{13}CO line distribution indicates that this region is expanding with a velocity of about 4 km s^{-1}. Kameya and Takakubo (1988) have studied a part of this region, and have shown that it is a shocked region due to the expansion of the HII region. If the whole area is shock compressed due to the expansion of the HII region, both column density and density of this area should be enhanced by a factor of four. Assuming the gas kinetic temperature of 25 K and relative abundance per velocity gradient of the HCO$^+$ molecule X(HCO$^+$)/(dV/dr) = 10^{-9} to 10^{-10} pc km^{-1} s, we estimated the molecular density of both a candidate of the shock compressed region (A in Figure 1) and that of the preshock region (B) under the large velocity gradient approximation. The estimated density is $10^{3.8-5.0}$ cm^{-3} for the shocked region, and $10^{3.3-4.3}$ cm^{-3} for the preshocked region. Therefore the density of the region A is about 3 times larger than that of region B, if the relative abundance of the HCO$^+$ molecule does not change between the two regions. Similarly the column density of the HCO$^+$ molecule was also estimated assuming the lines to be optically thin and in LTE. The enhancement of the column density is 3.5 to 6. These values of enhancement are consistent with the model that the shocked area is made by a swept-up effect of molecular gas due to the shock wave around the excitation star of the NGC 7538 HII region.

High-velocity flows and high density cores. Kameya et al. (1986, 1989) observed this region with J = 1 → 0 lines of CS and CO using the NRO 45m telescope, and showed that there are at least three high-velocity flows associated with the IRS1 complex, IRS9, and IRS11. The high-velocity flow associated with IRS1 complex shows clear bipolarity (e.g. Scoville et al., 1986). The other two flows probably show bipolarity (Kameya et al., 1989).

Figure 2 shows the integrated intensity map of the CO J = 3 → 2 line. The blue wing is integrated from -74 km s^{-1} to -64 km s^{-1}, and the red wing from -54 km s^{-1} to -44 km s^{-1}. The distributions of the blue and red wings are similar to those of CO J = 1 → 0 line. The small difference is an enhancement of blue wing at (60", -40"). The center of the flow associated with the IRS1 complex is shifted from IRS1 by 10" towards the south. This shift is consistent with observations done by Pratap et al. (1990) and Kameya et al. (1989).

Figures 1(b) and 1(c) show the distribution of high-density gas. There are four maxima around the IRS1 complex, IRS4, IRS9, and IRS11. It is noteworthy that the dense gas around the IRS1 complex and IRS9 elongate along the east-west direction, and these directions are perpendicular to the flow directions at the IRS1 complex and IRS9. The maximum velocity gradients are also east-west. Very high-density gas (n(H$_2$) > 10^6 cm^{-3}) traced by CS J = 7 → 6 line has maxima at the IRS1 complex, IRS9, and IRS11 (Kameya, 1990). Therefore these three regions are probably protostellar disks.

New H$_2$O masers, OH masers, and new ultracompact HII regions. In the NGC 7538 region masers were found at IRS1 and near IRS11 (e.g. Genzel and

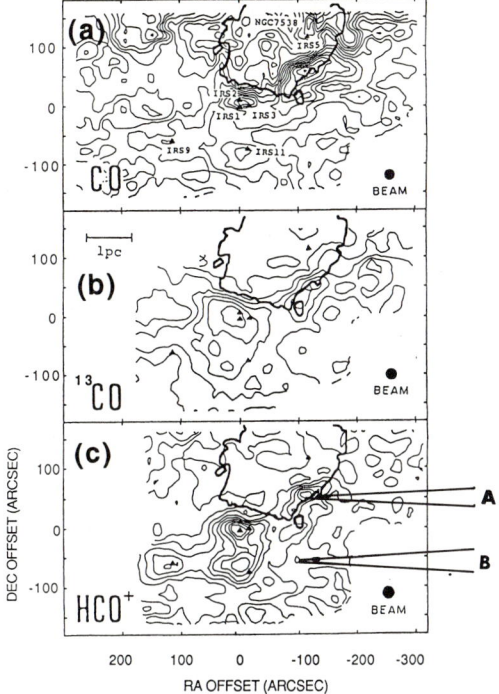

Figure 1: The peak antenna temperature maps of (a) CO J = 1 → 0, (b) ^{13}CO J = 1 → 0, (c) HCO$^+$ J = 1 → 0 lines. Contours are (1K,2K,...), (1K,2K,...) and (0.3K,0.6K,...) respectively. The origin of scale is at IRS1 (RA(1950) = $23^h\ 11^m\ 36.8^s$, Dec(1950) = 61° 11′ 48″). Triangles indicate infrared sources (Werner et al., 1979). The HII region NGC 7538 (Israel, 1977) is shown with a thick line.

Downes, 1977). No maser or ultracompact HII region has been reported in the other regions in the NGC 7538 region.

We made a sensitive survey of H$_2$O masers within the NGC 7538 region using the Nobeyama Millimeter Array, and we found five new H$_2$O masers within an 8′×8′ area (Kameya et al., 1990). One is associated with IRS9. The other four are not associated with any IR sources. These masers may be related to pre main sequence objects. We found two new ultra-compact continuum sources at IRS9 and at the H$_2$O position near IRS11 using VLA. A new OH maser was also found at IRS9.

IRS11 region. Figure 3 shows the detailed distribution of H$_2$O maser spots, OH position, IRS11, and the ultracompact radio source. The H$_2$O maser spots seem to be aligned in the NW–SE direction in a region of size 4″. This position of the maser spots coincides with the OH maser position within the observational error bar (Wynn-Williams et al., 1974). The ultracompact HII region has two maxima with the peak flux of 1.6 mJy/beam and 2.8 mJy/beam. The strongest position is very close to the cluster of H$_2$O maser spots (Y) and OH maser position. The weaker peak is 2″ northwest from the strongest position. The continuum

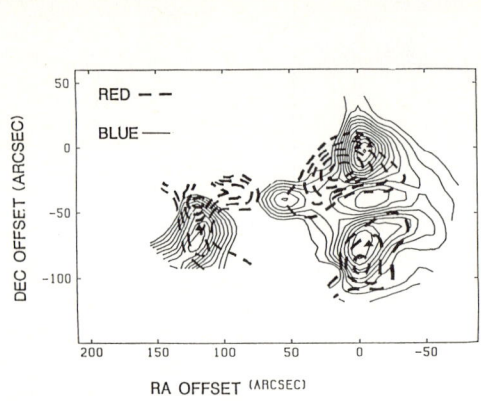
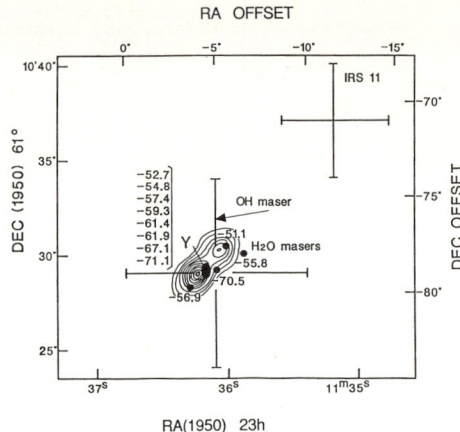

Figure 2: An integrated intensity map of CO J = 3 → 2 line wings. The blue wing is integrated from -74 km s^{-1} to -64 km s^{-1}, and the red wing from -54 km s^{-1} to -44 km s^{-1}. Contours are 21 K km s^{-1}, 26 K km s^{-1}, ... The origin of scale is same as Figure 1. The crosses indicate the observed positions.

Figure 3: A detailed distribution of H$_2$O maser spots (filled circles), OH position (error bar), IRS11 (error bar), and the ultra-compact radio source (thin contours). Contours are 0.15, 0.3, 0.6, 0.9, ..., 2.7mJy/beam.

distribution is very similar to that of H$_2$O maser spots. Campbell (1984) has found a "peanut" shaped structure of radio continuum at IRS1. We suggest that the radio continuum source near IRS11 is also in "peanut" shape. Werner *et al.* (1979) recognized that IRS11 is situated about 15″ northwest from the position of the H$_2$O/OH masers, and that IRS11 may be a reflection nebula. Therefore we suggest that star forming activity is centering at the continuum/H$_2$O position.

References

Campbell, B., 1984, Astrophys. J., **282**, L27.
Genzel, R. and Downes, D., 1977, Astron. Astrophys. Suppl., **30**, 145.
Israel, F.P., 1977, Astron. Astrophys., **59**, 27.
Kameya, O., 1990, in *Submillimetre Astronomy*, ed. G.D. Watt and A.S. Webster (Kluwer Academic Publishers), p. 181.
Kameya, O, Hasegawa, T.L., Hirano, N., Tosa, M., Taniguchi, Y. and Takakubo, K., 1986, PASJ, **38**, 793.
Kameya, O., Hasegawa, T.I., Hirano, N., Takakubo, K. and Seki, M., 1989, Astrophys. J., **339**, 222.
Kameya, O. and Takakubo, K., 1988, P.A.S.J., **40**, 313.
Kameya, O., Morita, K.I., Kawabe, R. and Ishiguro, M., 1990, Astrophys. J., **355**, 562.
Pratap, P., Batrla, W. and Snyder, L.E., 1990, Astrophys. J., **351**, 530.
Scoville, N.Z., Sargent, A.I., Sanders, D.B., Claussen, M.J., Masson, C.R., Lo, K.Y. and Phillips, T.G., 1986, Astrophys. J., **303**, 416.
Werner, M.W., Becklin, E.E., Gatley, I., Mathews, K., Neugebauer, G. and Wynn-Williams, C.G., 1979, Mon. Not. Roy. Astron. Soc., **188**, 463.
Wynn-Williams, C.G., Werner, M.W. and Wilson, W.J., 1974, Astrophys. J., **187**, 41.

Structure and physical properties of the molecular outflow in B335

N. HIRANO[1], O. KAMEYA[1], T. KASUGA[1], T. HASEGAWA[2], S.S. HAYASHI[3] and T. UMEMOTO[4]

Abstract We have made single-dish observations of the B335 outflow in the $J = 2 \to 1$ and $J = 1 \to 0$ transitions of ^{12}CO and ^{13}CO with angular resolutions of $16'' - 22''$, and the interferometric observations of the $J = 1 \to 0$ transition of ^{12}CO with angular resolution of $\sim 7''$. The single-dish maps show that the outflow has a clumpy shell-like structure with the higher excitation gas found inside the shell of the lower excitation gas. Two transitions of ^{12}CO and ^{13}CO are used to derive the optical depth and excitation temperature of the high-velocity molecular gas. Our results show that the much of the high-velocity ^{12}CO emission is optically thick and has an excitation temperature of 12 K. After correction for the opacity, the mass of the outflowing gas is estimated to be 0.6 $M_\odot$. The higher resolution interferometer maps show that the origin of the outflow is close to the millimeter continuum emission peak and that the outflow is focused within $\sim$ 1000 A.U. of its origin.

1 Introduction

Bok globule B335 is a typical low-mass star forming region and is known to have a distinct bipolar molecular outflow. B335 is one of the most suitable objects to study the molecular outflow, because of its proximity (250 pc; Tomita, Saito, and Ohtani, 1979), isolation from other star forming regions, and the quiet kinematics of the ambient material. Recent high resolution maps of the ^{12}CO ($J = 1 \to 0$) emission (Goldsmith et al., 1984, Hirano et al., 1988, Cabrit, Goldsmith, and Snell, 1988, Moriarty-Schieven and Snell, 1989) show that the outflow has biconical shape with a constant opening angle of $\sim 45°$ and is centered at the low-luminosity far-infrared source, IRAS 19345+0727. The axis of the outflow is nearly perpendicular to the line of sight, having an inclination angle of $\sim 10°$. Hirano et al. (1988) have detected strong high-velocity ^{12}CO emission toward the IRAS source, the N-S extent of which is not resolved with the 16″ beam. This result strongly suggests that the outflow is currently active and is focused within 4000 A.U. of the driving source.

In order to estimate the physical properties of the outflowing gas, we have made observations of the $J = 2 \to 1$ and $J = 1 \to 0$ transitions of ^{12}CO and ^{13}CO. We have also made higher resolution interferometric maps of ^{12}CO ($J = 1 \to 0$) emission around the central IRAS source which reveal small-scale structures of the outflow in the close vicinity of its origin.

[1] NOBEYAMA RADIO OBSERVATORY, NOBEYAMA, MINAMISAKU, NAGANO 384-13, JAPAN
[2] INSTITUTE OF ASTRONOMY, THE UNIVERSITY OF TOKYO, MITAKA, TOKYO, 181, JAPAN.
[3] JOINT ASTRONOMY CENTRE, 665 KOMOHANA STREET, HILO, HAWAII 96720, U.S.A.
[4] DEPARTMENT OF ASTROPHYSICS, NAGOYA UNIVERSITY, CHIKUSAKU, NAGOYA, 464-01, JAPAN.

2 Observations

Observations of the J = 2 → 1 transition of ^{12}CO and ^{13}CO were made with the 15m James Clerk Maxwell Telescope (JCMT) at Mauna Kea, Hawaii. The telescope had a beam size of 22" and the main beam efficiency of 0.84 at 230 GHz. The spectra of the J = 1 → 0 transition of ^{12}CO and ^{13}CO were taken with the 45m telescope of the Nobeyama Radio Observatory (NRO). The beam size and the main beam efficiency were 16" and 0.45, respectively. Interferometric observations of the ^{12}CO (J = 1 → 0) were made with the Nobeyama Millimeter Array (NMA; Ishiguro et al., 1984). The synthesized beam had a size of 8.1" × 5.0" (HPBW) with a position angle of 153.7°. The field centre was the position of IRAS 19345+0727, $\alpha = 19^h\ 34^m\ 35.3^s$, $\delta = +7°\ 27'\ 24"$ (1950).

3 Results and discussion

3.1 ^{12}CO (J = 2 → 1) map

Figure 1 shows the distribution of the blueshifted and redshifted ^{12}CO (J = 2 → 1) emission. The high-velocity ^{12}CO (J = 2 → 1) emission shows a fan-shaped bipolar distribution centered at IRAS 19345+0727. The red lobe has a shell structure with three prominent clumps in it, while the shell structure of the blue lobe is less clear. We can compare the J = 2 → 1 emission with the maps of the ^{12}CO (J = 1 → 0) emission in Hirano et al. (1988). The characteristics of our J = 2 → 1 maps are as follows:

1. The opening angle of the red lobe shell in the J = 2 → 1 map is ∼ 35°, which is smaller than that in the J = 1 → 0 map. This suggests that the higher excitation gas traced by the J = 2 → 1 emission is located inside the shell of the lower excitation gas.
2. The locations of the clumps seen in the J = 2 → 1 maps are shifted upstream in the outflow relative to the J = 1 → 0 maps. In addition, the intensity contrast between clumps and interclumps is ∼ 2.5 : 1 in the J = 2 → 1 maps, which is twice as large as that in the J = 1 → 0 maps.

3.2 Physical Parameters of the Outflow

Figure 2 shows the spectra of the J = 2 → 1 and J = 1 → 0 transitions of ^{12}CO and ^{13}CO at the position of the IRAS source. Using these spectra and assuming LTE, we can estimate the optical depth and the excitation temperature of the outflowing gas. In the inner wing region where ^{13}CO wing emission is seen, the opacity of each transition was determined from the ratio of ^{12}CO and ^{13}CO emission. We have obtained the ^{12}CO opacities of J = 1 → 0 and J = 2 → 1 transitions to be ∼ 9 and ∼ 20, respectively. From comparison between these opacities of two transitions, the excitation temperature is determined to be 12 K,

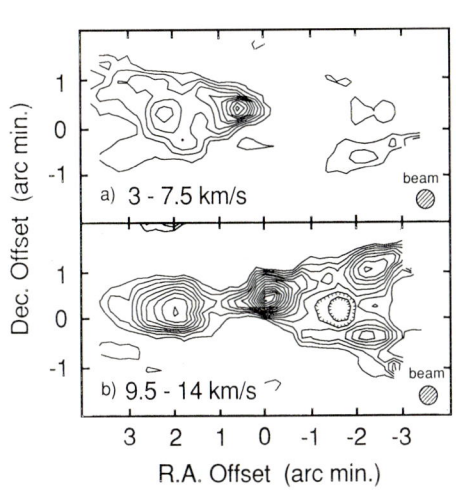

Figure 1: Maps of B335 showing the spatial distribution of the integrated intensity of the $^{12}\text{CO}(J = 2 \to 1)$ transition. The velocity intervals are (a) 3.0 – 7.5 km s^{-1} and (b) 9.5 – 14.0 km s^{-1}. The contours are spaced at every 0.6 K km s^{-1} with the lowest contours at 0.9 K km s^{-1}.

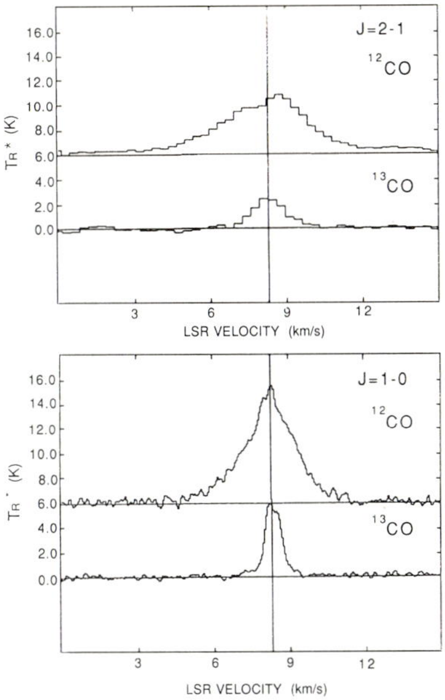

Figure 2: Spectra of the ^{12}CO and ^{13}CO emission in the $J = 2 \to 1$ transition (upper) and in the $J = 1 \to 0$ transition (lower) at the position of IRAS 19345+0727. The $J = 2 \to 1$ data were obtained with the 15m telescope of JCMT, and the $J = 1 \to 0$ data were obtained with the 45m telescope of NRO.

which is almost same as that of the ambient gas. In the outer wings region, where the ^{13}CO wings are invisible, the ^{12}CO opacities are obtained from the ratio of ^{12}CO $J = 2 \to 1$ and $J = 1 \to 0$ wing intensity (Plambeck, Snell, and Loren, 1983) with the assumption that the excitation temperature is $\sim$ 12 K. The ^{12}CO opacity of the $J = 1 \to 0$ transition decreases to a value of $\sim$ 3. Using the derived optical depths and excitation temperature we have calculated the total mass of the outflowing molecular material to be 0.6 M$_\odot$. The momentum and energy of the outflow corrected for the inclination angle of $\sim$ 10° are 4.6 M$_\odot$ km s^{-1} and 3.7×10^{44} ergs, respectively.

3.3 Small-Scale Structure of the Outflow

In Figure 3, we present channel maps of the ^{12}CO ($J = 1 \to 0$) emission taken by the interferometer. The map at the velocity of the line-center (Figure 3d), shows no significant emission. This implies that most emission from the ambient cloud is resolved out. We can consider that the emission seen in Figures 3a – 3c and Figures 3e – 3g is the wing emission which originates from the outflowing gas. One sees the double-lobed structure in Figures 3c and 3e,

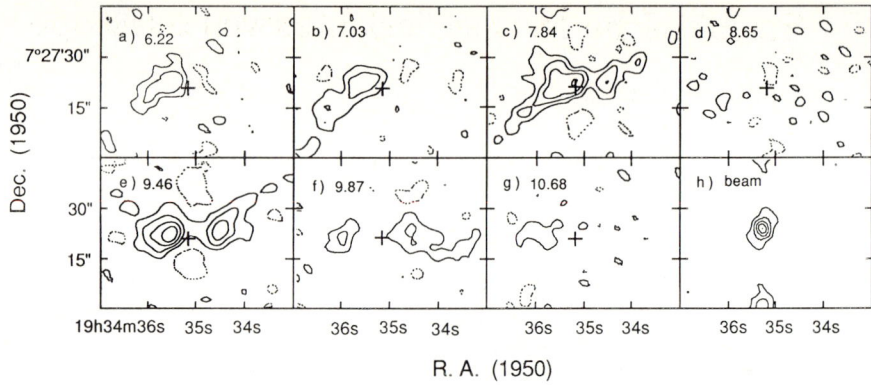

Figure 3: Channel maps of the ^{12}CO(J = 1 → 0) emission taken by the interferometer. The centre velocity of each channel is (a) 6.22 km s^{-1}, (b) 7.03 km s^{-1}, (c) 7.84 km s^{-1}, (d) 8.65 km s^{-1}, (e) 9.46 km s^{-1}, (f) 9.87 km s^{-1}, and (g) 10.68 km s^{-1}. Contours are every 0.97 Jy beam^{-1} with the lowest contours of 0.97 Jy beam^{-1}. The broken contours are for negative intensities. The position of the 2.7 mm continuum emission peak is marked with the cross. The synthesized beam is presented in panel (h).

which exhibit the low velocity (7.84 km s^{-1} and 9.46 km s^{-1}, respectively) wing emission. Two prominent peaks are separated by 10″ - 15″, which corresponds to 2400 – 3600 A.U. and a minimum occurs midway between them. Approximately 4″ southeast of this minimum point is an unresolved 2.7 mm continuum emission peak the flux of which is 0.08 Jy (Hirano et al., 1990). The double-lobed distribution of the outflowing gas suggests that the focusing of the outflow has occurred within ∼ 1000 A.U. of its origin and that there is a relatively small amount of high-velocity molecular gas between two peaks. Both eastern and western lobes contain the blueshifted and redshifted components. This implies that the axis of the outflow is kept to be nearly perpendicular to the line of sight up to the close vicinity of its origin. The outflow appears to settle its morphological characteristics at small radius where it is focused into two opposing jets.

References

Cabrit, S., Goldsmith, P.F., and Snell, R.L., 1988, Astrophys. J., **334**, 196.
Goldsmith, P.F., Snell, R.L., Hemeon-Heyer, M. and Langer, W.D., 1984, Astrophys. J., **286**, 599.
Hirano, N., Kameya, O., Nakayama, M., and Takakubo, K., 1988, Astrophys. J., **327**, L69.
Hirano, H. et al., 1990 (in preparation).
Ishiguro, M., Morita, K.I., Kasuga, T., Kanzawa, T., Iwashita, H., Chikada, Y., Inatani, J., Suzuki, H., Handa, K., Takahashi, T., Tanaka, H., Kobayashi, H. and Kawabe, R., 1984, in *Millimeter and Submillimeter Wave Radio Astronomy*, (URSI), p. 75.
Moriarty-Schieven, G.H. and Snell, R.L. 1989, Astrophys. J., **338**, 952.
Plambeck, R.L., Snell, R. L. and Loren, R. B., 1983, Astrophys. J., **266**, 321.
Tomita, Y., Saito, T. and Ohtani, H., 1979, Pub. Astr. Soc. Japan, **31**, 407.

Status of the James Clerk Maxwell telescope

GRAEME D. WATT

JOINT ASTRONOMY CENTRE, 665 KOMOHANA STREET, HILO, HAWAII 96720, U.S.A.

Abstract The *James Clerk Maxwell Telescope* has been in operation now for a period of well over two years. An increasing number of research papers are becoming available in the scientific press which refer to observations made with the *JCMT*. The aim of this extremely short report is to present what the observatory is capable of at present, what new projects lie on the immediate horizon and a brief summary of some of the science attempted. Hopefully this will gently persuade the reader to reach for a PATT Application form for submission by the next deadline.

The primary purpose for construction of this observatory was to explore the millimetre and submillimetre region of the electromagnetic spectrum. This region covers the range from around 2 mm down to 350 μm in wavelength. This segment of the electro-magnetic spectrum is sub-divided into a series of "windows" by molecules in the Earth's atmosphere (notably O_2 and H_2O) which absorb incoming radiation from space at certain wavelengths making ground-based detection around these wavelengths impossible. The *JCMT* is well equipped with bolometric instrumentation (to study the "heat" or continuum radiation) and with sensitive heterodyne receivers (to detect line spectra from excited states of interstellar molecules) designed to cover most of the frequency range that can possibly be of use. Teething troubles with antenna, carousel, encoders, drive motors, receivers, software reliability, staff training, etc. are not uncommon to new telescopes but have now been mostly overcome and the observatory is settling down to a more normal mode of operation. Over the next few years, major upgrades and additions to the suite of instrumentation together with significant, but subtle, improvements to the pointing/tracking, surface accuracy, software efficiency, etc. will push the *JCMT* well ahead of the competition in the class of ground-based submillimetre observatories.

The *JCMT* stands on the island of Hawaii, the largest island in the state of Hawaii. The telescope is located within the summit caldera of a dormant (but probably extinct!) volcano rising to 13,795 feet (4208 m) above sea level. The mid-ocean location, tropical latitude and the high altitude make this an excellent dry site for high frequency observations. Most of the time the thermal inversion layer lies around 10,000 feet keeping the clouds below and providing clear stable conditions above. The latitude of 20°N means that a considerable fraction of the Southern hemisphere sky is available for observing and, at the same time, important Northern sources are still available. Prime targets such as Orion and the Galactic Centre transit almost overhead.

The primary surface of the antenna is 15 m in diameter with an rms surface accuracy over the total area of better than 40 μm. The pointing of the antenna is currently around 3" rms and will certainly stay within this error if regular pointing checks are performed on a calibrator source within about 30° of the source position. Pointing (and tracking) checks are made on a regular basis (data is logged during each night's observing) and the model is improved when necessary. Most of the instrumentation resides in a Cassegrain cabin behind the primary surface and this tips in elevation along with the antenna. There are Nasmyth platforms located on either end of the elevation axis which are available to house the bulkier, heavier equipment or any instrument that does not like being tipped over.

Complete details of the current suite of instrumentation including relevant noise figures, tuning ranges, availability, etc. can be found in a document which is updated each semester and is available from either ROE or directly from the JAC (Matthews, 1990). Briefly, the heterodyne systems consist of two Schottky receivers and one InSb detector. RxA provides access to those lines in the 220–280 GHz range while RxB receives those in the 320–370 GHz band. These two frequency ranges are the "basebands" for *JCMT* operation. RxA is likely to be replaced by an SIS device by the summer of 1991 and RxB will be replaced first by another Schottky and then by an SIS system by late November 1990. These SIS instruments are expected to provide more stable and reliable instrumentation as well as significantly lower noise figures (greatly increased sensitivity). The InSb system known as RxC covers the CO transition at 461 GHz and the atomic Carbon line at 492 GHz with very little tuning capability. This instrument will be replaced in a few years by an SIS system with a much broader frequency coverage. For spectral line work the particular front end is connected to a single acousto-optical spectrometer (AOSC) which provides 500 MHz bandwidth with 1024 channels and an effective resolution of about 330 kHz. During 1991 this backend will be supplemented (or replaced) by a digital autocorrelation spectrometer (DAS) with 2048 channels divisible into 16 sections of 125 MHz providing up to 2 GHz of bandwidth.

The bolometer detector, UKT14, has a series of filters to cover each of the major atmospheric windows in the 2 mm to 350 μm range. There is a variable aperture iris in the instrument which enables the beam to be stopped down to diffraction limits at each wavelength. In 1992 this instrument will be replaced by a large array instrument, SCUBA, but UKT14 remains the workhorse instrument for the *JCMT* until then since it provides not only regular continuum data but the high sensitivity enables it to be used for pointing checks by observers using the spectral line systems. Moving between different instruments requires rotation of a tertiary mirror inside the receiver cabin. This is a relatively quick procedure and the collimation values for each instrument are well known.

The highest frequency spectral line work is achieved using a "guest" instrument. RxG (which belongs to the Max-Planck-Institut für Extraterrestrische Physik in Garching, Munich) is largely self-contained. It can cover several discrete frequencies around 690 GHz and 800 GHz and has its own AOS of 1 GHz bandwidth. The group from MPE are continually improving and upgrading the instrument and its interface with the *JCMT*.

The following is an extremely concise overview of some of the more exciting projects that are currently being studied by groups of observers at the *JCMT*.

A more detailed treatment of each topic mentioned as well as many other highly interesting projects can be found in Watt and Webster (1990) and the references therein.

Spectral line surveys in the submillimetre have concentrated on OMC-1, SgrB2 and IRC+10216. Detailed spectra have been obtained from 257–273 GHz, several segments below 330 GHz, and from 339–361 GHz. Analyses of these data are well under way. Fewer lines are evident in the 350 GHz band than in the 230 GHz range. The IRC+10216 spectrum is relatively sparse and contains features from only the more abundant (expected) circumstellar species. Many existing molecules have transitions within these bands (in SgrB2 a great number of lines are attributed just to SO_2) although several unidentified features stand out clearly in both frequency ranges. These surveys are taken to about the 0.1 K level and deeper searches await the arrival of more state-of-the-art equipment.

Searches for individual lines from previously unidentified species do not in themselves constitute evidence for the existence of that species unless other transitions at different frequencies are also evident. However this does not prevent observers from searching and species such as HOC^+, H_3O^+, H_2D^+, MgH, $^{16}O^{18}O$ and SH^+ have all been hunted with varying success. To claim any new molecules at this stage would be a little presumptious.

Considerable work has taken place in the higher frequency windows which require much clearer observing conditions. Preliminary studies (some spectra and maps) of CO and atomic C have been made in the 460–490 GHz window and work continues on CO, HCN, HCO^+ around 690 GHz and up in the 800 GHz window. These observations are highly sensitive to the weather conditions at the *JCMT*, to the accuracy of the antenna surface and (since the beam is small – around 8″ at 690 GHz) to the pointing and tracking accuracy. A bonus from this is that the high frequency work is useful for the commissioning team as a method of carefully characterizing the antenna beam patterns.

Considerable studies have been performed using the more common interstellar molecules (CO, CS, HCN, HCO^+, etc.) with low level transitions that lie in the base (230/345 GHz) windows. The size and surface accuracy of the antenna, the high quality pointing and tracking, and the sensitivity of the receivers have allowed considerable detailed mapping of such regions as shock front photoionization/dissociation edges, wind structures in outflow sources, detection of embedded energy sources in dense cores, and line emission from molecules in dust rings around stars.

Continuum studies have been made in all of the windows available for ground based observations. Topics to date include detection of warm dust particles (radius around 200 μm around Vega-like stars), studies of circumstellar disks (including variability of IRC+10216), monitoring of blazars and quasars (especially during flaring phases) as well as searches for cosmological features (such as gravitational lens effects) around 1 mm wavelength. A large database of planetary observations exists since this data is used primarily for calibration purposes.

High resolution studies have been made of starburst galaxies in attempts to determine whether the emission arises from thermal radiation or from internal synchrotron sources. Mapping of galaxies has shown that intensity distributions and hence the gas masses derived from cold dust continuum observations (eg. 450 μm) disagree considerably with those masses determined from CO line measures. The cause of the discrepancy is not yet clearly understood.

Mapping of extended regions in the continuum takes a considerable fraction of the observing time. The small beam of the *JCMT* at 450 μm and 350 μm and a rapid "on-the-fly" data collection technique allow observers to make detailed high resolution maps of dust emission in quite short periods of time. Maps from a large variety of galactic sources have now been completed at 2 mm, 1.3 mm, 1 mm, 850 μm, 800 μm, 600 μm and (on days of excellent weather) 450 μm and 350 μm. A calibration unit has now been attached to UKT14 and regular skydipping is to be initiated in order to keep a close watch on the weather conditions. The calibration of submillimetre data can be extremely difficult when there are no bright sources (primarily planets) of known strength visible. The recent addition of a polarimeter to UKT14 has opened a new area for study and preliminary work on a few regions including the Galactic Centre complex and the Crab nebula has already begun.

Some of the most exotic projects include mapping of the solar disk in the continuum. Such work has been used to show the chromospheric super-granulation network, detailed sunspot structure, prominences and flares. Much of this work is done combined with Calcium K-line data from Hawaii Solar Observatories (on Hawaii and on Maui) and is being used to develop some of the energy transport theories for the Sun. Several other Solar System objects are the subject of study by visiting groups. There have been recent submillimetre detections of Jovian moons (Ganymede and Callisto), of Saturn's largest moon (Titan), several bright asteroids (Ceres, Pallas, Juno, Vesta and others) and several comets (Okazaki-Levy-Rudenko, Brorsen-Metcalf).

References

Matthews, H.E., 1990, *The James Clerk Maxwell Telescope: A Guide for the Prospective User.*
Watt, G.D. and Webster, A.S., (editors), 1990, *Submillimetre Astronomy*, (Kluwer, Dordrecht).

Further CS observations of selected star-forming regions previously mapped in ammonia

G. ANGLADA[1], J. BUJ[1], R. ESTALELLA[1], R. LÓPEZ[1,2], J. PASTOR[1] and P. PLANESAS[3]

1 Introduction

It is generally assumed that dense cores in molecular clouds are the sites where low-mass stars are forming. The morphology and kinematics of these regions can be explored by means of molecules like NH_3 and CS, which trace the gas with density higher than $\sim 10^4\,\mathrm{cm}^{-3}$. However, although their critical densities for detectable emission are similar, the emissions of these two molecules show noticeable discrepancies in several sources. The origin of these differences is not well understood at present. In order to give some insight into the problem, we have mapped the CS (J = 1 → 0) emission around six outflow sources: L1524 (Haro 6-10), AFGL 5142, AFGL 5157, NGC 2068 (HH 19-27), L43 (RNO 91) and HHL 73. These sources were previously mapped in the (J,K)=(1,1) inversion transition of NH_3 at Haystack Observatory (Anglada *et al.*, 1989, Verdes-Montenegro *et al.*, 1989).

2 Results and Discussion

The CS (J = 1 → 0) observations were carried out using the 14 m radio telescope at the Centro Astronómico de Yebes (Spain). At the frequency of this transition, the beam size is 1'.9 and the spectral resolution is 0.31 km s^{-1}. In order to make the comparison between the CS and the NH_3 emissions more meaningful, CS observations were made using a grid of 1'.4 (the beamwidth of the Haystack telescope at the NH_3 frequency). We have detected and mapped the CS emission in all of the observed sources. As a general trend, we found that the CS emission is more extended than the NH_3 emission. In two of the observed sources (AFGL 5142 and AFGL 5157) the CS emission is barely resolved, coinciding with the NH_3 emission. These high density structures are located around the center of symmetry of the CO outflows and include, in each case, an IRAS source. In the sources that are well resolved by the beam, our maps reveal small clumps of enhanced column density inside the regions emitting in CS. These enhancements coincide, in general, with the NH_3 cores, giving support to the statistical correlation between the location of the NH_3 emission peaks and the

---

[1] DEPARTAMENT D'ASTRONOMIA I METEOROLOGIA, UNIVERSITAT DE BARCELONA, AND GRUP D'ASTROFÍSICA, SOCIETAT CATALANA DE FÍSICA, I.E.C., SPAIN.
[2] CENTRE D'ESTUDIS AVANÇATS DE BLANES, C.S.I.C., SPAIN.
[3] CENTRO ASTRONÓMICO DE YEBES, I.G.N., SPAIN.

exciting sources of the outflows found by Anglada et al. (1989). Furthermore, the CS emission seems well suited to reveal the lower intensity connections between the NH_3 cores present in a given cloud. This fact is particularly noticeable in the complex regions of NGC 2068 (HH 19-27) and HHL 73. In NGC 2068, the CS emission extends over $\sim 20'$ in the N–S direction, connecting the regions of the CO outflows and showing two main enhancements towards the IR sources HH 26 and HH 24. In the case of HHL 73, the CS E-W elongated structure connects the different tracers of star formation and also, the three main enhancements of the emission are spatially associated with three IRAS sources present in this region. However, in the case of L1524, the CS and NH_3 maps are very different and the emission peaks are $\sim 2'.5$ displaced. In L43, the CS emission exhibits two maxima, separated $\sim 3'$ and located on either side of the NH_3 peak.

Acknowledgements

We thank the staff of the Centro Astronómico de Yebes, and specially A. Fuente and J. Alcolea, for their support during the observations. G.A. acknowledges a fellowship from DGU and CIRIT de Catalunya.

References

Anglada, G., Rodríguez, L.F., Torrelles, J.M., Estalella, R., Ho, P.T.P., Cantó, J., López, R. and Verdes-Montenegro, L., 1989, Astrophys. J., **341**, 208.

Verdes-Montenegro, L., Torrelles, J.M., Rodríguez, L.F., Anglada, G., López, R., Estalella, R., Cantó, J. and Ho, P.T.P., 1989, Astrophys. J., **346**, 193.

A 257 – 273 GHz spectral survey of the OMC-1 cloud core

JANE S. GREAVES AND GLENN J. WHITE

DEPARTMENT OF PHYSICS, QUEEN MARY AND WESTFIELD COLLEGE, UNIVERSITY OF LONDON, MILE END ROAD, LONDON E1 4NS, U.K.

1 Introduction

A spectral survey was made of the core of OMC1, the high-mass star-forming cloud in Orion. The core contains a luminous embedded object IRc2 ($L \sim 10^5 L_\odot$), the source of a high-velocity outflow ($v \sim 100$ kms^{-1}). Close to IRc2 are clumps rich in oxygen- and nitrogen-bearing species, making this region one of great chemical complexity. It is therefore necessary to obtain a uniformly calibrated database of spectral lines on which to base future studies of the gas in the core.

The 257 – 273 GHz region of the OMC1 spectrum was observed using the James Clerk Maxwell Telescope (JCMT). This is the highest frequency unbiassed survey of this source made so far. The spatial resolution was also higher than that used for earlier surveys (19″ beam, corresponding to 0.05 pc at a distance of ~ 500 pc). The data were obtained in double-sideband mode, and separated to single-sideband spectra using a new algorithm (Greaves and White, 1990).

2 Results

152 spectral lines were detected (Figure 1), including ~ 7 from unidentified molecules. The rest of the lines were from between 17 and 20 known species (Table 1).

Table 1

2-atom	3-atom	4-atom	5-atom	6-atom	7-atom	8-atom	9-atom
SO	SO_2	H_2CS	HC_3N	CH_3CN		$HCOOCH_3$	C_2H_5CN
SiO	OCS	H_2CO ?	CH_2CO	CH_3OH			CH_3OCH_3
	HCN	NH_2D ?	C_3H_2 ?				
	HNC						
	CCH						
	HCO^+						
	HDO						

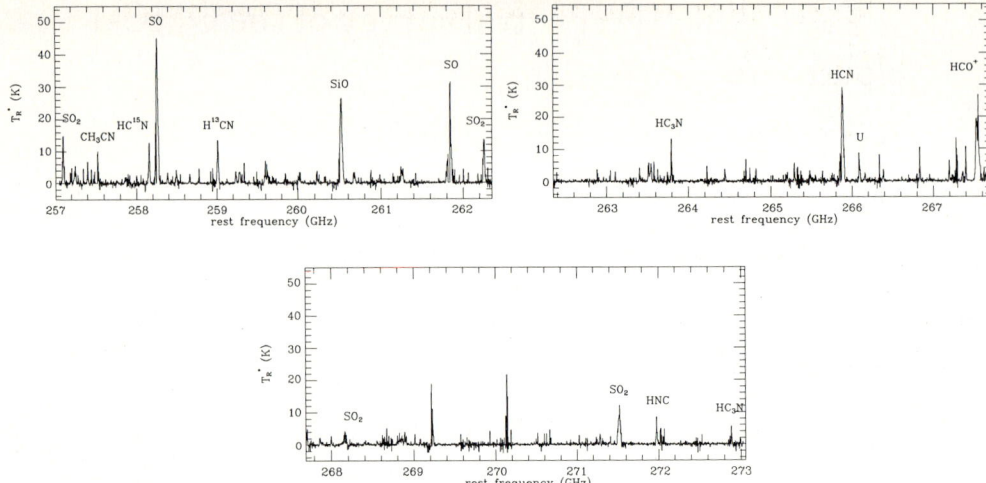

Figure 1: 257–273 GHz spectrum of the OMC1 core, with the strongest transitions identified.

3 Conclusions

No new molecules have been identified in OMC1 in this survey, but there are several unidentified features (U-lines). Their frequencies do not match the transitions of possible interstellar species studied in the laboratory, indicating that the OMC1 chemistry is not yet understood.

Very high spatial resolution is needed to study OMC1, since the antenna temperatures measured increase with decreasing beam size. For example, the lines in this survey are (on average) a factor of 2 stronger than when observed in the 257 – 263 GHz region of the OVRO survey (Blake *et al.*, 1986), which had twice the beam area.

The distribution of number of lines detected against peak antenna temperature is $N(T_R^* \to T_R^* + \Delta T_R^*)$ proportional to $T_R^{*-1.2}$ (for $\Delta T_R^* = 1$ K). If this distribution can be extrapolated to weaker lines, then the confusion limit (blended lines right across the passband) will be reached at $T_R^* \sim 0.05$ K. Since the noise level of this survey was ~ 0.5 K, there is a long way to go before before the confusion limit is reached.

References

Blake, G.A., Sutton, E.C., Masson, C.R. and Phillips, T.G., 1986,
    Astrophys. J. (Suppl.) **60**, 357.
Greaves, J.S. and White, G.J., 1990, Astr. Astrophys., (submitted).

Cloud collapse and low mass star formation in Orion-KL

V. MIGENES[1], K. J. JOHNSTON[2], T. A. PAULS[2] and T. L. WILSON[3]

Introduction. The nature of the observed turbulent motion in dense cores within molecular clouds and its role in star formation is not quite clear. The formulation of empirical relations among the internal velocity dispersion, number density and the size of the clouds has helped us clarify their nature (Larson, 1981). Basically, all the models which incorporate the observed turbulence predict the $\sigma \propto R^{0.5}$ and the $n \propto R^{-1.3}$ relations observed. The simplest explanation for the power laws among these physical quantities is that the clumps are in virial equilibrium.

In this paper we present evidence which supports that: (1) it is possible to extrapolate these power laws to the size scales represented by our data (Migenes et al., 1989) and (2) the "clumpy" nature of the OH masers and NH_3 emission in Orion-KL not only offer a consistent picture of the star formation process taking place, but also show that both species are part of the same molecular envelope. The clumps discussed in this paper seem to be in the latest stages of star formation and lend support to the star-forming scenario described by Hoyle (1953).

Discussion. The authenticity and scale of the "clumpiness", in OH maser emission, can be studied by forming a two-point spatial correlation function for the maser components. The distribution for features with $\Delta v \leq 1\,\mathrm{km\,s^{-1}}$ demonstrates the magnitude of the clumping, $\approx 1.0''$ ($10^{15} - 10^{16}$ cm). The larger clumps can be formed by a superposition of clumps or by clump collisions. The velocity distribution within the clumps can also be studied. The distribution of masers with a separation less than $1.0''$ can be well fitted with two Gaussians. The main component has a FWHM of $4.5\,\mathrm{km\,s^{-1}}$ which corresponds to the combined effects of the Zeeman splitting ($\approx 0.5\,\mathrm{km\,s^{-1}}$ in Orion-KL) and clumping. More extensive analysis indicates that the maser features seem to delineate "cloudlets" of gas which support maser emission.

The NH_3 clumps, detected with the (3,2) transition, have the same size, velocity and spatial distributions as the OH maser emission. The mass of these fragments is within an order of magnitude of the mass expected when the collapse and fragmentation process reaches its end. The empirical correlations seem to hold for self-gravitating clumps with size scales between 0.04 and 100 pc but large deviations seem to occur for very compact objects. The ammonia and maser clumps observed in our data are significantly smaller than this size scale (10^{-4} to 10^{-2} pc). As an example, in Figure 1, we have plotted the velocity dispersion $\sigma(v)$ of the maser and ammonia clumps with respect to their size. Even though the data is largely scattered there is a clear trend (followed by the dotted line

---

[1] UNIVERSITY OF MANCHESTER, NUFFIELD RADIO ASTRONOMY LABORATORIES, JODRELL BANK, MACCLESFIELD, CHESHIRE, SK11 9DL, U.K.
[2] NAVAL RESEARCH LABORATORIES, WASHINGTON, D.C., U.S.A.
[3] MAX-PLANCK-INSTITUT FÜR RADIOASTRONOMIE, BONN 1, F.R.G.

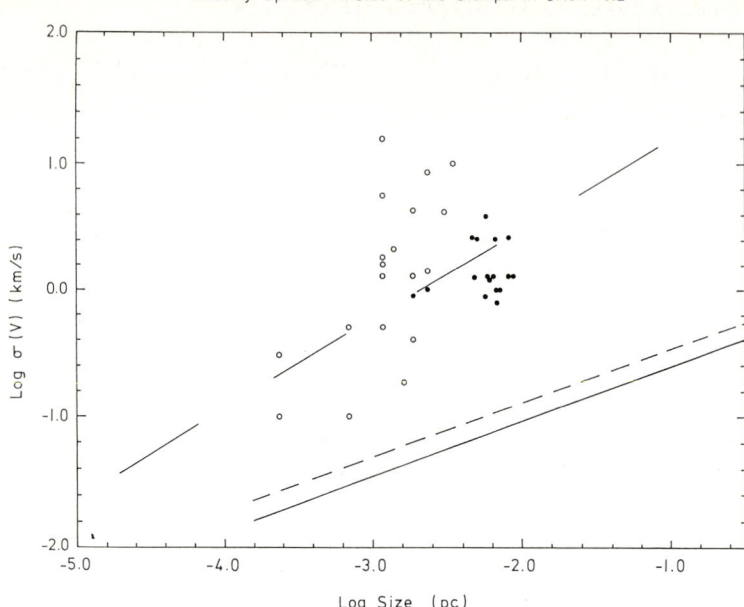

Figure 1: The velocity dispersion (vs) size relation for the ammonia (solid circles) and OH maser (open circles) clumps. The linear sizes have been determined using 480 pc as the distance to Orion-KL. The general interpretation, given by $\sigma(v) \propto L^{0.5}$, has been drawn in the lower right with its 1 σ error.

through the center of the scatter) which suggests the existence of a correlation with a slope of 0.7±0.2. The largest source of error in this relation arises from the size estimates for the NH_3 and OH clumps.

Our data supports two of the models which explain the origin of the empirical power laws: (1) the effect of magnetic support in these clumps, so that the clumps are in both virial equilibrium and magnetic balance against gravity (Myers and Goodman, 1988), (2) angular momentum transfer, from the increasing rotational energy during collapse, into the supersonic "random" motion observed (Henriksen and Turner, 1984). The link between maser emission and the NH_3 clumping, and the fragmentation process could be the triggering of velocity gradients in the medium, due to the supersonic relative motion of the NH_3 clumps, which will create density enhancements that could favor maser emission.

References

Henriksen, R.D. and Turner, B.E., 1984, Astrophys. J., **287**, 200.
Hoyle, F., 1953, Astrophys. J., **81**, 37.
Larson, R.B., 1981, Mon. Not. Roy. Astron. Soc., **194**, 809.
Migenes, V., 1989, Ph.D Thesis, Univ. of Pennsylvania.
Myers, P.C. and Goodman, A.A., 1988, Astrophys. J., **326**, L27.

The TMC-1 mapped in HC_5N

C. CODELLA[1,2], G. BRUNI[1], G. COMORETTO[3], G. MACCAFERRI[4],

G.G.C. PALUMBO[2] and F. SCAPPINI[1]

The Medicina (Bologna, Italy) 32m dish radiotelescope, operated by the Istituto di Radioastronomia-CNR, has been used for the first time, to obtain a map of a region of sky in molecular lines.

The molecular line chosen was the $J = 4 \rightarrow 3$ rotational line of HC_5N at $\nu_0 = 10.650643$ GHz, and the region scanned was the Taurus Molecular Cloud 1 (TMC-1), which is a small portion of the Heiles Cloud 2 (HCL2) complex. In the same region this line was already detected and partially mapped by MacLeod et al. (1979), with a 10 KHz resolution.

Our observations were done in different periods during 1988 and 1989. The HPBW was 3' at 10.65 GHz and the efficiency 0.45. The receiver, which was operated in the total power mode, is a cryogenically cooled HEMT-FET. The system temperature was typically 60 K. Spectral information was obtained with a 512 channel autocorrelation spectrometer (Comoretto, 1983), operating with a resolution of 6 KHz (0.17 km s^{-1}) and using position switching at 5 minutes on/off intervals.

A grid of 39 points was obtained in steps of half beamsize covering a region of 14'×6'. Figure 1 shows a typical spectrum of the $J = 4 \rightarrow 3$ transition of HC_5N resolved into the three strongest hyperfine components. A contour map of the antenna temperature, summed over the three components, is shown in figure 2. The peak antenna temperature is $\Sigma T_A = 0.213$ K, corresponding to a column density of $N = 2.6 \times 10^{13}$ cm^{-2}. The lines are very narrow with a FWHM of about 0.4 km s^{-1} and a radial velocity of 5.4 ± 0.2 km s^{-1}.

The map obtained is in fair agreement with that of MacLeod et al. (1979); however it covers a more extended region of $14' \times 6'$ (these dimensions refer to the 3σ detection limit) compared to the previous $10' \times 4'$. The present result has proven that, particularly for extended sources, the Medicina radiotelescope is suitable for molecular observations.

References

Comoretto G., 1983, Journal of Physics E: Scientific Instruments, **16**, 836.
MacLeod J.M., Avery L.W. and Broten N.W., 1979, Astrophys. J., **233**, 584.

[1] ISTITUTO DI SPETTROSCOPIA MOLECOLARE DEL CNR, 40126 BOLOGNA, ITALY
[2] DIPARTIMENTO DI ASTRONOMIA, UNIVERSITÀ DI BOLOGNA, ITALY
[3] OSSERVATORIO ASTROFISICO DI ARCETRI, L.GO. E. FERMI 5, 50125 FIRENZE, ITALY
[4] ISTITUTO DI RADIOASTRONOMIA DEL CNR, BOLOGNA, ITALY

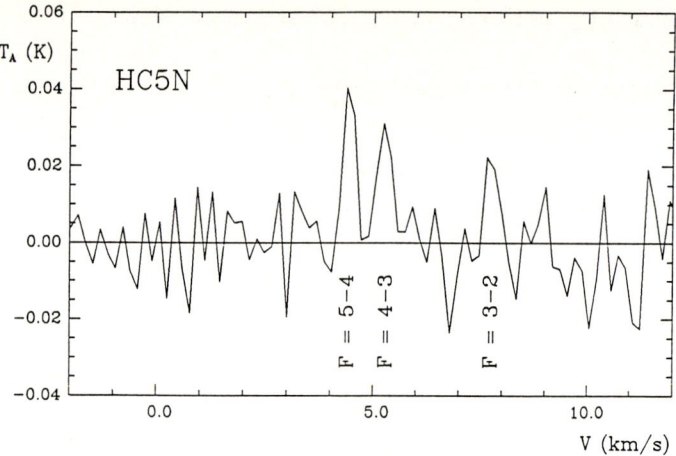

Figure 1: HC$_5$N in TMC-1. 6 kHz spectrum of the J = 4 → 3 rotational transition at α (1950) = 4^h 38^m 46.0^s, and δ (1950) = 25° 31′ 00″. The strongest hyperfine components are indicated by their quantum number F.

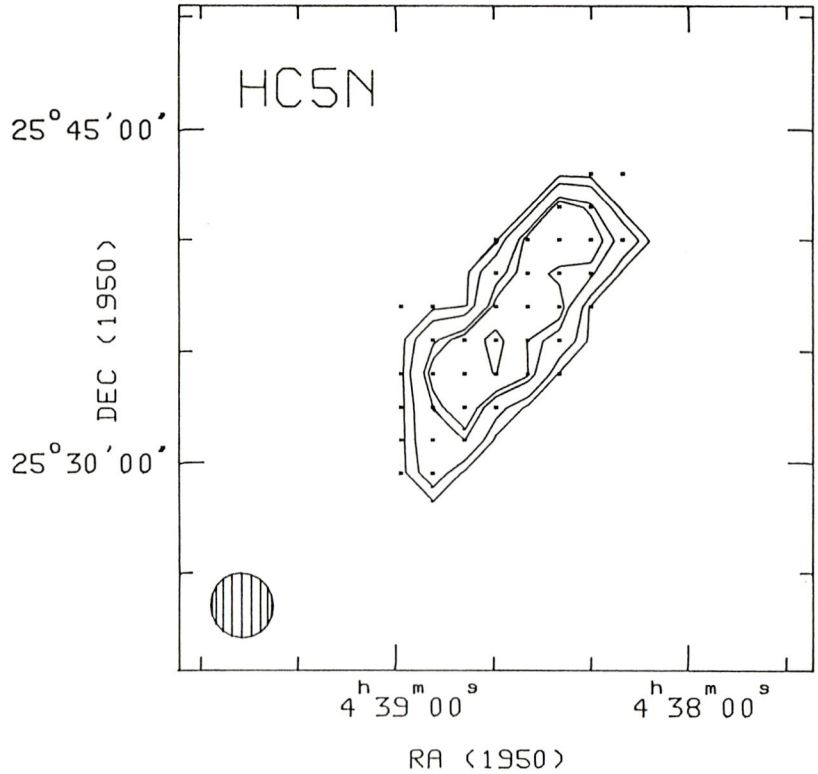

Figure 2: Antenna temperature contour map of the J = 4 → 3 rotational transition of HC$_5$N in TMC-1. Dots indicate the positions observed. Contours are at 5, 20, 50, 60 and 90% of the peak antenna temperature $\sum T_A = 0.213$ K, respectively.

Water vapour in the Orion molecular cloud

R. F. KNACKE[1] and   H. P. LARSON[2]

We report infrared observations of interstellar gas-phase H_2O in the spectrum of the BN object in Orion. There are absorptions (S/N = 2 – 5) at the positions of four of the strong lines in the $O_{00} - O_{01}$ ν_3 (2.66 µm) vibration rotation band. The new observations corroborate the detection reported earlier (Knacke, Larson and Noll, 1988). With an estimated excitation temperature of 150 K, the column density of gaseous H_2O to BN in the OMC-1 cloud is $(2 \pm 1) \times 10^{17}\,cm^{-2}$. The intensities of the lines imply an ortho/para ratio of 1 ± 0.5 indicating recent sublimation of H_2O from low temperature grains.

The results give gas-phase abundance ratios of $H_2O/CO = 0.03 \pm 0.02$ and $HDO/H_2O = 10^{-3}\text{–}10^{-4}$ toward BN. The velocities of the H_2O absorptions agree with those of the Ridge Source and CO outflow, but the position along the line of sight is not well constrained. The gas/solid ratio is $H_2O_{gas}/H_2O_{ice} \leq 0.05$. Less than 1% of the oxygen is in H_2O gas (assuming total cosmic abundance). Most of the H_2O in the line of sight to BN, and by inference in quiescent regions of molecular clouds generally, is frozen on grains.

The complete paper will be published in The Astrophysical Journal.

Reference

Knacke, R.F., Larson, H.P. and Noll, K.S., 1988, Astrophys. J., **335**, L27.

[1] NASA MARSHALL SPACE FLIGHT CENTER
[2] LUNAR AND PLANETARY LABORATORY, UNIVERSITY OF ARIZONA.

Magnetic fields in interstellar clouds

PHILIP C. MYERS

HARVARD-SMITHSONIAN CENTER FOR ASTROPHYSICS, 60 GARDEN STREET, CAMBRIDGE, MASSACHUSETTS 02138, U.S.A.

Abstract Physical properties of diffuse, dark and giant interstellar clouds and their constituent dense cores are summarized. Magnetic field strengths in from a few μG to a few $\times$ 10 mG in these clouds are reviewed, based on observations of the Zeeman effect in atomic and molecular spectral lines. For the intercloud medium and diffuse clouds, the magnetic energy density is similar to the kinetic energy density in clouds and in the intercloud medium. These three energy densities are much greater than the gravitational energy density of the typical diffuse cloud. For dense clouds and dense cores, the magnetic, kinetic and gravitational energy densities are similar to each other, but much greater than the kinetic energy density of the intercloud medium. Magnetic field directions, as indicated by optical polarization are reviewed. They appear coherent on scales of order 5 pc or greater, where the visual extinction is about 1 magnitude. They show a wide range of angles between the long axis of a dark cloud and the local projected field direction. A model of the field directions based on combination of uniform and nonuniform field components indicates that many clouds have comparable energy density in uniform and nonuniform components.

1   Cloud and Core Properties

Table 1 shows cloud size R, density n, mass M, FWHM line width Δv, kinetic temperature T and constituent stars and dense cores, based mainly on observations of the 2.6 mm fundamental rotational line of ^{12}CO (Myers, 1990). Table 2 presents a similar summary of dense core properties, based mainly on observations of the 1.3 cm (J, K) = (1, 1) rotation-inversion line. Among massive dense cores, we distinguish between "large" and "small", based respectively on maps made with filled-aperture telescopes and interferometers. An important property of clouds and cores not included in Tables 1 and 2 is their projected elongation. A study of the shapes of 16 dense cores as mapped in lines of NH_3, CS and $C^{18}O$ indicates that the typical core map is elongated, with axial ratio about 0.5; and that a significant number of cores with elongated maps are prolate, rather than oblate, in three dimensions (Myers *et al.*, 1991).

2   Magnetic Field Strength Measurements

Table 3 summarizes most available measurements of magnetic field strength in the intercloud medium and in interstellar clouds, in terms of the cloud types described in Tables 1 and 2.

In Table 3, the meaning of the measured field strength B_{obs} depends on the method of measurement. For rotation measure data, B_{obs} is the magnitude of

Table 1. *Molecular Cloud Properties Based on the $J = 1 \to 0$ Line of ^{12}CO.*

Type	R (pc)	n (cm^{-3})	M ($M_\odot$)	Δv (km s^{-1})	T (K)	Cores and stars
Diffuse	0.3–3	30–500	0.5–100	0.7–1.5	10?	Low-mass
Dark	3–10	10^{2-3}	10^{3-4}	1–3	10	Low-mass
Giant	20–100	10–300	10^{5-6}	5–15	10–20	Massive (and low-mass)

Table 2. *Dense Core Properties Based on the $(J, K) = (1,1)$ Line of NH_3.*

Type	R (pc)	n (cm^{-3})	M ($M_\odot$)	Δv (km s^{-1})	T (K)	Stars
Low-mass	0.05–0.2	10^{4-5}	0.3–10	0.2–0.4	10	T Tauri
Massive...						
Large	0.3–0.6	10^{4-5}	30–10^4	1–2	10–30	OB
Small	0.01–0.03	10^{6-7}	0.3–300	1–3	30–100	OB

the line-of-sight component of the field strength which gives rise to the Faraday rotation of the linear polarization. For the Zeeman data, B_{obs} is again the magnitude of the line-of-sight component of the field strength, when the frequency shift between right- and left-circularly polarized spectral lines is small compared to the line width. This applies to all Zeeman data in Table 3 except for the "small" massive cores. For the Zeeman data where the frequency shift is large compared to the line width, as in the OH masers in the small massive cores, the reported field strength is the three-dimensional field strength.

Field strength measurements via the Zeeman effect have several sources of uncertainty and complexity in addition to those associated with spectral line observations (see, e.g. Heiles *et al.*, 1991). For the Zeeman shift of the HI line associated with a dark cloud, one can sometimes identify the Zeeman shift of the one or more emission components and also of a self-absorption dip against the broader emission profile. In L204, the emission component gives about 5 μG while the self-absorption dip gives about 10 μG (Heiles, 1988). Observations with finer angular resolution tend to give larger field strengths in some cases: in B1, OH Zeeman observations with the 43m telescope of NRAO in Green Bank (18' beam) give 19 ± 3 μG, while observations with the 305m telescope of the NAIC in Arecibo (3' beam) give 27 ± 4 μG (Goodman *et al.*, 1989). In S106, observations with the 18' beam of the 43m telescope give 137 μG (Kazes *et al.*, 1988) while observations with the 10" beam of the VLA give field strengths of 0.5 to 1 μG, with the direction of the line-of-sight component varying across the map (Loushin and Crutcher, 1989).

Table 3. *Magnetic Field Strength Measurements.*

Region	Cases	Method	Example	Reference	B_{obs} (μG)
Intercloud Medium	> 100	rotation measure	....	1	5
Diffuse	~ 10	Zeeman HI	NCP Loop	2	7
Dark	~ 5	Zeeman HI	L204	3	5–10
		Zeeman OH	B1	4	20–30
Giant	1	Zeeman HI	L1641	5	10
Massive core...					
Large	~ 7	Zeeman OH	Orion A	6	125
Small	~ 15	Zeeman OH	Orion KL	7	3×10^3
H$_2$O Maser	~ 5	Zeeman H$_2$O	Orion KL	8	3×10^4

References:
1. Rand and Kulkarni, 1989.
2. Heiles, 1989.
3. Heiles, 1988.
4. Goodman *et al.*, 1989.
5. Heiles, 1987.
6. Troland, Crutcher and Kazes, 1986.
7. Norris, 1984.
8. Fiebig and Güsten, 1989.

3 Equipartition Field Strengths

We present useful expressions for magnetic field strength B in terms of observational variables and energy densities for a uniform sphere of radius R, mean column density N, mean particle mass m, FWHM line width Δv of the molecule with mass m, and uniform magnetic field strength B, surrounded by an intercloud medium exerting constant pressure P. The quantities R, N and Δv are obtained from observed line and map parameters, as for example in Myers and Goodman (1988a). The magnetic energy density is

$$\mathbf{M} = \frac{B^2}{8\pi} \tag{1}$$

the kinetic energy density is

$$\mathbf{K} = \frac{9mN\Delta v^2}{64 \ln 2 \, R} \tag{2a}$$

for clouds and

$$\mathbf{K} = \frac{3P}{2} \tag{2b}$$

for the intercloud medium. The gravitational energy density is

$$\mathbf{G} = \frac{9\pi m^2 G}{20} N^2 \tag{3}$$

where G is the gravitational constant.

Equations (1), (2) and (3) can be used to write expressions for the magnetic field strength B in terms of the observables Δv, N and R, and ratios of either one or two pairs of the energy densities $\mathbf{M}$, $\mathbf{G}$ and $\mathbf{K}$. We denote these identities for B with subscripts indicating the energy densities on which they depend: $B_{\mathbf{MG}}$, $B_{\mathbf{MK}}$ and $B_{\mathbf{MGK}}$:

$$B_{\mathbf{MG}} = 3\pi m \left(\frac{2G}{5}\right)^{0.5} N \left(\frac{\mathbf{M}}{\mathbf{G}}\right)^{0.5} = 2.6 \frac{m}{m_H} \frac{N}{\text{mag}} \left(\frac{\mathbf{M}}{\mathbf{G}}\right)^{0.5} \mu\text{G}, \tag{4}$$

$$B_{\mathbf{MK}} = \left(\frac{9\pi m N \Delta v^2}{8 \ln 2 \, R}\right)^{0.5} \left(\frac{\mathbf{M}}{\mathbf{K}}\right)^{0.5}$$

$$= 5.2 \frac{\Delta v}{\text{km s}^{-1}} \left(\frac{m}{m_H} \frac{N}{\text{mag}} \bigg/ \frac{R}{\text{pc}}\right)^{0.5} \left(\frac{\mathbf{M}}{\mathbf{K}}\right)^{0.5} \mu\text{G} \tag{5a}$$

for clouds and

$$B_{\mathbf{MK}} = (12\pi P)^{0.5} = 7.2 \left(\frac{P}{10^4 k \text{ cm}^{-3} \text{ K}}\right)^{0.5} \mu\text{G} \tag{5b}$$

for the intercloud medium with pressure P. Here k is Boltzmann's constant. Also,

$$B_{\mathbf{MGK}} = \frac{3}{8 \ln 2} \left(\frac{5}{2G}\right)^{0.5} \frac{\Delta v^2}{R} \left(\frac{\mathbf{M}}{\mathbf{K}} \frac{\mathbf{G}}{\mathbf{K}}\right)^{0.5}$$

$$= \frac{B_{\mathbf{MK}}^2}{B_{\mathbf{MG}}}$$

$$= 11 \left(\frac{\Delta v}{\text{km s}^{-1}}\right)^2 \left(\frac{R}{\text{pc}}\right)^{-1} \left(\frac{\mathbf{M}}{\mathbf{K}} \frac{\mathbf{G}}{\mathbf{K}}\right)^{0.5} \mu\text{G} \tag{6}$$

for clouds and cores. Here m_H is the atomic mass unit.

As equation (6) indicates, these field strengths for clouds are interrelated so that only two of them are independent. When each energy density ratio in equation (4), (5) or (6) is unity, then these expressions give the corresponding equipartition field strengths. Equations (4), (5) and (6) are thus general formulas for field strength, of which the equipartition field strengths given by Myers and Goodman (1988a, b) are special cases. The numerical coefficients in equations (4), (5) and (6) differ slightly from those in Myers and Goodman (1988a, b) because the latter models use equality of energy density terms of the virial theorem, while the present expressions use simple equality of energy density.

4 Comparison of Observed and Equipartition Field Strengths

When the observed field strengths in Table 3 are compared with the equipartition field strengths in equations (4), (5) and (6), two general conclusions emerge: (a) for the diffuse clouds and for the intercloud medium, $B_{obs} \approx B_{MK}$; and (b) for denser clouds and for dense cores, $B_{obs} \approx B_{MGK}$. Thus for diffuse clouds,

$$\mathbf{M} \approx \mathbf{K} \approx \mathbf{P} \gg \mathbf{G} \qquad (7)$$

while for clouds and dense cores,

$$\mathbf{M} \approx \mathbf{K} \approx \mathbf{G} \gg \mathbf{P} \qquad (8)$$

The evidence for equation (7) is still somewhat sparse and more cloud data needs to be brought into the comparison. Evidence for equation (8) was presented by Myers and Goodman (1988a) for 14 dense clouds and cloud cores and it will be important to examine new measurements in these terms as they become available. Some magnetic field measurements and upper limits in dark clouds reported by Crutcher et al. (1991) appear consistent with equation (8).

It is sometimes suggested that the tendency for molecular cloud line width Δv to increase with the cloud map size R, first pointed out by Larson (1981), is also evidence for equipartition between magnetic, kinetic and gravitational energy, provided the relative variation in field strength is small compared to the relative variation in scale size. Here we note that this latter condition is highly dependent on the sample of clouds under consideration. For example, the 14 clouds considered by Myers and Goodman (1988a) show no evidence of correlation between Δv and R, but they show strong evidence for equipartition among $\mathbf{M}$, $\mathbf{G}$ and $\mathbf{K}$. Furthermore, if the clouds under consideration have equal gravitational and kinetic energy (or, equivalently, "virial" equilibrium) and if they have a relatively narrow range of column density N compared to the range of R, then equations (4), (5) and (6) show that the sample will exhibit a correlation between Δv and $R^{0.5}$, even if the narrow range of N does not arise from equipartition between magnetic and kinetic energy and a narrow range of B. For example, a narrow range of N could arise from selection. Therefore, the appearance of a correlation between Δv and $R^{0.5}$ is by itself neither necessary nor sufficient to indicate equipartition among **M**, **G** and **K**, but the correlation can be consistent with such equipartition. A more detailed discussion is given in Myers and Goodman (1988b). See also the review by Heiles *et al.* (1991).

5 Magnetic Field Directions

Several efforts have been made to understand the patterns of projected magnetic field directions associated with dark clouds, indicated by observations of the polarization of starlight toward background stars shining through the more transparent parts of the dark clouds. Vrba, Strom and Strom (1976) noted that the direction of polarization in the Ophiuchus complex is nearly aligned with some of the filamentary dark clouds in the complex. Moneti *et al.* (1984) showed that the direction of polarization toward the Taurus complex varies slightly across the complex, while the directions of the individual cloud axes varies more. Heyer *et al.* (1987) showed that the polarization is nearly perpendicular to the long axis of the B216–B217 cloud in Taurus. Goodman *et al.* (1990) compared directions of polarization and of cloud axes in Ophiuchus, Perseus, Taurus and other clouds. They found that the polarization near a cloud usually shows a well-defined direction, but that the direction has no obvious relation to the cloud symmetry axis, from one cloud to the next.

The dispersion in polarization direction toward dark clouds is almost always greater than instrumental uncertainty, suggesting that this dispersion should contain some information about the degree of variation of the magnetic field in the cloud, or in the cloud environment. Chandrasekhar and Fermi (1953) modeled this dispersion on the scale of galactic spiral arms as arising from wavelike distortions in the galactic field, due to interaction with the turbulent motions of interstellar clouds. Zweibel (1990) extended this treatment to a model of clumps and an interclump medium within a cloud, including effects of ambipolar diffusion. Myers and Goodman (1991) compared dispersions in many clouds and found that regions with prominent embedded star clusters have significantly greater dispersion in the direction of polarization than do their neighboring, more quiescent dark clouds. They modeled the dispersion as arising from a magnetic field with a uniform component, and with a nonuniform component having an isotropic probability distribution of direction, a Gaussian distribution of amplitude, and with N correlation lengths along the line of sight. They found that the model probability distribution of electric field direction is approximately a Gaussian, with dispersion

$$s = \frac{\sigma_B}{N^{0.5} B_{0x}} \tag{9}$$

where σ_B is the dispersion in the distribution of field amplitudes and B_{0x} is the projection of the uniform component of the field on the plane of the sky.

This model can be used to estimate the three-dimensional field magnitude and direction when the polarization map is combined with a map of the line-of-sight component of the field strength, from Zeeman observations. For many clouds, the model indicates that the nonuniform and uniform components of the field energy density are comparable.

Acknowledgements

I thank Richard James, Tom Millar and the other organizers of the Seventh Manchester Meeting for their support of my attendance, for their kind hospitality and for their invitation to speak. I thank my collaborators A. Goodman, R. Crutcher, C. Heiles, I. Kazes, T. Troland, P. Bastien and F. Menard.

References

Chandrasekhar, S. and Fermi, E., 1953, Astrophys. J., **118**, 113.
Crutcher, R.M., Troland, T.H., Goodman, A.A., Kazes, I. and Myers, P.C., 1991, (in preparation).
Fiebig, D. and Güsten, R., 1989, Astron. Astrophys., **214**, 133.
Goodman, A.A., Crutcher, R.M., Heiles, C., Myers, P.C. and Troland, T.H., 1989, Astrophys. J., **338**, L61.
Goodman, A.A., Bastien, P., Myers, P.C. and Menard, F., 1990, Astrophys. J., **359**, 363.
Heiles, C., 1987, in *Interstellar Processes*, ed. D.J. Hollenbach and H.A. Thronson (Reidel:Dordrecht), p. 171
Heiles, C., 1988, Astrophys. J., **324**, 321.
Heiles, C., 1989, Astrophys. J., **336**, 808.
Heiles, C., Goodman, A.A., McKee, C.F. and Zweibel, E., 1991, in *Protostars and Planets III*, ed. E.H. Levy and J.I. Lunine, (University of Arizona Press, Tucson, in press).
Heyer, M.H., Vrba, F.J., Snell, R.L., Schloerb, P.F., Strom, S.E., Goldsmith, P.F. and Strom, K.M., 1987, Astrophys. J., **321**, 855.
Kazes, I., Troland, T.H., Crutcher, R.M. and Heiles, C., 1988, Astrophys. J., **335**, 263.
Larson, R.B. 1981, Mon. Not. Roy. Astron. Soc., **194**, 809.
Loushin, R. and Crutcher, R.M., 1990, in *Galactic and Intergalactic Magnetic Fields*, ed. R. Beck, P.P. Kronberg and R. Wielebinski, (Kluwer: Dordrecht).
Moneti, A., Pipher, J.L., Helfer, H.L., McMillan, R.S. and Perry, M.L., 1984, Astrophys. J., **282**, 508.
Myers, P.C. and Goodman, A.A., 1988a, Astrophys. J., **326**, L27.
Myers, P.C. and Goodman, A.A., 1988b, Astrophys. J., **329**, 392.
Myers, P.C., 1990, in *Molecular Astrophysics, A Volume Honouring Alexander Dalgarno*, ed. T. Hartquist (Cambridge University Press, Cambridge) p. 328.
Myers, P.C., Fuller, G.A., Goodman, A.A. and Benson, P.J., 1991, (submitted to Astrophys. J.)
Myers, P.C. and Goodman, A.A., 1991, (submitted to Astrophys. J.)
Norris, R.P., 1984, Mon. Not. Roy. Astron. Soc., **207**, 127.
Rand, R.J. and Kulkarni, S.R., 1989, Astrophys. J., **343**, 760.
Troland, T.H., Crutcher, R.M. and Kazes, I., 1986, Astrophys. J., **304**, L57.
Vrba, F.J., Strom, S.E. and Strom, K.M., 1976, Astron. J., **81**, 958.
Zweibel, E., 1990, Astrophys. J., **362**, 545.

Star formation in dark globules

S.M. SCARROTT

PHYSICS DEPARTMENT, UNIVERSITY OF DURHAM, DURHAM DH1 3LE, U.K.

Abstract The dense accumulation of matter in cometary and Bok globules suggests that they should be suitable locations for star formation yet there are few examples of either form of globule playing host to such activity. Optical polarization studies of nebulosities apparently associated with dark globules have revealed systems in which reflection nebulae are illuminated by stars still embedded within the dark material (e.g. The "Bruck Bipolar", Scarrott *et al.*, 1987, HH46/47, Scarrott and Warren-Smith, 1988). These observations indicate that limited star formation is taking place in dark globules and the previous lack of such evidence lies in failing to recognise that the small nebulosities, frequently found with hidden illuminators, are the earliest visible signs of newly formed stars.

In this contribution we present further evidence for star formation in dark globules.

The CG30 Cometary Globule: A complex of related cometary globules, CG30, CG31ABCDE and CG38 are located in the extensive Gum Nebula. CG30 has the characteristic mass (~ 70 M$_\odot$) and size (0.3 pc) of a Bok globule and seen projected against the central dark region of CG30 is a small patch of optical nebulosity (dimension $\sim 15''\sim 0.03$ pc $= 7000$AU at a distance of 450 pc, Reipurth, 1981). The bright knot within the diffuse nebulosity was shown to be an HH object (HH120; Pettersson, 1984). Although there is no known nearby optical source capable of exciting the CG30 nebulosity an IR source, CG30IRS4, located to the south of the optical nebula, is assumed to be the exciting and illuminating object (Pettersson, 1984, Graham, 1988).

Figure 1 shows intensity contour and optical polarization maps of the CG30 nebulosity. The contour map shows a generally amorphous distribution of nebulosity with a central brightness knot which corresponds to the proposed HH object HH120. The polarization map has a basically circular pattern indicating that the system is a reflection nebula illuminated by a point-like source but at the position of this source there is no optical feature. Although our illuminator is completely hidden from direct view we know that it must be radiating in the optical part of the spectrum to produce the visible nebula – the source must still be deeply embedded in the globule. The position of our illuminator and the source CG30IRS4 are coincident, they are one and the same object. Further details concerning this object are given in Scarrott, Gledhill *et al.* (1990).

The Re10 nebulosity: The small and compact optical nebulosity known as Re10 is located within a dark cloud in the direction of the Coalsack complex. Although there is no optical stellar-like feature associated with the nebulosity there is a strong 2.2 μm source in the region (Reipurth, 1981) which is coincident, within positional uncertainties, with IRAS 12542-6115.

The intensity contour and polarization maps of Figure 2 show that Re10 is a small bipolar nebula with a central dust lane, presumably in form of a circumstellar disc, which totally obscures the illuminator from direct view at optical wavelengths. The observation that Re10 is a bipolar nebula with the character-

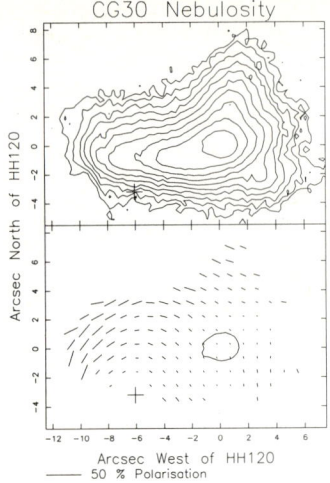
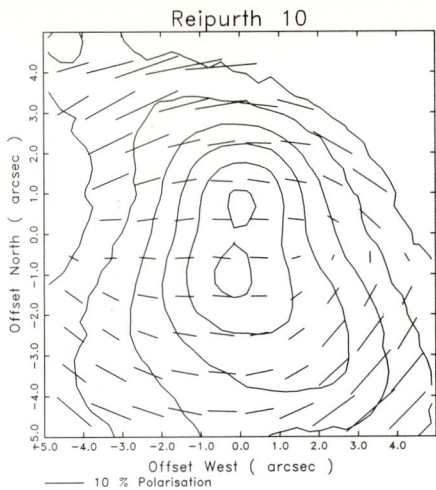

Figure 1: Intensity contour and linear polarization maps of the CG30 nebulosity. The circular pattern of vector orientations indicates that the system is a reflection nebula with a point-like illumiantor at the position marked +.

Figure 2: Intensity contour and linear polarization maps of the Re10 nebulosity. The system is a small bipolar nebula with the illuminating source hidden within the central dust lane.

istics of other pre-main sequence reflection nebulae strongly suggests that star formation is taking place in this dark cloud. If the cloud is part of the Coalsack then this is the first evidence for star formation in this region. These results are discussed in more detail by Eaton et al. (1990).

The L810 dark cloud: The L810 globule is a relatively large system of dark material, faint extensive nebulosity and compact and relatively bright clumps of nebulosity. Evidence for multiple star formation in L810 has been presented by Neckel et al. (1985) and Neckel and Staude (1987, 1990) who claim to have found up to 15 young stars embedded in L810, five of them associated with, and illuminating, individual cometary reflection nebulae. A visible star has been identified as both the dominant illuminating object and prime heating source of the dark cloud.

Figure 3 shows a polarization map of the extended nebulosity in the L810 dark cloud. The relatively simple and circular pattern of vectors indicates that there is a single dominant illuminating source in the neighbourhood of stars #7 and #17. Star #7 has been assumed to be the sole illuminator of the cloud. Figure 4 gives a high spatial resolution polarization map of the region in which the source is located. The pattern clearly centres some few arcseconds from star #7 in an area where there is no obvious optical feature, yet again we find the source of illumination of optical nebulosity to be completely hidden from direct view – it is deeply embedded in the cloud. Our map definitely indicates that star #7 is not an illuminator of nebulosity. Figure 5 shows the polarization in a more extended region about our illuminator. Stars #7, #17, #18, #8 and #16, which are

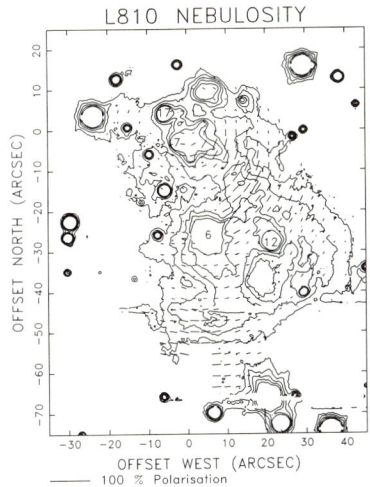 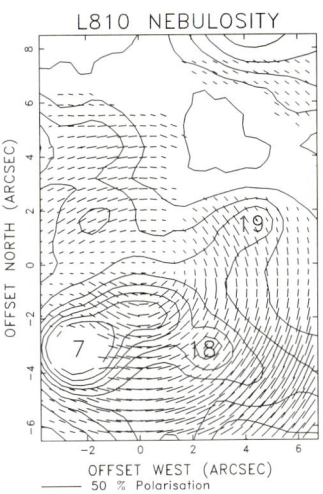

Figure 3: Intensity contour and linear polarization maps of the extended nebulosity in the L810 dark cloud. The illuminator appears to be in the region of stars #7 and #17.

Figure 4: A high resolution polarization map of the central source region. There is no optical feature at the centre of the pattern of vectors indicating that the source is deeply embedded in the dark cloud. Stars #7 and #17 play no part in the illumination of the reflection nebulosity.

supposed to illuminate individual cometary nebulae, show no indication in our data of illuminating any nebulosity whatsoever; in fact the supposed cometary nebulae are just enhanced regions of scattering centres illuminated by our single and central hidden source. With the exception of star #12, which is faintly nebulous and may be foreground to the cloud, we find no evidence whatsoever in our polarimetric data for multiple star formation in the L810 dark cloud. Our polarization data for the L810 dark cloud are given in greater detail in Scarrott, Rolph and Tadhunter (1990).

Conclusion: The polarization data presented here and cited earlier indicate that the small nebulosities associated with dark globules are each illuminated by stars which, although radiating in the optical waveband to create the visible features, are hidden from direct view. The stars are deeply embedded in the clouds but light can indirectly escape from them to create reflection nebulae. This means that there must be routes through the clouds along which light can escape without suffering too much extinction, we assume these routes just trace out the cavities blown in the cloud by outflow activity from the embedded objects. The reflection nebulosities are then formed by the illumination of the cavity walls.

Although our data indicate that star formation is taking place in dark globules they also show that the scale is limited, in each case investigated, we only find evidence for the creation of a single star. This is particularly significant in the case of the L810 system where multiple star formation had previously been supposed.

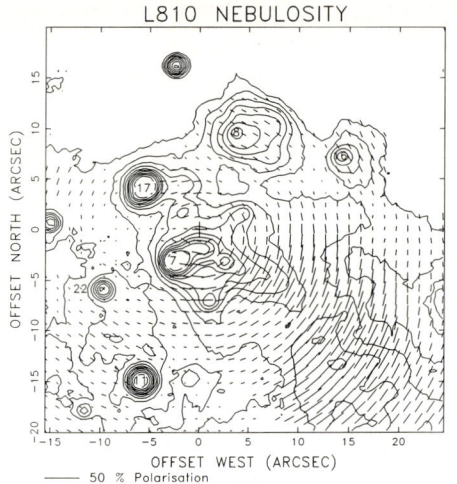

Figure 5: A polarization map in the regions of the supposedly individual cometary nebulae illuminated by stars #7, #17, #18, #8 and #16 respectively. All the nebulosity in this region is illuminated by our hidden source; there are no additional illuminators.

It appears that there is some mechanism inhibiting multiple star formation; possibly, the creation of the initial star disrupts secondary star formation; perhaps, the lifetime of the globule is such that only limited star formation can occur; or are there other effects such as magnetic fields within the globules that inhibit collapse leading to star formation?

References

Eaton, N., Rolph, C.D. and Scarrott, S.M., 1990, Mon. Not. Roy. Astron. Soc., **244**, 527.
Graham, J.A., 1988, in *Progress and Opportunities in Southern Hemisphere Optical Astronomy*, ed. V.M. Blanco and M.M. Phillips, Pub. ASP, p. 1.
Neckel, Th., Chini, R., Güsten, R. and Wink, J.E., 1985, Astron. Astrophys., **153**, 253.
Neckel, Th. and Staude, H.J., 1987, IAU Symposium No. 122 *Circumstellar Matter*, ed., I. Appenzeller and C. Jordan, (Reidel), p. 183.
Neckel, Th. and Staude, H.J., 1990, Astron. Astrophys., (in press).
Pettersson, B., 1984, Astron. Astrophys., **139**, 135.
Reipurth, B., 1981, Astron. Astrophys. Suppl., **44**, 379.
Scarrott, S.M., Warren-Smith, R.F., Draper, P.W., Bruck, M.T. and Wolstencroft, R.D., 1987, Mon. Not. Roy. Astron. Soc., **225**, 17P.
Scarrott, S.M., Gledhill, T.M., Rolph, C.D. and Wolstencroft, R.D., 1990, Mon. Not. Roy. Astron. Soc., **242**, 419.
Scarrott, S.M., Rolph, C.D. and Tadhunter, C.N., 1990, Mon. Not. Roy. Astron. Soc., (in press).
Scarrott, S.M. and Warren-Smith, R.F., 1988, Mon. Not. Roy. Astron. Soc., **232**, 725.

The relationship of optical to molecular outflows

T. P. RAY[1], R. POETZEL[2] and R. MUNDT[2]

HH objects and jets from molecular outflow sources. Most of the early studies of Herbig-Haro (HH) objects and highly collimated HH-like jets (see Mundt 1988, Edwards, Ray and Mundt 1990) concentrated on low luminosity stars with a few notable exceptions (*e.g.* Axon and Taylor, 1984). Naturally the question then arises as to whether such outflows are common amongst high luminosity young stellar objects (HLYSOs) and if so what, if any, is their relationship with the molecular outflows (see Lada, 1985)?

In the past few years several examples of HH objects have been found to be associated with HLYSOs. For a full list the reader is referred to Edwards, Ray and Mundt (1990). Even more recently jets have been discovered emanating from LkHα 234, AFGL 4029 (Ray *et al.*, 1990) and Z CMa (Poetzel *et al.*, 1989). Many of the HLYSOs with optical outflows are highly obscured but almost all of those that are visible (for example LkHα 198, LkHα 234, MWC 1080 and V645 Cyg) show a Herbig Ae/Be spectrum (Finkenzeller and Mundt, 1984, Goodrich, 1986).

Generally speaking the optical outflows from HLYSOs have higher velocities (up to $400 - 700$ km s^{-1} in the line wings) than those from low luminosity young stars (Axon and Taylor, 1984, Hartigan *et al.*, 1986, Reipurth, 1989, Poetzel, 1990, Poetzel *et al.*, 1990, Ray *et al.*, 1990). Although jets have been seen, the HH emission can in some cases consist of an isolated knot (*e.g.* LkHα 198, see Ray, Poetzel and Mundt, 1990) while in others, *e.g.* V645 Cyg (Zou and Solf, 1990), it may be spread over a range of angles.

As an example of an optical outflow from a HLYSO with an associated molecular outflow we shall consider the case of Cep A (see Figure 1 and Ray, Poetzel and Mundt, 1990). Cep A is interesting in that many of the HH objects to the west of this source (the GGD 37 complex, see Hartigan *et al.*, 1986) have either a comma- or a bow-shaped appearance. All of them are blueshifted and several possess large line widths indicative of high shock velocities. Hartigan *et al.* (1986) have successfully modelled the GGD 37 HH objects in terms of individual bow shocks driven into the ambient medium by "bullets" ejected from the central YSO. Turning now to the other side of this source, Lenzen (1988) reported the discovery of a HH object with a high proper motion two arcminutes northeast of Cep A. That object (HH-NE) can be seen in Figure 1 along with another newly-discovered HH object (HH-NE2) some 0.75′ further west. Hereafter we refer to HH-NE as HH-NE1. As in the case of the GGD 37 complex both HH-NE1 and HH-NE2 display a bow-shaped morphology. In particular HH-NE2 has what appears to be "wings" to the west and southwest, the central axis of which point toward the Cep A source. Looking at Figure 1, we also see some very diffuse bands of nebulosity to the west and the south of the HH objects. The western nebulosity has already been seen by Lenzen (1988) who

---

[1] DUBLIN INSTITUTE FOR ADVANCED STUDIES, SCHOOL OF COSMIC PHYSICS, 5, MERRION SQUARE, DUBLIN 2, IRELAND.
[2] MAX PLANCK INSTITUTE FOR ASTRONOMY, HEIDELBERG, F.R.G

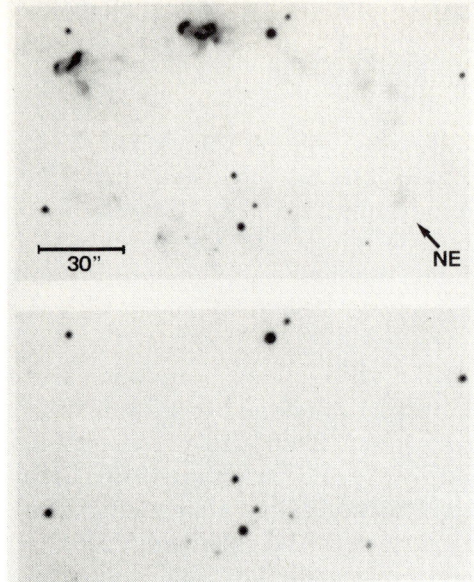

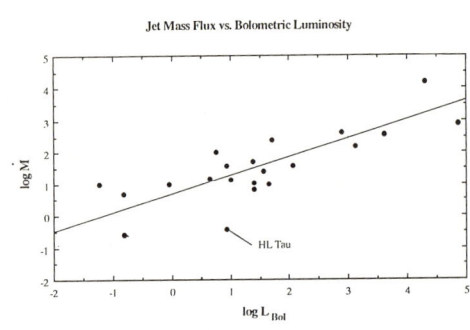

Figure 1. Eastern portion of Cep A through a [SII] filter (top) and a continuum filter (bottom). The details regarding the filters are as given in the caption for Fig. 1. The [SII] image shows the already-known HH object HH-NE1 (centre-top) along with the newly-discovered HH object HH-NE2 (upper left corner).

Figure 2. A plot of jet mass flux (in units of 10^{-9} $M_\odot$ yr^{-1}) versus bolometric luminosity of the parent source. The same power-law dependence on L_{bol} is seen as for the mass flux required to drive molecular outflows.

also pointed out that the northeastern edge of this nebulosity coincides with the string of radio knots (HW 1-6) discovered by Hughes and Wouterloot (1984) and Hughes (1985). Lenzen (1988) concluded that the emission from this nebulosity was continuum emission since it was also seen in his I band images. In fact, the structures observed in I and HH emission lines (*e.g.* [SII]$\lambda\lambda$6716,6731) differ in their morphology (Ray, Poetzel and Mundt, 1990) so the possibility of some line emission cannot be automatically excluded. Moreover deep narrowband continuum images (close to but excluding the HH [SII] doublet) show no trace of these nebulous bands (see Figure 1 and Ray, Poetzel and Mundt, 1990) implying that they are in fact shock-excited. Thus not only does the somewhat scattered complex of HH objects to the west of Cep A suggest the presence of a poorly-collimated high velocity wind from this source but the HH objects and diffuse emission to the east do as well.

What is the connection between optical and molecular outflows? As emphasised for example by Mundt (1988), HH objects and HH-like jets map the highest velocity and most highly collimated components of the outflows from YSOs. Molecular outflows by comparison are much more poorly collimated with considerably lower velocities (Lada, 1985, Fukui, 1988). Moreover, whereas jets may

Table 1.

Source	L_{bol} ($L_\odot$)	$\dot{M}^a$ ($10^{-10} M_\odot \, yr^{-1}$)	Sp. Type[b]	Ref.
AFGL 2591	90000	600	FUOR	1,2
AFGL 4029	20000	15000	H Ae/Be	3
AS 353A	5.8	100	cTTS	4
DG Tau	8.5	37.5	cTTS	4
DG Tau B	0.9	10	cTTS	4
Haro 6-5 B	0.06	10	cTTS	4
HH 7-11 (SSV 13)	115	38	emb	4
HH 24 (SSV 63)	24.7	10	emb	1,5
HH 30	0.15	2.5	cTTS	1,6
HH 34 IRS	45	10	emb	4,7
HH 46/47-IRS	24	50	emb	4
HH 83	10	13	emb	1,7
HH 111	25	6.5	emb	1,7
HL Tau	8.5	4.0	cTTS	1,6
L1551 IRS5	36.3	25	FUOR	4
LKHα 234	1300	150	H Ae/Be	3
R Mon	785	430	H Ae/Be	4
Th 28	0.15	5	cTTS	4
T Tau S	4.4	15	emb	4
VLA-1 HH1/HH2	50	250	emb	4
Z CMa	3500	350	FUOR	1,8

a) The jet internal shock velocities are assumed to be 70 km s^{-1} except for HH30 where 50 km s^{-1} is used because of this jet's lower excitation.
b) Spectral types are either that of a classical T Tauri (cTTS), FU Orionis (FUOR) or Herbig Ae/Be (H Ae/Be) star where optically visible. However many of the stars are embedded (emb).
References:
(1) This paper, (2) Poetzel *et al.*, 1990, Ray *et al.*, 1990, (4) Mundt, Brugel and Bührke, 1987, (5) Raga, Mundt and Ray, 1990, (6) Mundt *et al.*, 1990, (7) Reipurth 1989, (8) Poetzel, Mundt and Ray, 1989.
Luminosities are taken from the above references and from Cohen and Schwartz (1987).

be identified with, or at least considered to be a component of, the winds from YSOs, the high velocity molecular gas seems to be a shell of swept-up material driven by the wind (Snell *et al.*, 1988). Leverault (1988) has shown that the mass loss rate in the wind driving a molecular outflow scales roughly with $L_{bol}^{0.6}$ where L_{bol} is the bolometric luminosity of its source. If, following Levreault (1988), we plot mass loss rates of optical jets versus L_{bol} (see Figure 2 and Table 1) it is immediately seen that the mass loss rate similarly scales with $L_{bol}^{0.6}$. Note however, as emphasized by Mundt, Brugel and Bührke (1987), that the mass loss rates in the jets are typically one to two orders of magnitude less than that of their associated molecular outflows so that the jets cannot be the driving wind. Combining the results of Levreault (1988) with those presented here, it seems very likely that the jets and the driving wind must come from the same source since their mass loss rates both scale roughly with the same power of L_{bol}. Edwards, Ray and Mundt (1990) have argued that both the winds driving the molecular outflows and jets originate in the disk surrounding the parent YSO. The jet can then be looked-upon as perhaps the central component of a largely neutral wind. Such winds have been observed directly for example by Lizano (1988) in the case of HH 7-11 with velocities up to several hundred km s^{-1} and Lizano et al. (1988) and Koo (1989), have shown that these winds contain sufficient momentum to drive their associated molecular outflows.

References

Axon, D.J. and Taylor, K., 1984, Mon. Not. Roy. Astron. Soc., **207**, 241.
Catala, C., 1989, in ESO Workshop on *Low Mass Star Formation and Pre-main Sequence Objects*, ed. B. Reipurth, (ESO, Garching), p. 471.
Cohen, M. and Schwartz, R.D., 1987, Astrophys. J., **316**, 311.
Edwards, S., Ray, T.P. and Mundt, R., 1990, in *Protostars and Planets III*, ed. E. Levy and J. Lunine, (University of Arizona Press, in press).
Finkenzeller, U. and Mundt, R., 1984, Astron. Astrophys. Suppl., **55**, 109.
Fukui, Y., 1988, Vistas in Astron., **31**, 217.
Goodrich, R.W., 1986, Astrophys. J., **311**, 882.
Hartigan, P., Lada, C.J., Stocke, J. and Tapia, S., 1986, Astron. J. **92**, 1155.
Hughes, V.A., 1985, Astrophys. J., **298**, 830.
Hughes, V.A. and Wouterloot, J.G.A., 1984, Astrophys. J., **276**, 204.
Koo, B.-C., 1989, Astrophys. J., **337**, 318.
Lada, C.J., 1985, Ann. Rev. Astron. Astrophys., **23**, 267.
Lenzen, R., 1988, Astron. Astrophys., **190**, 269.
Levreault, R.M., 1988, Astrophys. J., **330**, 897.
Lizano, S., Heiles, C., Rodriguez, L.F., Koo, B.-C., Shu, F.H., Hasegawa, T., Hayashi, S.S. and Mirabel, I.F., 1988, Astrophys. J., **328**, 763.
Mundt R., 1988, in *Formation and Evolution of Low Mass Stars*, ed. A.K. Dupree and M.T.V.T. Lago, (Kluwer, Dordrecht), p. 257.
Mundt, R., Brugel, E.W. and Bührke, T., 1987, Astrophys. J., **319**, 275.
Mundt, R., Ray, T.P., Bührke, T., Raga, A. and Solf, J., 1990, Astron. Astrophys., (in press).
Poetzel R., 1990, *Optical Investigations of Outflows from Luminous Young Stars*, Ph. D. Thesis, University of Heidelberg.
Poetzel, R., Mundt, R. and Ray, T.P., 1989, Astron. Astrophys., **224**, L13.
Poetzel, R., Mundt, R., Ray, T.P. and Solf, J., 1990, (in preparation).
Ray, T.P., Poetzel, R., Solf, J. and Mundt, R., 1990, Astrophys. J. **357**, L45.
Reipurth, B., 1989, in ESO Workshop on *Low Mass Star Formation and Pre-main Sequence Objects*, ed. B. Reipurth, (ESO, Garching), p. 247.
Snell, R.L., Huang, Y.-L., Dickman, R.L. and Claussen, M.J., 1988, Astrophys. J., **325**, 853.
Zou, H. and Solf, J., 1990, (in preparation).

Cha T1 luminosity function

P.R. WESSELIUS[1], T. PRUSTI[1], D.C.B. WHITTET[2], R. ASSENDORP[1]

Abstract The luminosity function of the star formation region Cha T1 is derived. It is compared with that of ρ Oph. There are no significant differences, indicating the same mix of stars and the same stage of evolution.

1 Introduction

There are two good reasons to study the stellar content of nearby star formation regions (SFR's) in great detail using all possible wavelength regions: deriving an initial mass function (IMF) and finding the lowest mass stars.

The IMF is a fundamental parameter for evolution studies of galaxies. One way to derive an IMF is to study the stellar contents of a SFR. Studying just one SFR is insufficient: all stars need not have been born yet. By studying several star formation regions in different stages of evolution an IMF may be derived.

The low-mass part of the IMF ($< 0.1\,M_\odot$) is poorly known. Although present scarce data suggest a downward trend, this might taper off at even lower masses, and there might be hidden mass in very low-mass "stars". Very low mass stars could be found in SFR's.

Determining the IMF for a SFR is not possible yet. It is possible to determine the total luminosities of members of a SFR, provided that multispectral observations have been done and the distance is known. A luminosity function (LF) can then be constructed. In a SFR several of the members may be in an early stage of their evolution, on the Hayashi track, and consequently be superluminous. Converting the LF of a star formation region into an IMF, the function that is really of interest, is non-trivial. The position of an object in the H-R diagram and evolutionary tracks are needed for such a conversion (see e.g. Cohen and Kuhi, 1979). Putting a pre-main-sequence (PMS) star in a H-R diagram is problematic, because finding the temperature of a star is usually done through its spectral type: many of our objects do not have a spectral type or have a deviating one.

2 Cha T1 Members

First, a reliable list of members has to be compiled. Using IRAS colours alone, like Hughes and Emerson (1988, HE), is insufficient, because not only

---

[1] KAPTEYN INSTITUTE, P.O. BOX 800, 9700 AV GRONINGEN, THE NETHERLANDS.
[2] SCHOOL OF PHYSICS AND ASTRONOMY, LANCASHIRE POLYTECHNIC, PRESTON PR1 2TQ, U.K.

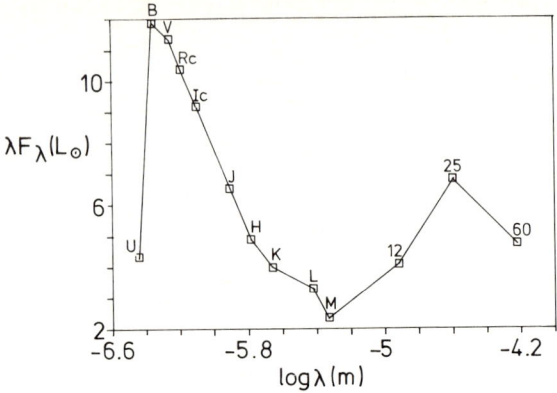

Figure 1: The SED of T32.

PMS stars occupy the relevant region of the IRAS colour-colour diagram. Many bona-fide members of low-mass star formation regions are not so easily detectable with IRAS in the infrared: both confusion and lack of sensitivity play a role.

We have compiled a list of members from several sources. The primary one was an article of Whittet et al. (1987); they carefully studied all objects observed towards Cha T1 and made a list of members containing 56 entries. We found 73 members: 58 optically-selected (Hα, variability, near reflection nebulae), 12 through near-infrared observations and three others. Feigelson and Kriss (1989) describe Einstein observations towards Cha T1. They find 22 soft X ray sources, 17 associated with PMS stars. Follow-up optical studies on 15 stars close to the X ray positions have resulted in 6 to 7 possible new members, "naked" T Tauri stars. These have not been included.

41 of the 73 sources could be identified in the IRAS point source catalogues published. Evidently, the IRAS catalogues do not contain fluxes for all members of even such a nearby SFR. The same applies for ρ Oph (Wilking et al., 1989). Thus to select members of a SFR through IRAS colour-colour plots will likely lead to an incomplete LF for nearly all SFR's. Even worse, non-members can be included in the LF, because the IRAS colours are insufficient discriminators.

At least for the nearest SFR's it might be possible to find IRAS fluxes for nearly all members by going back to the raw IRAS data, both survey and additional observations (AO: repeatedly scanning a small region of sky and thus increasing the signal-to-noise and spatial resolution), and combining these data before point source detection in a coadded image. That can increase the signal-to-noise by a factor 3 to 10. Such techniques have been developed by us for Cha T1, where a large number of AO's is available (of order 100). Wilking et al. (1989) used the published images of special AO's that mainly enhance the spatial resolution and were able to find IRAS fluxes for many more members.

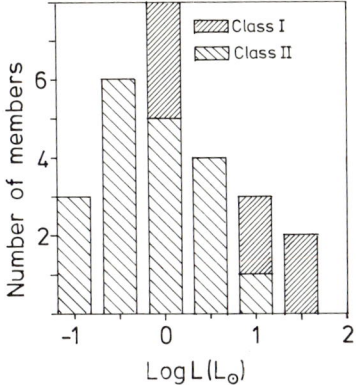

Figure 2: The LF of Cha T1 (Present Paper).

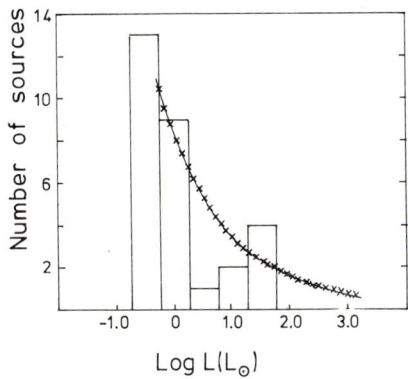

Figure 3: The LF of Cha T1 (HE)

3 Cha T1 LF

Figure 1 shows a very complete spectral energy distribution (SED) of the member T32. 41 out of the 73 members had rather complete SED's and their luminosities have been derived. A distance of 140 pc was used (Whittet *et al.*, 1987).

The luminosity function of these 41 members (Figure 2) can be compared with that of HE (Figure 3) and that of ρ Oph (Figure 4) All these LF's are incomplete. HE's sample suffers most from selection effects as discussed above. The SFR Cha T1 is advantageous to study because it is rather small (about 4 square degrees) and well-defined. At least two selection effects play a role for Cha T1: clearly not all members have yet been identified, and we have only included objects observed over a sufficiently broad spectral range. ρ Oph suffers from an extra selection effect: only part of the whole cloud has been described by Wilking *et al.* (1989) because it is so large, and it is not clear whether there is really one cloud or it concerns a complex of clouds.

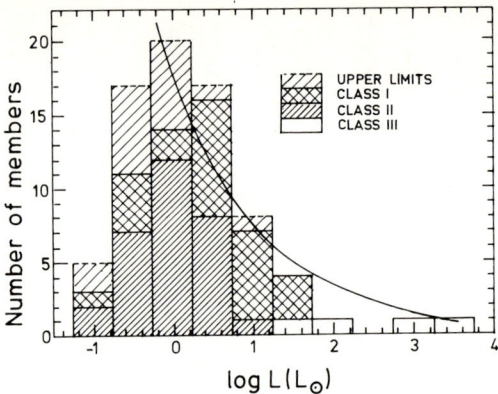

Figure 4: The LF of ρ Oph

Bearing these restrictions in mind we can compare the three LF's. HE and our Cha T1 LF differ considerably, especially in the range 1–10 $L_\odot$. This is because HE only used the far-infrared fluxes and some extrapolation to derive the LF. They therefore find much too low fluxes for visible members. Figure 1 (equal area = equal luminosity) illustrates that most of the luminosity may be in the optical spectral region and that the optical luminosity cannot be extrapolated from the far-infrared one.

The LF's of ρ Oph and Cha T1 look quite similar. Also the distributions of class I (the youngest) and class II (T Tauri star) sources are not very different between these two SFR's. We may conclude that ρ Oph and Cha T1, both low-mass SFR's, have the same mix of objects and are in a similar stage in their evolution.

References

Cohen, M. and Kuhi, L.V., 1979, Astrophys. J., **41**, 743.
Hughes, J.D. and Emerson, J.P., 1988, in *Comets to Cosmology*, ed. A. Lawrence, (Springer-Verlag, Berlin), p. 159.
Feigelson, E.D. and Kriss, G.A., 1989, Astrophys. J., **338**, 347.
Whittet, D.C.B., Kirrane, T.M., Kilkenny, D., Oates, A.P., Watson, F.G. and King, D.J., 1987, Mon. Not. Roy. Astron. Soc., **224**, 497.
Wilking, B.A., Lada, C.J. and Young. E.T., 1989, Astrophys. J., **340**, 823.

Modelling the 2.2 micrometre polarisation from the reflection nebula GL 2591

N. R. MINCHIN AND J. H. HOUGH

DIVISION OF PHYSICAL SCIENCES, HATFIELD POLYTECHNIC, HATFIELD, HERTFORDSHIRE AL10 9AB, U.K.

Young stellar objects embedded in molecular clouds are characterized by mass outflow, usually observed as a bipolar high-velocity flow of molecular gas. One of the most famous examples of this phenomenon is GL 2591. Lada et al. (1984) detected both blue-shifted and red-shifted high-velocity CO gas in the vicinity of GL 2591 IRS, along an approximate east-west direction.

Forrest and Shure (1986) have presented near-IR images of GL 2591 showing the presence at 1.6 μm and 2.2 μm of a bright 'loop' which is aligned with the direction of the high-velocity blue-shifted outflow and which they interpret as the outline of a 'bubble' or cavity with GL 2591 IRS at its eastern edge. Radiation associated with the loop is assumed to arise from the scattering of photons from GL 2591 IRS off dust grains at the surface of the cavity. More recently, Burns et al. (1989) have published an image of GL 2591 at 1.6 μm that shows the presence of a second loop of radiation beginning at the western edge of the first.

We made new, high spatial resolution, near-IR polarisation measurements of the blue-shifted outflow from GL 2591 IRS. Observations were made at the 3.8 metre United Kingdom IR Telescope (UKIRT), using the common-user near-IR camera (IRCAM) and the IR Polarimeter (IRPOL), on Mauna Kea, Hawaii. Observations were made using standard J (1.25 μm), H (1.65 μm) and K (2.2 μm) filters and at the high resolution 0.62 arcsec/pixel plate scale.

Figure 1 displays polarisation vectors overlayed upon surface brightness contours for GL 2591 at K. The polarisation vectors form a strong centro-symmetric pattern around the IR source along the western (blue-shifted) outflow, indicating that GL 2591 IRS is illuminating an IR reflection nebula. Our near-IR surface brightness maps show the existence of two loops of radiation, confirming the (1.65 μm) observations of Burns et al. (1989).

We devised computer models to simulate two different scattering geometries for the outflow region. First, scattering from a spherical or 'bubble' geometry with an illuminating star at its eastern edge (as proposed by Forrest and Shure, 1986) and secondly from a flat 'slab' at the end of an outflow of arbitrary shape. We assume single Rayleigh scattering and that the flux from the star drops as r^2. We examined the radial dependence of K polarisation values along cuts of constant azimuthal angle from GL 2591 IRS and compared these with values from our models. The results are shown in figure 2. The rate of increase of the polarisation values closely approximates the observations for an outflow inclined between 50° and 55° from the plane of the sky up to 6″ from the illuminator, where the observed percentage polarisation remains approximately constant. The radial

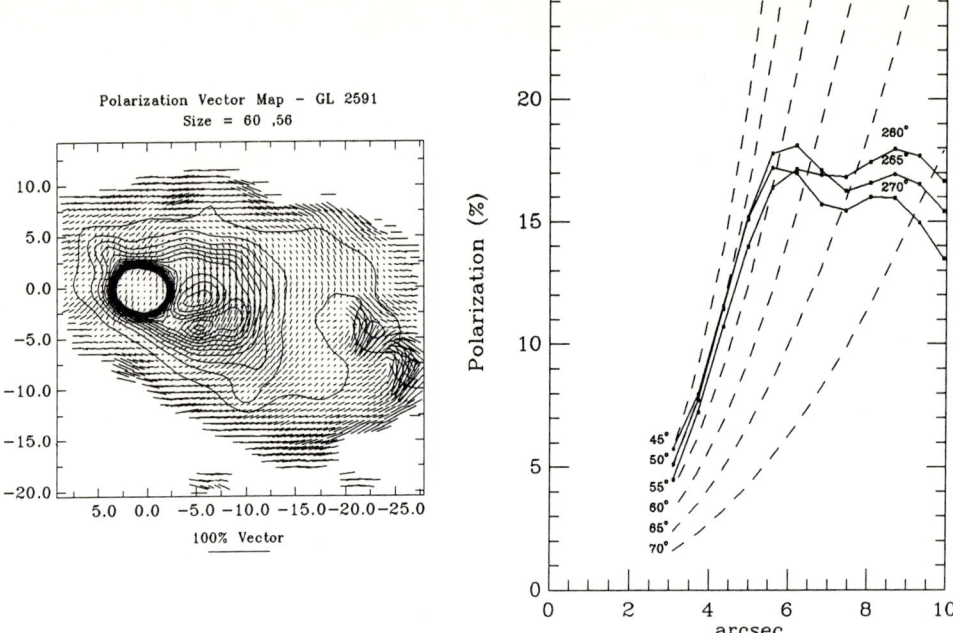

Figure 1. High resolution (0.62 arcsec/pixel) K (2.2 μm) imaging polarimetry of GL 2591. The polarisation vectors are superimposed upon the K surface brightness contours. The base level contour is at $1\,\mathrm{mJy\,arcsec^{-2}}$ and the contour step is $1\,\mathrm{mJy\,arcsec^{-2}}$. Offsets are in arcsec.

Figure 2. Polarisation values along the centre of outflows for various inclinations from the plane of the sky (45° to 70°) using the slab model (dashed lines) overlayed upon the observed percentage polarisation values at K along the centre of the western (blue-shifted) outflow (azimuthal angles 260°, 265° and 270°).

dependences of polarisation values calculated from the bubble model were found to be incompatible with the observations.

We propose that the two loops of radiation observed along the direction of the blue-shifted outflow are the result of two discrete mass outbursts from GL 2591 IRS.

A more detailed presentation and discussion of the observations and modelling will be given elsewhere (Minchin *et al.*, 1990).

References

Burns, M.S., Hayward, T.L., Thronson, H.A. and Johnson, P.E., 1989, Astron. J., **98**, 659.
Forrest, W.J. and Shure, M.A., 1986, Astrophys. J., **311**, L81.
Lada, C.J., Thronson, H.A., Jr., Smith, H.A., Schwartz, P.R. and Glaccum, W, 1984, Astrophys. J., **286**, 302.
Minchin, N.R., Hough, J.H., McCall, A., Yamashita, T., Aspin, C. and Hayashi, S.S., 1990, Mon. Not. Roy. Astron. Soc., (submitted).

Molecular clouds in the optical: dark clouds

JOACHIM A. STÜWE

ASTRONOMISCHES INSTITUT DER RUHR-UNIVERSITÄT BOCHUM,

POSTFACH 10 21 48, D-4630 BOCHUM 1, F.R.G.

Abstract This paper presents a method of obtaining physical parameters of dark clouds without the need of performing new observations, solely based on material already gathered. For this purpose Schmidt-photographs of two fields of the ESO/SRC Southern Sky Survey were digitized in both available colours (SRC-J and ESO(R)). An automatic starcount technique was applied to this data. The Bahcall and Soneira (1980) model of our Galaxy, in comparison with the results of the automatic starcounts yields distances for 43 dark clouds in these fields accurate to within 30%. Standard techniques of extinction mapping by starcount data give radii of these clouds in the range from 0.2 pc to 1.8 pc and masses of several ten to several hundred times the mass of the sun. The internal density structure $n(r) \sim r^{-\beta}$ of the clouds shows the same power-law index ($\beta = 1.0 \pm 0.4$) as the relation between the overall mean density of the clouds and their size $\langle n \rangle \sim R^{-\tilde{\beta}}$, which amounts to $\tilde{\beta} = 1.3 \pm 0.3$.

Introduction. In recent years several models of our Milky Way have been published, which in principle compute star numbers by solving the following equation:

$$N(m;l,b) = \omega \int_{-\infty}^{m} \int_{0}^{\infty} D(\mathbf{r}) \Psi(M,\mathbf{r}) r^2 \, dr \, dm' \qquad (1)$$

They adopt a space density distribution D for the stars in the Galaxy as well as a luminosity function Ψ, which are fitted to yield the observed stellar distribution in the Milky Way. With the aid of computers the cumulative star numbers as a function of apparent brightness can be computed for a certain galactic direction in a reasonable time.

Also recently the progress in computer techniques allows us to let the boring and time consuming work of star counts on photographic plates to be done automatically. This paper presents a method how these two developments brought together make it possible to determine distances as well as sizes and masses of dark clouds.

Principle of the method. An automatic star count and brightness classification algorithm (P. Grosbøl, 1983), applied on digitalized data from photographic Schmidt plates yields positions and a certain brightness parameter p for each detected star. If this is done in an unobscured region of the sky, cumulative star numbers as a function of this parameter are directly comparable to the model numbers as a function of apparent brightness m.

$$N_{obs}(p) = N_{mod}(m) \qquad (2)$$

This comparison yields a calibration relation $m(p)$, which then allows the computation of the apparent brightness for each detected star.

For a distance determination of a dark cloud it is necessary to infer a dark cloud of certain extinction A and distance d in the Galaxy model. These parameters are then determined via a "least square" analysis between the model numbers and the observed star counts.

The procedure of extinction mapping is straight forward and is described for example by Dickman (1987a, b), only that it is also done by the computer now. An imaginary counting grid is placed at the area to be analysed and for each grid point all stars up to a certain magnitude m_{obs} are counted. The galaxy model yields the star numbers in the same region of the sky for the case of no dark cloud and the comparison $N(m_{obs}) = N(m_{mod})$ directly gives the amount of extinction A (in magnitudes).

$$A = m_{obs} - m_{mod} \qquad (3)$$

Results. The method outlined above has been applied to 15 regions in the southern sky. In total 43 distinct small dark clouds have been detected in the extinction maps of these regions. Due to the parallel determination of the distance, it is possible to derive absolute sizes R for each cloud. The empirical relation between the extinction and the total column density of hydrogen N_H (Dickman, 1987b) also permits the computation of masses for the clouds

$$M = \sum N_H\, m_H\, \Omega_{grid}\, d^2 \qquad (4)$$

where m_H is the hydrogen mass, d the cloud distance and Ω_{grid} the solid angle of one counting grid point.

A detailed description of the method, as well as the exact results has been published in Astronomy and Astrophysics (Stüwe, 1990).

References

Dickman, R.L., 1987a, Astron. J., **83**, 363.
Dickman, R.L., 1987b, Astrophys. J. Suppl. Ser. **37**, 407.
Grosbøl P., 1983, SEARCH command in *Image Handling And Processing (IHAP)*, ed. F. Middleburg, (ESO).
Stüwe, J.A., 1990, Astron. Astrophys., **237**, 178.

Extragalactic molecules

R.S. BOOTH

ONSALA SPACE OBSERVATORY, CHALMERS UNIVERSITY OF
TECHNOLOGY, S-43900 ONSALA, SWEDEN

Abstract Molecular line observations of external galaxies are reviewed with emphasis on results of CO-line surveys, including new results for the Large Magellanic Cloud, and their ramifications for the conversion of CO integrated line intensities to molecular hydrogen masses. New results of surveys of CO in distant luminous far infrared galaxies are presented and the physical conditions in mergers are discussed. A list of known extragalactic molecules is given in the form of a table showing that the chemistry of external galaxies is probably not too different from that in the Milky Way.

1   Introduction

The first extragalactic molecule to be observed was OH, detected in absorption against the nuclear continuum sources in M82 and NGC 253 by the late L. Weliachew (1971). More recent searches for OH have revealed extragalactic masers of extreme strength, the mega masers, e.g. Baan *et al.* (1982). To date some 18 extragalactic molecules have been detected and there is every reason to believe that external galaxies will exhibit the same chemical complexity as the Milky Way. Detection of more complex species awaits further sensitivity improvements in the new generation of millimetre telescopes and their receivers.

Despite the evident interest in the chemical complexity of external galaxies, the most important results to date have come through observations of the relatively simple CO molecule. CO is believed to be a good tracer of the most abundant molecule, molecular hydrogen, which itself is only observed under somewhat extreme excitation conditions through near-infrared vibrational/rotational transitions. Observations of the CO molecule, which is excited at relatively low densities, have been used to determine global molecular properties of galaxies and to investigate the relationship between the molecular gas and other important galactic parameters like star formation rate and efficiency.

However, although CO is a good tracer of the molecular distribution and dynamics, not everyone is convinced of the efficacy of CO as a tool to determine the column density of H_2 and therefore galactic molecular masses. There has thus been a lot of discussion and no small controversy over the determination of a suitable conversion factor between the observed CO intensity and the H_2 column density. This is normally defined as $X = N(H_2)/\int T(^{12}CO)dv$ mols cm^{-2} K^{-1} km^{-1}s. The conversion factor normally used is determined solely from observations of Milky Way molecular clouds and this may result in misleading results. For example, there are indications that factors like metallicity and ambient UV radiation fields have a substantial effect on the CO abundance. In this context observations of the Magellanic Clouds offer an unique possibility to investigate the ratio, X in another galaxy. The new Swedish-ESO Submillimetre Telescope, SEST, has made it possible for the first time to map individual molecular clouds in the LMC to study the effect of metallicity on the CO-H_2 conversion factor.

Because of its obvious importance, a large fraction of time on millimetre telescopes has been devoted to studying CO in nearby galaxies. Maps of the CO distribution have been obtained for the nearer galaxies both with single antennas and with the recently available millimetre interferometer arrays at Berkeley, Caltech and Nobeyama, and will soon be possible with the recently commissioned IRAM interferometer.

In addition to these detailed CO observations of individual galaxies, statistical studies of large samples of galaxies have been undertaken to investigate molecular content as a function of morphological type. In this work a frequently used selection criterion has been strong flux in the far IR as determined by IRAS. Some new work done at Onsala and SEST will be presented and compared with earlier work by Devereux and Young (1990a), Young (1990) and others. In addition, some new results on CO detections in more exotic IRAS galaxies, also from SEST, (Mirabel et al. 1989, 1990) will be given. These galaxies have infrared luminosities in the range $10^{12} - 10^{13}$ $L_\odot$ and are suggested to be an evolutionary step towards quasars.

In principle, measurements of several CO transitions and those of the isotopically substituted species ^{13}CO should yield information on the physical parameters of the molecular clouds, e.g. excitation temperature, optical depth, etc. (see e.g. Israel, 1988). However, simple interpretations may lead to confusion in the case of galaxies because of the possible lack of uniformity in size and composition of the clouds filling the beam. Some new information on the LMC and on NGC 3256, a merging galaxy, will be presented to illustrate this point.

Other molecules will be discussed only briefly, but a current list of extragalactic molecules is given and some recent observations are described.

2 CO as a tracer of molecular mass

Despite the high optical depth of the 2.6 mm ground rotational line of CO, with certain assumptions, its emission may be shown to trace the molecular mass of giant molecular clouds and their molecular hydrogen masses may be determined from a standard conversion ratio X.

This is based on the following assumptions (Young and Scoville, 1982, Maloney and Black, 1988):

1. The ensemble average of excitation temperatures and mean densities is constant across the galactic disc.
2. The peak antenna temperature is a measure of the beam filling factor and hence the number of clouds in the beam.
3. All clouds are approximately in virial equilibrium in the sense that CO line widths are directly related to cloud mass and hence to the cloud area-averaged column density.

The last assumption is the reason for the strong correlation between the velocity width (σ_v) and (size)$^{0.5}$ found for Galactic Clouds (Solomon et al., 1987).

The theoretical basis for the constant conversion ratio, based on these assumptions has been reviewed by Young (1990) and following her paper and references cited therein we define the luminosity of a cloud as

$$L_{CO} = d^2 I_{CO}$$

where d is the distance to the cloud and $I_{CO}(= \int T dv)$, the integrated brightness temperature of the CO emission in the telescope beam.

For a cloud of radius R and line width Δv, the CO luminosity is given by

$$L_{CO} = \pi R^2 T_{CO} \Delta v$$

where T_{CO} is the peak brightness temperature in the CO line.

For a cloud of mass M and density ρ, in virial equilibrium, $\Delta v = (GM/R)^{0.5}$ and

$$L_{CO} = (3\pi G/4\rho)^{0.5} T_{CO} M$$

Thus, the CO luminosity is proportional to the molecular mass and to the ratio $T_{CO}/\rho^{0.5}$. If we can show that $T_{CO}/\rho^{0.5}$ is constant, on average, from galaxy to galaxy and if the other assumptions are correct, we can take the CO luminosity to be directly proportional to the mass of the molecular clouds.

An attempt to determine the N_{H_2}/I_{CO} ratio on a theoretical basis was made by Dickman et al. (1986). They considered statistical averages of an ensemble of clouds within a uniform antenna beam. They calculated the CO integrated intensity from the ensemble and compared it with the actual column density of H_2 in the beam. They assumed, as in the discussion above, that all clouds were virialized and that cloud-cloud shielding was negligible. For a mean H_2 number density of 200 cm^{-3} and a kinetic temperature of 6 K they obtained a value for X of 2.8×10^{20} cm^{-2} (K km s^{-1})$^{-1}$. The assumption of virialized molecular clouds may not be valid for the least massive ones (Maloney, 1990), which leads to an overstimate of the molecular mass for these clouds. Such clouds may contribute a significant fraction of the beam averaged CO emission from an external galaxy (see also Polk et al., 1988), even though they account for a small fraction of molecular mass.

There have been a number of observational attempts to determine empirically the constant of proportionality X between the H_2 column density and CO integrated intensity (cf. Frerking et al., 1982, Young and Scoville, 1982, Bloemen, 1988, Wolfendale, 1988). The techniques that have been applied are:

1 Correlation of ^{12}CO and ^{13}CO integrated intensities of, as well as column densities for ^{13}CO and C^{18}O in nearby molecular clouds, with the visual extinction A_v.

2 Correlation of virial masses, derived from molecular cloud sizes and ^{12}CO, and ^{13}CO, linewidths, with masses derived from CO integrated intensities.

3 Comparison of the distribution of CO intensity with the distribution of diffuse γ-ray emission, which traces the total hydrogen column density in the Milky Way.

The derived values of X cover the range $(1.5\text{--}4.8) \times 10^{20}$ cm^{-2} (K km s^{-1})$^{-1}$ (cf. Scoville and Sanders, 1987). The relations between I_{CO} and visual extinction A_v have been estimated for diffuse clouds of relatively low optical extinction. These values have then been extrapolated to A_v typical for dense molecular clouds, which can be an order of magnitude larger.

Estimates of the clouds temperatures and densities have recently been derived from CO multi-transition studies of M82 (Turner et al., 1989), and IC 342 (Eckart et al., 1990). These results show the existence of two or more molecular cloud components with different temperatures and densities: a cool one with $T_k \sim 10\text{--}20$ K and H_2 densities of $< 10^3$ cm^{-3}, and a warm component with $T_k \sim 75$ K and $n(H_2) \sim 5 \times 10^4$ cm^{-3}. These components turn out to give similar values for the ratio $T/\rho^{0.5}$. Thus, even in M82 which is one of the more extreme galaxies, the assumption of a constant conversion between CO luminosity and H_2 mass leads to a global molecular mass that is accurate to a factor of 2.

It may be that nature conspires to make X more or less constant over a relatively wide range of gas temperatures and densities, which would be very fortuitous for molecular line observers. However, until the constancy of X has been verified by observations, all results regarding the mass and distribution of molecular material in external galaxies should be viewed with caution and some scepticism. As has been shown by Maloney and Black (1988), radial as well as galaxy-to-galaxy variations of parameters such as gas kinetic temperatures, mean H_2 densities and CO abundance, can lead to considerable over- or underestimates of the molecular mass. This is particularly important when considering galaxy types different from our own Milky Way galaxy: e.g. dwarf irregulars which usually have a very low metallicity, lenticular and elliptical galaxies where the heating rate of the molecular gas may be less than in the Milky Way, and starburst galaxies where the physical properties of the interstellar medium are probably completely different from normal spirals.

3   The effect of metallicity on X – CO observations of the Large Magellanic Cloud

Maloney and Black (1988) and Elmegreen (1989) have argued that metallicity and UV radiation flux may affect the conversion of CO intensity to H_2 mass. In the case of a metal poor system, the lower abundance of C and O will give fewer CO molecules per hydrogen, which will be less able to shield themselves in the intense radiation field due to star formation. Molecular hydrogen, on the other hand, will continue to form rapidly on dust grains and will remain self-shielding. Thus the volume occupied by *detectable* CO, as a fraction of the total molecular

Figure 1: The Swedish – ESO Submillimetre Telescope, SEST.

volume (defined by the H_2), may be smaller in a system of low metallicity. In extreme cases, CO lines may not even be optically thick in typical clouds. Thus, the conversion X may be expected to be higher in the Magellanic clouds.

Arnault et al. (1988) from a small sample of metal poor systems found that CO luminosity varies with the metallicity squared but it is not clear from their analysis whether the conversion ratio X is affected.

Carbon monoxide in the Large Magellanic Cloud (LMC), a known low metallicity galaxy, has been studied by Israel et al. (1986) and by Cohen et al. (1988) with small telescopes (beam width 2′ and 8.8′, respectively corresponding to linear dimensions of 33 pc and 140 pc). In both cases, for cloud ensembles on the scale of the beam, virial mass was found to be greater than the mass derived from the CO profile and Cohen et al. have argued that LMC molecular clouds have six times more mass per unit CO luminosity than Milky Way clouds. Are these results somehow related to the coarse resolution of the measurements?

The 15m SEST telescope (Booth et al., 1989) shown in Figure 1 has enabled us, for the first time to observe individual molecular clouds in the LMC (Johansson and Booth, 1988, Johansson et al., in prep.). The dramatic step in resolution is shown in Figure 2 which presents a portion of the contour map of integrated CO in the LMC, near 30 Dor, observed by Cohen et al. (1988) compared with SEST data in the N 159 region south of 30 Dor. The linear resolution of the SEST observation is about 10 pc (c.f. sizes of GMC in the Milky Way which are 10–300 pc).

The preliminary results of Johansson et al. are presented in Figure 3. They are based on data collected during the commissioning phase of SEST. Since that

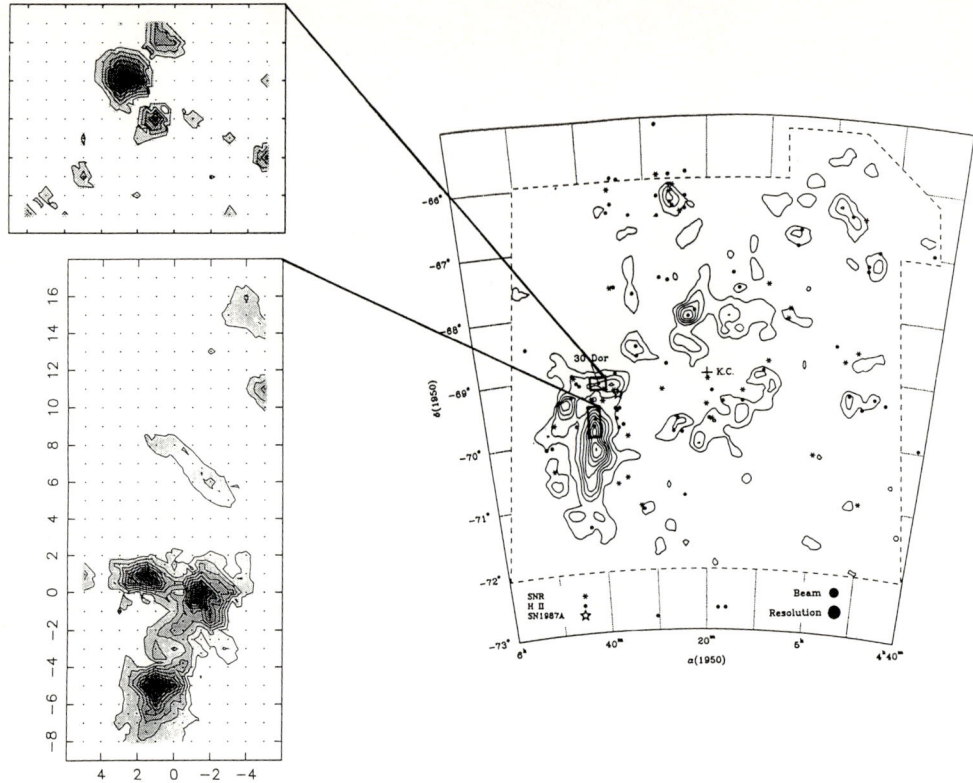

Figure 2: A map of the CO in the Large Magellanic Cloud (Cohen *et al.* 1988) showing the regions observed by Johansson *et al.* with the SEST telescope. (see text).

time, CO observations of the Magellanic Clouds have been designated as a SEST Key Project and a joint programme is being conducted by astronomers from the Swedish and ESO communities.

Figure 3a shows a plot of Log L_{CO} against Log M_{vir} derived from the SEST observations where some 20 clouds are identified and Figure 3b shows the size-velocity dispersion relation for the same clouds. The solid lines drawn through the data of Figure 3 show the equivalent relations derived for the Milky Way (Solomon *et al.*, 1987). They indicate (Figure 3b) that the LMC giant molecular clouds are virialized but Figure 3a clearly implies that the conversion, from L_{CO} to M_{H_2} is different, on average, by only a factor of ~ 2, from the Milky Way. While at first sight, this result favours a constant value of X from galaxy to galaxy, Johansson *et al.*, through observations of both the $(1 \rightarrow 0)$ and $(2 \rightarrow 1)$ rotational lines of ^{12}CO and ^{13}CO, find that the ^{12}CO in the LMC has an optical depth as low as 2. They therefore suggest that the CO clouds may indeed have suffered photodissociation and not be completely (spatially or dynamically) representative of the molecular hydrogen. Hence the ratio X may be somewhat greater than 2.

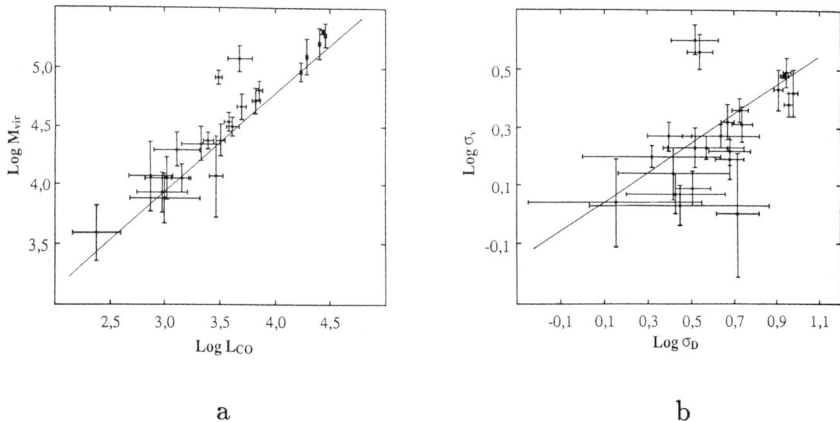

Figure 3: a) A plot of L_{CO} against virial mass for individual CO clouds in the 30 Dor region of the Large Magellanic Cloud. The Galactic relation (Solomon et al., 1987) is shown as a solid line. b) Plot showing the relationship between size and velocity width of the same clouds, again showing the Galactic correlation of Solomon et al.

4 CO in spiral galaxies

4.1 Global properties

Much of the work to date on the molecular properties of spiral galaxies has been done by Young with the 14m telescope of the U.Mass. group. She has presented an excellent review of this and other related work at the recent Wyoming Conference (Young, 1990). She points out the inherent difficulties of making comparisons among the available data both on observational grounds (different telescopes with different beam sizes, the importance of sampling the total CO, calibration inconsistencies, etc.) and due to the fact that galaxies, even of the same morphological type, may have very different absolute physical dimensions. However, Kenney and Young (1988) have carried out a careful analysis of the data available and claim that the flux uncertainties are within 32% so long as the CO has been sampled at several positions, at least along the major axis.

The major aim of studying the global properties of galaxies is to understand their evolution by relating the molecular data to other observable parameters. In this way several important apparent correlations have been found but their further translation into real physics must still be treated with caution, especially when a derived molecular mass is involved.

I will now summarise these correlations:

4.1.1 CO v radio continuum

The earliest relationship to be identified was a correlation between the CO line and radio continuum emission in nearby galaxies (Israel and Rowan-

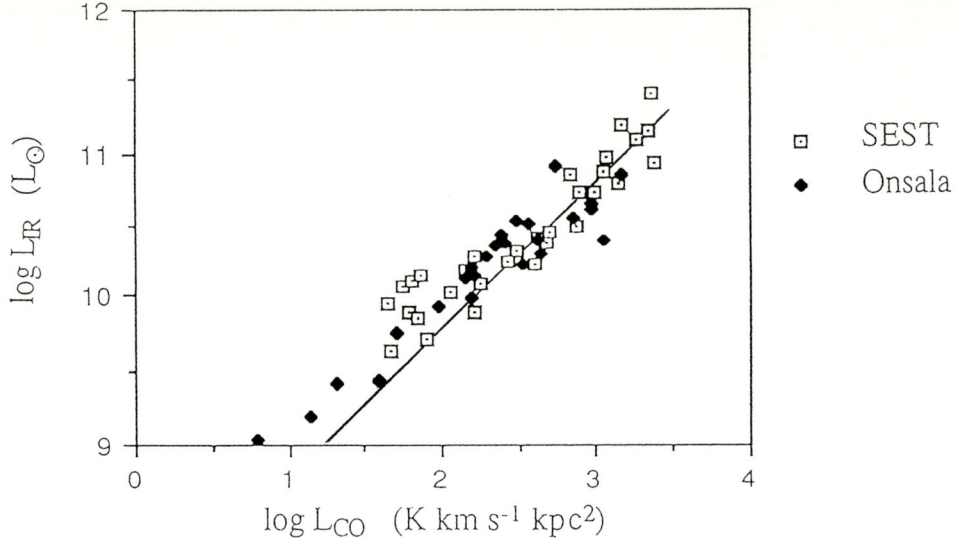

Figure 4: The correlation between CO luminosity and FIR luminosity for a sample of IRAS galaxies observed with the Onsala 20m telescope and with SEST. The correlation determined by Young (1990) is shown on the diagram as a solid line.

Robinson, 1984) These investigators interpreted their result in terms of the efficiency of OB star formation. The correlation has since been consolidated through IRAS and the tight global infrared – radio continuum correlation which extends over six orders of magnitude with little scatter (Dickey and Salpeter, 1984, de Jong et al., 1985, Helou et al., 1985). This relation strongly suggests that the radio emissivity is closely linked to the warm dust component in galaxies and hence to the recent (10^8 yr) star formation rate. In fact, Hummel (1990) has suggested that the (non-thermal) radio emission is one of the better indicators of recent star formation. This is very interesting since it may imply experimental evidence for the role of magnetic fields in the star formation process.

4.1.2 *CO versus IRAS flux*

The CO – IRAS flux correlation implied by the discussion above has been shown by a number of investigators e.g. Young et al., (1986a, 1986b), Sanders et al., (1986), Kenney and Young, (1988). Figure 4 shows a recent plot by Booth et al. (unpublished) derived from CO observations of IRAS galaxies with the Onsala and SEST telescopes. The slope derived from the plot of Young (1990) is shown as a solid line through the data.

If the molecular mass could be reliably derived from the CO luminosity, it would be possible to derive the star formation efficiency per unit molecular mass. Such studies, using the "standard" CO – H2 conversion factor imply that some very luminous infrared galaxies have extremely high star formation efficiencies (see section 8).

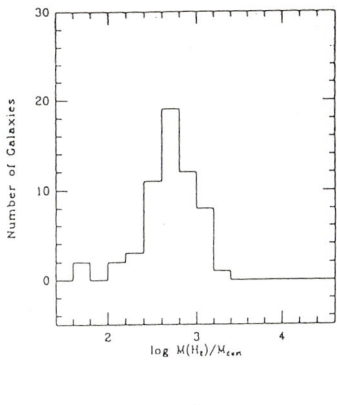 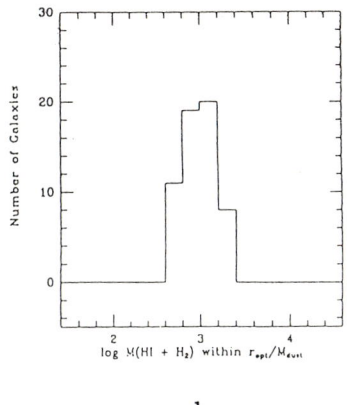

Figure 5: Histograms of the gas to dust ratio in a sample of spiral galaxies (Young, 1990). a) Molecular hydrogen/dust. b) (Atomic hydrogen within $R_{opt}/2$ + molecular hydrogen)/dust.

4.1.3 CO and dust

Remarkably, one of the best correlations to be found is that between molecular masses and IRAS-derived dust masses (Young *et al.*, 1986, Stark *et al.*, 1986) although the implied gas to dust ratios are typically 570 ± 50, rather than the value of nearer 150 which is generally taken as the ratio in the Milky Way (e.g. Spitzer, 1978). I say the correlation is remarkable because, again, the "standard" value of X has been used to derive the H2 mass for each galaxy. Turning this around, one might use the good correlation as an indication that the "standard" value of X is, after all, a reasonable conversion ratio, at least in the context of spiral galaxies! A histogram of the gas to dust ratio for molecular gas in spirals is shown in figure 5a.

Devereux and Young (1990b) have made a detailed study of the gas to dust ratio in a sample of 58 galaxies whose H_2 and HI distributions are known. They find that the addition of the atomic component does not change the scatter but if they add only that part of the atomic gas associated with the inner disk, ($R < R_{opt}/2$) of the galaxies, then the dispersion is dramatically reduced (see Figure 5b). Thus, the gas/warm dust ratio is well determined, with a value of 1080 ± 70. As Young (1990) points out, if the true gas to dust ratio is at or near the value of 150 measured for the Milky Way, then the high value derived from the IRAS data indicates that some 80–90% of the dust mass in spiral galaxies is radiating at wavelengths longer than 100 μm and is colder than about 30 K. It seems unlikely that the use of the "standard" value of X, would be the cause of such a discrepancy.

4.2 Detailed CO studies of spiral galaxies

Detailed studies of spiral galaxies are of great interest and may resolve several questions and ambiguities posed by observations of our own Milky Way galaxy. Important questions in this context relate to the nature of central condensations of gas in galaxies, the existence and nature of molecular spiral arms, their formation and lifetime, the role of density waves, etc.

Detailed studies of the spatial distribution of CO in the nearest galaxies have been going on using the larger millimetre telescopes for several years. These are inevitably long term projects since the resolution (beam size) needed to detect spiral arm enhancements unambiguously must be very small (typically $< 30''$), while the galaxies are many minutes in angular diameter, and the CO signals, particularly in the outer regions of a galaxy may be very weak, requiring long integrations. Thus, a galaxy mapping project may take several years to complete since it is unlikely that a large millimetre antenna will be given over entirely to a single project. The difficulties of such long term projects are compounded by the pointing and calibration problems inherent in most of the world's millimetre antennas and care and attention to detail is the hallmark of a successful mapping experiment.

Nevertheless, careful work is underway and anyway some of the questions can be answered by partial mapping and by interferometers. Techniques to enhance the resolution of single telescope observations of galaxies have been developed (Rydbeck *et al.*, 1988) and procedures to include the zero spacing (single telescope) data in spectral line interferometer maps have been elaborated by Mundy *et al.* (1988), but not yet applied to galaxy data.

5 Structure and dynamics of spiral galaxies

5.1 Spiral arm structure

Perhaps the best studied galaxy, with respect to molecular spiral arms is M51. The results of detailed work by Rydbeck *et al.*, using a maximum entropy deconvolution procedure to enhance the resolution of the Onsala observations to an equivalent $10''$ (500 pc at M51) show that molecular arms are clearly delineated and that the arm-interarm contrast ratio is at least 4 (see Figure 6). Interferometer maps by Lo *et al.* (1987) and more recently by Vogel *et al.* (1989) and Rand and Kulkani (1990) also show molecular clouds and cloud complexes confined to the spiral arms as outlined by prominent dust lanes and continuum ridges. Their estimate of the contrast is of order 3, but it is difficult to quantify since the interferometer resolves out 75% of the (extended) structure. Clearly, it is desirable to include zero spacing data in these maps. Finally, direct high resolution observations with the IRAM 30m antenna in the CO ($J = 2 \rightarrow 1$) rotational transition with a resolution of $11''$ show an arm-interarm contrast of 3–5:1.

What is clear is that a large fraction of the CO lies in the interarm region (Hjalmarson and Rydbeck, 1988) but arm-interarm variations of molecular density and temperature could be masking the true molecular hydrogen situation

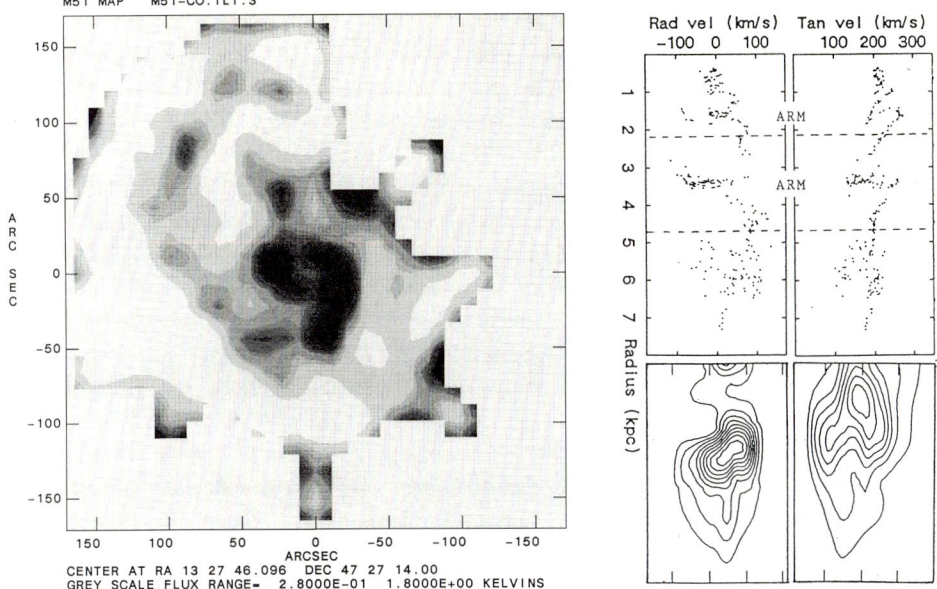

Figure 6: A grey scale plot of the integrated CO line temperature in M51. Plotted alongside are velocities of point mass particles from simulations with the corresponding observed velocity-position graphs.

in view of the $T_{CO}/\sqrt{\rho}$ dependence of the conversion factor X. Lord and Young (1990) and Tacconi and Young (1989) have compared their CO observations of M51 and NGC 6946 with Hα observations and find that the ratio, Hα/CO is enhanced on the arms. They suggest that this implies that there is a nonlinear dependence of star formation rate on the gas surface density. One mechanism which could explain this is cloud-cloud collisions (Noguchi and Ishibashi, 1986, Scoville et al., 1986). Cloud-cloud collisions are also considered by Rand and Kulkarni to explain the individual cloud complexes "superclouds" revealed in their beautiful interferometer map.

5.2 Velocity structure

Early observations of M51 by Rydbeck et al. (1985) revealed large velocity shifts across the spiral arms. Such shifts, tangential as well as radial, are very clear in the higher resolution Onsala data of Rydbeck et al., 1988 (Figure 6) and in the Owens Valley interferometer data (Vogel et al., 1989, Rand and Kulkarni, 1990). These gradients are in agreement with the density wave simulations of Roberts and Stewart (1987) which show a boost in the tangential velocity in the direction of rotation and a shift in radial velocity towards the nucleus.

6 CO in early-type galaxies

Although we are sometimes taught that early type galaxies are gas-free and inert, Hubble, in 1936, had noted the presence of dark patches on some dwarf ellipticals which he attributed to dust, and Mayall surveyed ionized gas in early type galaxies in 1939. After the detection of the 21 cm line of neutral hydrogen (see Wardle and Knapp, 1986), in 22 ellipticals and 90 S0 galaxies and the recent optical data on the presence of dust (e.g. Ebneter and Balick, 1985), it is perhaps not surprising that CO has been detected in such galaxies (Huchtmeier et al., 1988). Surveys of early-type galaxies detected by IRAS have been conducted by Sage and Wrobel (1989), Thronson et al. (1989) and by Wiklind and Henkel (1989). These and other observations have been collated by Wiklind (1990). Of 40 detections of CO, some 80% are S0 galaxies and the rest are ellipticals.

Wiklind and Henkel (1989) have compared the properties of their sample of early-type galaxies with those of more than 120 spiral galaxies, compiled from the literature. They find that the CO intensity is weaker in the early types by about a factor of 10, which they take to imply an order of magnitude less molecular hydrogen (i.e. they use the same conversion factor, X, which they take as 2×10^{20}, although they caution that this value may be too low for the early type galaxies). The far-IR luminosities are also an order of magnitude less than those of the spirals and so by using L_{FIR}/L_{H_2} as a measure of recent star formation efficiency, they derive a similar value for the early-types as for the spiral galaxies. Based on the LMC CO results of Johansson et al. reported in this review and on the fact that small lenticulars and S0 galaxies are almost certainly metal deficient relative to normal spirals, it is more probable that Wiklind and Henkel have underestimated the H_2 masses for the early-type galaxies whose star formation rates are then somewhat smaller than for the spirals

7 Centaurus A

The relatively nearby peculiar galaxy NGC 5128 exhibits a prominent dust lane, contains a large quantity of atomic and molecular gas, and harbours a strong radio continuum source, Centaurus A. The compact nuclear source is so strong at cm and mm wavelengths that molecular lines of OH, H_2CO, C_3H_2, ^{12}CO, and ^{13}CO can be seen in absorption against it (Gardner and Whiteoak 1976a, b, Bell and Seaquist 1988, Israel et al., 1990, Eckart et al., 1990). This means, for example, that information about the global content of molecular gas can be extracted from the CO emission in the usual ways, while the properties of individual giant molecular clouds can be derived from the absorption measurements. The ^{12}CO J $= 1 \rightarrow 0$ emission line mapping indicates a total molecular mass in a disk component 3' across of approximately 2×10^8 $M_\odot$ (Eckart et al., 1990) for an adopted distance of 3 Mpc (Ford et al., 1989). A comparable mass is inferred from the thermal dust emission if the contribution of cool dust is taken into account (Cunningham et al., 1984, Eckart et al., 1990). The ^{12}CO and ^{13}CO absorption features are evidently measured against the unresolved nuclear con-

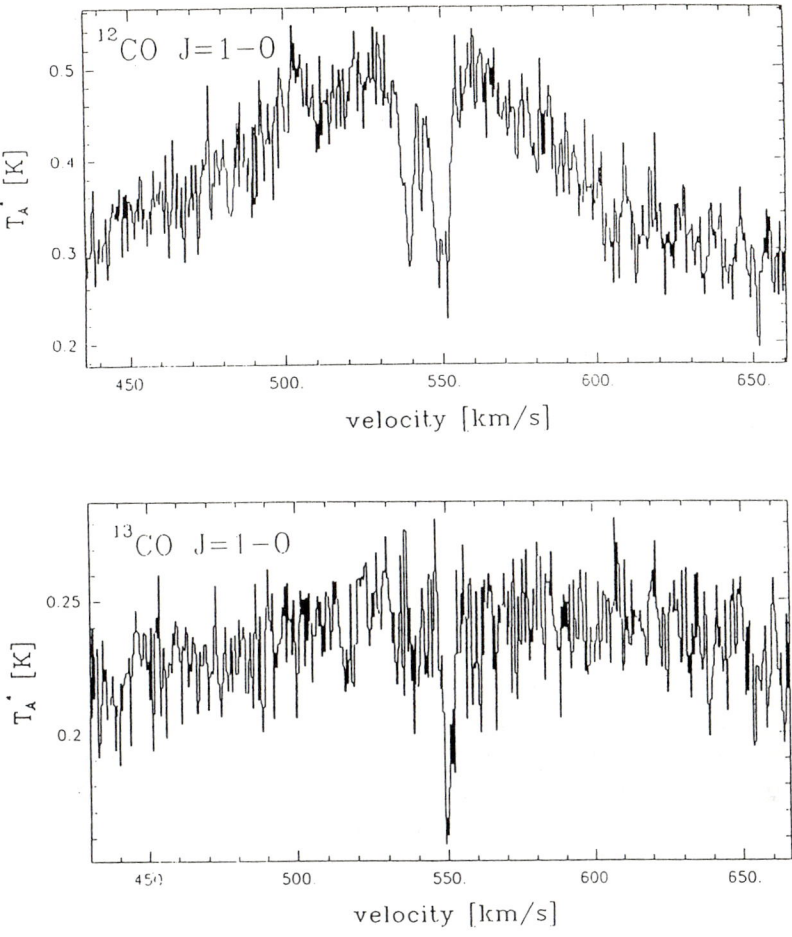

Figure 7: ^{12}CO and ^{13}CO spectra towards the centre of Centaurus A showing the absorption against the nuclear continuum source.

tinuum source and probably arise in a few individual clouds located closer to the nucleus than the average emitting gas. The absorption features (Figure 7) have narrow widths, 4 – 10 km s^{-1}, which are characteristic of individual giant molecular clouds in the Milky Way (Eckart *et al.*, 1990, Israel *et al.*, 1990). In principle, the measurement of two or more transitions, eg. ^{12}CO J = 1 → 0, 2 → 1, 3 → 2, etc., would yield both optical depths and excitation temperatures for the absorbing cloud(s).

Preliminary analysis suggests that the absorbing gas is hotter than the emitting gas, as might be expected for gas more closely confined near the central core. The observed infrared line emission of hot H$_2$ is unresolved on a scale of approximately 4″ centered on the nucleus and thus probes an even more compact component of molecular gas (Israel *et al.*, 1990).

8 Merging galaxies

8.1 CO in luminous far-IR galaxies or mergers

Among the many exciting results from the IRAS survey was the discovery of galaxies with luminosities dominated by far-infrared emission.(Soifer et al., 1984). In most of these galaxies the far-infrared flux appears to come from dust heated as a consequence of intense star formation. However, as the far-IR emission rises above $L_{IR} > 10^{11}$ $L_\odot$, in addition to star bursts, these objects are often powered by an active nucleus. Sanders et al. (1988) have proposed that galaxies that radiate in the infrared as much energy as quasars, i.e. $L_{IR} > 10^{12} L_\odot$ in the 8 μm – 1000 μm wavelength range, really are quasars enshrouded by dust. Since the ultimate source of fuel for intense star formation and/or an active nucleus resides in the interstellar medium, studies of the atomic and molecular gas are of importance for our understanding of these systems.

A broad observational attack has been led by Mirabel and Sanders who have studied the atomic hydrogen (Mirabel and Sanders, 1988), and CO emission in a sample of galaxies derived from the IRAS data. (Mirabel et al., 1988, 1990, Sanders et al., 1990). CO is detected in a high proportion of the sample, e.g. SEST observations of the Southern hemisphere part of the data (Mirabel et al., 1990) have resulted in 28 detections out of 33 sources observed, and even the non detections were probably due to inaccurate redshift data. (Many of these objects have relatively high redshifts, $z > 0.1$). A main result of this work is the demonstration that a common property of the luminous far-IR galaxies is an abundant supply of molecular gas. This is also confirmed by detections of OH maser emission in most of the galaxies with infrared luminosities above $10^{12} L_\odot$. In addition to a large supply of molecular gas, these megamasers require a continuum input, assumed to be the active galactic nucleus (Martin et al.,1989, Henkel and Wilson, 1989, Diamond et al., 1989).

Using the "standard" CO–H_2 conversion factor, Mirabel et al., 1990, determine molecular masses in the range of $0.6 - 6.0 \times 10^{10}$ $M_\odot$ for the luminous far infrared galaxies. These masses are 2 – 20 times the mass of molecular gas in the Milky Way, and on average substantially larger than the derived H_2 masses in the nearby luminous spiral galaxies like M101. The high abundance of molecular gas is taken to indicate that we are observing pairs of giant spiral galaxies in the process of merging — a fact confirmed by high resolution optical images of some of the galaxies (see Melnick and Mirabel, 1990) which show double nuclei, wisps, or long slender tails.

If we accept the values of derived molecular masses from the CO data on the luninous far infrared galaxies, we find a trend of increasing molecular mass with increasing L_{IR}. Furthermore, taking the far infrared luminosity as a measure of star formation rate, the star formation efficiency as determined by the ratio $L_{FIR}/M(H_2)$ has values in the range 10 – 80 $L_\odot/M_\odot$. This is larger than the average ratio found in "classical" starbursts like M82. These results must again be treated with caution because we do not understand completely the physical conditions in the galaxies and so the conversion of CO to H_2 could be in error.

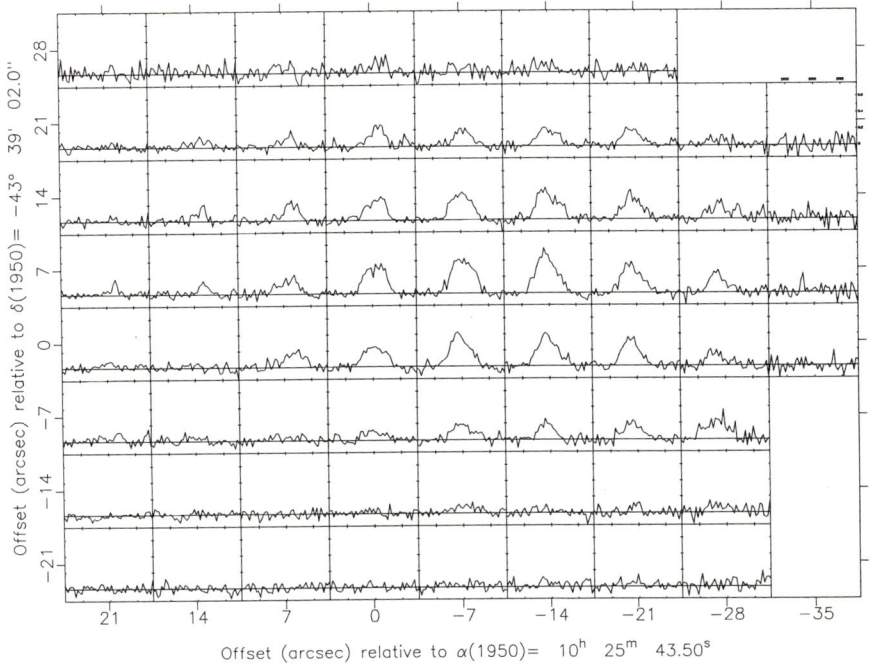

Figure 8: ^{12}CO (2 → 1) map of the merging galaxy NGC 3256.

8.2 Physical conditions in mergers — observations of NGC 3256

The bursts of star formation that occur in the centre of merging galaxies must lead to unusual conditions in the interstellar medium with large streaming motions and high temperatures. In such conditions it will be especially difficult to derive the total molecular mass from observations of the CO lines. Remember the conversion factor used is derived from observations of the Milky Way optically thick molecular clouds in virial equilibrium. It scales as $T/\sqrt{\rho}$, and for clouds in mergers that are both hotter and denser than Galactic clouds, it should not, in principle, lead to large uncertainties. In confirmation of the high density, Solomon *et al.* (1990) have detected CS in Arp 220 (the prototypical CO merger) showing that a large fraction of the molecular gas in this system may have densities higher than 10^5 cm^{-3}, 1000 times the density of ordinary clouds. However, if taken at face value, observations of the CO rotational lines do not always indicate such high densities showing that observations of ^{12}CO alone are not sufficient to understand the physical state of the molecular gas in the centre of merging galaxies. To illustrate this fact, I refer to observations of NGC 3256 (using SEST) by the Onsala group (Aalto *et al.*, unpublished) and by the Meudon group (Casoli *et al.*, private communication).

NGC 3256 is a southern merger galaxy, as shown by its optical appearance, experiencing a strong starburst. Its infrared luminosity is $L_{IR} = 3 \times 10^{11}$ $L_\odot$. We have mapped the galaxy in the two lowest rotational transitions of ^{12}CO with

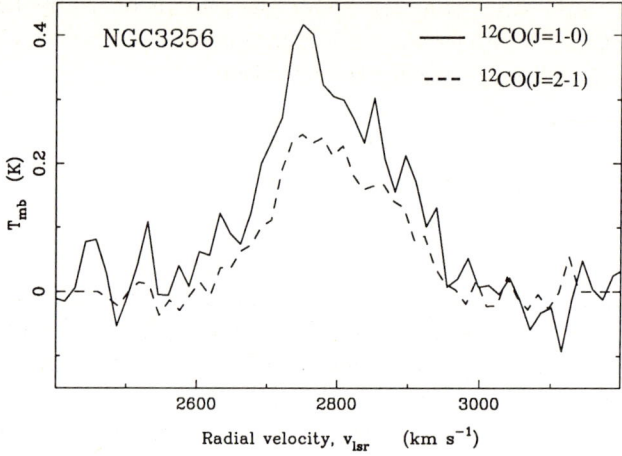

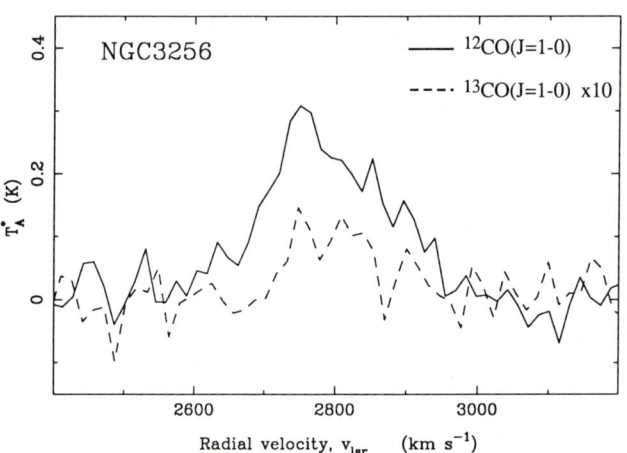

Figure 9: CO spectra towards the centre of NGC 3256 showing the line ratios.

better than half beamwidth sampling in each case. The $(2 \rightarrow 1)$ map is shown in Figure 8. The ratio $CO(2 \rightarrow 1)/CO(1 \rightarrow 0)$ is about unity when convolved to the same resolution. However, when observed in the lines of ^{13}CO, we find that the signals are relatively weak and that the ratio $^{12}CO(1 \rightarrow 0)/^{13}CO(1 \rightarrow 0)$ is more than 25 (see Figure 9). This would, conventionally, be taken to imply that a large fraction of the gas is of low optical depth, provided that the beam filling factors are the same for the two species. Alternatively, solutions can be found when the optical depths are high, but that requires a rather large isotope ratio. If the emission is optically thin then a lower mass would be obtained, compared to that derived assuming optically thick virialised clouds. However, the story may be even more complicated. Observations of ^{13}CO and $C^{18}O$ in the star burst galaxy M82 by Wild (Ph.D. thesis, 1990) give a ratio $^{13}CO/C^{18}O$ which in some places is as low as 3, indicating that the ^{13}CO gas has an optically thick component even though other CO observations indicate optically thin ^{12}CO. Clearly, more observations are needed to clarify the situation.

9 Observations of distant sources

The detection of CO in the ultraluminous far infrared galaxies, some of which have red shifts z > 0.1 together with the accumulating evidence that gas-rich galaxies may play an important role in the origin and evolution of quasars has led to a number of attempts to detect CO in more distant galaxies. Sanders *et al.* (1988) have begun a search for CO in a sample of "warm" ultraluminous infrared galaxies These galaxies were selected on grounds of both their 25 μm and 60 μm flux densities, using criteria similar to those used by de Grijp *et al.* (1988) and Low *et al.* (1988) to identify active galaxies from the IRAS data base. They have detected CO from a UV-excess quasar, Mrk 1014 at a red shift, z = 0.163. Mirabel *et al.* (1989) have detected CO in the powerful radio galaxies Perseus A (3C 84) and 4C 12.50 and suggest that large amounts of molecular gas, and therefore, high rates of star formation are involved in the genesis of the central engines that power extragalactic radio sources.

10 Other molecules in galaxies

The detection of the first extragalactic molecule, OH, was greeted with great excitement in the world press with talk of extragalactic life forms! There are now some 18 known extragalactic molecules, some detected in various isotopically substituted forms. However, I am not aware of any equivalent press hype.

I have listed the molecules associated with external galaxies in Table 1, together with their year of discovery. Nearly half the number have been detected in the past 4 years with the IRAM 30m telescope on Pico Veleta. This remarkable achievement reflects the sensitivity and power of that instrument.

10.1 High density tracers

Since most studies of the molecular gas in galaxies has been confined to the rotational transitions of CO, which trace the lower density gas ($\rho < 10^3$ cm^{-3}), a number of groups have begun studies of high density tracers like CS, (Mauersberger and Henkel, 1989, Solomon *et al.*, 1990), and the 2 cm transition of formaldehyde, (Baan *et al.*, 1990). Solomon *et al.* (1990) find that the CS(3 → 2) luminosity of Arp 220 is half the ^{12}CO luminosity of the Milky Way, and that since CS emission traces much higher density gas than that traced by CO emission, conclude that Arp 220 contains many more high density, star forming molecular cloud cores than normal galaxies. They claim that the existence of so many high

Table 1 *The extragalactic molecules known in March 1990*

Molecule	OH	H_2CO	CO	H_2O	HCN	H_2	NH_3	HCO^+	CH
Year	1971	1974	1975	1975*, 1977	1977	1978	1979	1979	1980
Reference	1	2	3, 4	5,6	7	8	9	10	11
Molecule	CS	C_3H_2	CH_3OH	CN	C_2H	HNC	HC_3N	N_2H^+	CH_3C_2H
Year	1985	1986	1987	1988	1988	1988	1990	1989	1989
Reference	12	13	14	15	15	15	15,16	17	18

* Tentative detection reported

References:

1. Weliachew, 1971
2. Gardner and Whiteoak, 1974
3. Rickard *et al.*, 1975
4. Solomon and de Zafra, 1975
5. Andrew *et al.*, 1975
6. Churchwell *et al.*, 1977
7. Rickard *et al.*, 1977
8. Thompson *et al.*, 1978
9. Martin and Ho, 1979
10. Stark and Wolff, 1979.
11. Whiteoak *et al.*, 1980.
12. Henkel and Bally, 1985.
13. Seaquist and Bell, 1986
14. Henkel *et al.*, 1987.
15. Henkel *et al.*, 1988.
16. Mauersberger *et al.*, 1990.
17. Mauersberger *et al.*, 1989a.
18. Mauersberger *et al.*, 1989b.

density gas clouds supports a starburst rather than an accretion disc - black hole origin for the extremely high far infrared luminosity of Arp 220.

Observations of the $J = 2 \rightarrow 1$ rotational line of CS and the $2_{11} \rightarrow 2_{12}$ transition of formaldehyde have been made in the galaxies M82 and NGC 253 by Baan *et al.* (1990). These observations show that large quantities of dense gas are present also in these galaxies. A comparison of observations with the Bell Labs antenna (Henkel and Bally, 1985) and the new observations with the 30m IRAM antenna on Pico Valeta indicates that the high density gas in M82 is confined to the nuclear region. The CS has a double peaked structure, similar to those observed in CO, ^{13}CO and CN (Nakai *et al.*, 1987, Loiseau *et al.*, 1988, Henkel *et al.*, 1988) and has been interpreted as a rotating torus of gas at a radius of 80 – 400 pc (Baan *et al.*). The presence of 6 cm absorption (Graham *et al.*, 1978) and the 2 cm emission in M82 is explained by Baan *et al.* in terms of a two component model. In this model, a compact and dense (10^6 cm^{-3}) source is embedded in a more extended low density ($< 10^4$ cm^{-3}) gas which covers the whole continuum source. In their paper, Baan *et al.* also report the first detection of extragalactic para-formaldehyde.

11 Concluding remarks

This paper is an attempt to review the current status of the observational data on extragalactic molecules. The work is at a very exciting stage with many interesting results. The volume of the Universe amenable for molecular observations is very much greater than we could have expected a few years ago, and as the search to distant galaxies is intensified, it becomes clear that molecular gas has a significant role as a source of fuel for the powerful radio galaxies and quasars.

There are, inevitably, some possible sources of error in our interpretation of the CO observations in terms of H_2 masses. It would seem however that as long as we are dealing with dense, optically thick gas that these may not be too serious, although the implied variations of density in mergers should engender caution. In galaxies of low metallicity, on the other hand, there is now enough evidence for us to conclude that the CO to molecular hydrogen conversion factor is probably greater than for the Galaxy

The chemistry of external galaxies is poised for exciting new developments as the large millimetre telescopes are tuned to new line frequencies; even today we can infer with some confidence that the chemical history of the Milky Way is mimicked in its distant neighbours.

Acknowledgements

I am grateful to several people for help and support during the preparation of this manuscript. I want to thank John Black for helpful suggestions and critical comments, Lars E. B. Johansson and Susanne Aalto for providing material before publication and for help with diagrams and Christian Henkel for information on the "other" molecules.

References

Andrew, B.H., Bell, M.B., Broten, N.W. and MacLeod, J.M., 1975, Astron. Astrophys., **39**, 421.
Arnault, P., Casoli, F., Combes, F. and Kunth, D., 1988, Astron. Astrophys., **205, 41**.
Baan, W.A., Wood, P.A. and Haschick, A.D., 1982, Astrophys. J. 260, L49.
Baan, W.A., Henkel, C., Schilke, P., Mauersberger, R. and Güsten, R., 1990, **Astrophys. J., 353, 132**.
Bell, M.B. and Seaquist, E.R., 1988, Astrophys. J., **329**, L17.
Bloemen, H., 1988, In *Molecular Clouds in the Milky Way and External Galaxies*, ed. **R.L. Dickman**, R.L. Snell, and J.S. Young, (Berlin:Springer), p. 71.
Booth, R.S., Delgado, G., Hagström, M., Johansson, L.E.B., **Murphy, D.C., Olberg, M.,** Whyborn, N.D., Greve, A., Hansson, B., Lindström, C.O. **and Rydberg, A.,** 1989, Astron. Astrophys., **216**, 315.
Churchwell, E., Witzel, A., Huchtmeier, W., Pauliny-Toth, I., **Roland, J. and Sieber, W.,** 1977, Astron. Astrophys., **54**, 969.
Cohen, R.S., Dame, T.M., Garay, G., Montani, J., **Rubio, M. and Thaddeus, P.,** 1988, Astrophys. J., **331**, L95.

Cunningham, C.T., Ade, P.A.R., Robson, E.I. and Radostitz, J.V., 1984, Mon. Not. Roy. Astron. Soc., **211**, 543.
Devereux, N.A. and Young, J.S., 1990a, Astrophys. J., **350**, L25.
Devereux, N.A. and Young, J.S., 1990b, Astrophys. J., **359**, 42.
Diamond, P.J., Norris, R.P., Baan, W.A. and Booth, R.S., 1989, Astrophys. J., **340**, L49.
Dickey, J. and Salpeter, E.E., 1984, Astrophys. J., **284**, 461.
Dickman, R.L., Snell, R.L. and Schloerb, F.P., 1986, Astrophys. J., **309**, 326.
Ebneter, K. and Balick, B., 1985, Astron. J., **90**, 183.
Eckart, A., Cameron, M., Rothermel, H., Wild, W., Zinnecker, H., Rydbeck, G., Olberg, M. and Wiklind, T., 1990, Astrophys. J., **363**, 451.
Eckart, A., Downes, D., Genzel, R., Harris, A.I., Jaffe, D.T., and Wild, W., 1990, Astrophys. J., **348**, 434.
Elmegreen, B.G., 1989, Astrophys. J., **338**, 178.
Ford, H.C., Ciardullo, R., Jacoby, G.H. and Hui, X., 1989, in *Planetary Nebulae*, IAU Symposium 131, ed. S. Torres-Peimbert, p.335.
Frerking, M.A., Langer, W.D. and Wilson, R.W., 1982, Astrophys. J., **262**, 590.
Gardner, F.F. and Whiteoak, J.B., 1974, Nature, **247**, 526.
Gardner, F.F. and Whiteoak, J.B., 1976a, Proc. Australian Astron. Soc., **3**, 63.
Gardner, F.F. and Whiteoak, J.B., 1976b, Mon. Not. Roy. Astron. Soc., **175**, 9P.
Graham, D.A., Emerson, D.T., Weiler, K.W., Wielebinski, R. and de Jager, G., 1978, Astron. Astrophys., **70**, L69.
Helou, G., Soifer, B. and Rowan-Robinson, M., 1985, Astrophys. J., **298**, L7.
Henkel, C. and Bally, J., 1985, Astron. Astrophys., **150**, L25.
Henkel, C., Jacq, T., Mauersberger, R., Menten, K. and Steppe, H., 1987, Astron. Astrophys., **188**, L1.
Henkel, C., Mauersberger, R. and Schilke, P., 1988, Astron. Astrophys., **201**, L23.
Henkel, C., Whiteoak, J.B., Nyman, L.-A. and Harju, J., 1990, Astron. Astrophys., **230**, L5.
Henkel, C. and Wilson, T.L., 1990, Astron. Astrophys., **229**, 431.
Hjalmarson, Å. and Rydbeck, G., 1988, in *Molecular Clouds in the Milky Way and External Galaxies*, ed. R.L. Dickman, R.L. Snell, and J.S. Young, (Berlin:Springer), p. 432.
Huchtmeier, W.K., Bregman, J.N., Hogg, D.E. and Roberts, M.S., 1988. Astron. Astrophys., **198**, L17.
Hummel, E., 1990, in *Windows on Galaxies*, ed. G.Fabbiano *et al.*, Kluwer Academic Publishers.
Israel, F.P. and Rowan-Robinson, M., 1984, Astrophys. J., **283**, 81.
Israel, F.P., de Graauw, Th., van de Stadt, H. and de Vries, C.P., 1986, Astrophys. J., **303**, 186.
Israel, F.P., 1988, in *Millimetre and Submillimetre Astronomy*, ed. R.D. Wolstencroft and W.B. Burton, Kluwer Academic Publishers, p. 281.
Israel, F.P., van Dishoeck, E.F., Baas, F., Koornneef, J., Black, J.H. and deGraauw, T., 1990, Astron. Astrophys., **227**, 342.
Johansson, L.E.B. and Booth, R.S., 1990, in *Recent Developments of Magellanic Cloud Research*, ed. K.S. de Boer, F. Spite and G. Stasinska, O.P., p. 147
de Jong, T., Klein, U., Wielebinski, R. and Wunderlich, E., 1985, Astron. Astrophys., **147**, L6.
Keel, W.C., de Grijp, M.H.K., and Miley, G., 1988, Astron. Astrophys., **203**, 250.
Kenney, J.D.P and Young, J.S., 1989, Astrophys. J., **344**, 171.
Lo, K.Y., Ball, R., Masson, C., Phillips, T., Scott, S. and Woody, D., 1987, Astrophys. J., **317**, L63.
Loiseau, N., Reuter, H.-P., Wielebinski, R., and Klein, U., 1988, Astron. Astrophys., **200**, L1.
Lord, S. and Young, J.S., 1990, Astrophys J. **356**, 135.
Low, F.J., Cutri, R.M., Kleinmann, S.G. and Huchra, J.P., 1989, Astrophys. J., **340**, L1.
Maloney, P., 1990, Astrophys. J., **348**, L9.
Maloney, P. and Black, J.H., 1988, Astrophys. J., **325**, 389.
Martin, R.N. and Ho, P.T.P., 1979, Astron. Astrophys., **74**, L7.
Martin, J.M., Bottinelli, L., Dennefeld, M., Gougnenheim, L., and LeSqueren, A.M., 1989, Astron. Astrophys., **208**, 39.
Mauersberger, R. and Henkel, C., 1989, IAU Circular 4889.

Mauersberger, R., Sage, L.J., and Henkel, C., 1989 (a, b), IAU Circulars 4906 and 4914.
Mauersberger, R. and Henkel, C., 1989, Astron. Astrophys., **223**, 79.
Mauersberger, R., Henkel, C., Wilson, T.L. and Harju, J., 1989, Astron. Astrophys., **226**, L5.
Mauersberger, R., Henkel, C., and Sage, L., 1990, Astron. Astrophys., **236**, 63.
Melnick, J. and Mirabel, I.F., 1990, Astron. Astrophys., **231**, L19.
Mirabel, I.F., Booth, R.S., Garay, G., Johansson, L.E.B. and Sanders, D.B., 1988, Astron. Astrophys., **206**, L20.
Mirabel, I.F. and Sanders, D.B., 1988, Astrophys. J., **335**, 104.
Mirabel, I.F., Sanders, D.B. and Kazes, I., 1989, Astrophys. J., **340**, L12.
Mirabel, I.F., Booth, R.S., Garay, G., Johansson, L.E.B. and Sanders, D.B., 1990, Astron. Astrophys., **236**, 327.
Mundy, L.G., Cornwell, T.J., Masson, C.R., Scoville, N.Z., Bååth, L.B. and Johansson, L.E.B., 1988, Astrophys. J., **325**, 382.
Nakai, N., Hayashi, M., Handa, T., Sofue, Y., Hasegawa, T. and Sasaki, M., 1987, Pub. Astron. Soc. Japan, **39**, 685.
Noguchi, M. and Ishibashi, S., 1986, Mon. Not. Roy. Astron. Soc., **219**, 305.
Polk, K.S., Knapp, G.R., Stark, A.A. and Wilson, R.W., 1988, Astrophys. J., **332**, 432.
Rand, R.J. and Kulkarni, S.R., 1990, Astrophys. J., **349**, L43.
Rickard, L.J., Palmer, P., Morris, M., Zuckerman, B. and Turner, B.E., 1975, Astrophys. J., **199**, L75.
Rickard, L.J., Palmer, P., Turner, B.E., Morris, M. and Zuckerman, B., 1977, Astrophys. J., **214**, 390.
Roberts, W.W. and Stewart, G.R., 1987, Astrophys. J., **314**, 10.
Rydbeck, G., Hjalmarson, Å. and Rydbeck, O.E.H., 1985, Astron. Astrophys., **144**, 282.
Rydbeck, G., Hjalmarson, Å., Wiklind, T., and Rydbeck, O.E.H., 1988, in *Molecular Clouds in the Milky Way and External Galaxies*, ed. R.L. Dickman, R.L. Snell, and J.S. Young, (Berlin:Springer), p.446.
Sage, L.J. and Wrobel, J.M., 1989, Astrophys. J., **344**, 204.
Sage, L.J., Shore, S.N. and Solomon, P.M., 1990, Astrophys. J., **351**, 422.
Sanders, D.B., Scoville, N.Z., Young, J.S., Soifer, B.T., Schloerb, F.P., Rice, W.L. and Danielson, G.E., 1986, Astrophys. J., **305**, L45.
Sanders, D.B., Soifer, B.T., Elias, J.H., Madore, B.F., Matthews, K., Neugebauer, G. and Scoville, N.Z., 1988, Astrophys. J., **325**, 74.
Sanders, D.B., Scoville, N.Z. and Soifer, B.T., 1988, Astrophys. J., **335**, L1.
Scoville, N.Z., Sanders, D.B. and Clemens, D.P., 1986, Astrophys. J., **310**, L77.
Scoville, N.Z. and Sanders, D.B., 1987, in *Interstellar Processes*, ed. D.J. Hollenbach and H.A. Thronson, (Reidel, Dordrecht), p. 21.
Seaquist, E.R. and Bell, M.B., 1986, Astrophys. J., **303**, L67.
Soifer, B.T., Rowan-Robinson, M., Houck, J.R., de Jong, T., Neugebauer, G., Aumann, H.H., Beichman, C.A., Bogess, N., Clegg, P.E., Emerson, J.P., Gillett, F.C., Habing, H.J., Hauser, M.G., Low, F.J., Miley, G. and Young, E., 1984, Astrophys. J., **278**, L71.
Solomon, P.M. and de Zafra, R., 1975, Astrophys. J., **199**, L79.
Solomon, P.M., Rivolo, A.R., Barrett, J. and Yahil, A., 1987, Astrophys. J. **319**, 730.
Solomon, P.M., Radford, S.J.E., and Downes, D., 1990, Astrophys. J., **348**, L53.
Spitzer, L., 1978, in *Physical Processes in the Interstellar Medium*, (New York:Wiley).
Stark, A.A. and Wolff, R.S., 1979, Astrophys. J., **229**, 118.
Stark, A.A., Knapp, G.R., Bally, J., Wilson, R.W., Penzias, A.A. and Rowe, H.E., 1986, Astrophys. J., **310**, 660.
Tacconi, L. and Young, J.S., 1989, Astrophys. J., **352**, 595.
Thompson, R.I., Lebofsky, M.J. and Rieke, G.H., 1978, Astrophys. J., **222**, L49.
Thronson, H.A., Tacconi, L., Kenney, J., Greenhouse, M.A., Margulis, M., Tacconi-Garman, L. and Young, J.S., 1989, Astrophys. J., **344**, 747.
Turner, J.L., Martin, R.N. and Ho, P.T.P., 1990, Astrophys. J., **351**, 418.
Vogel, S.N., Kulkarni, S.R., Scoville, N.Z. and Hester, J., 1988, in *Molecular Clouds in the Milky Way and External Galaxies*, ed. R.L. Dickman, R.L. Snell, and J.S. Young, (Berlin:Springer), p.437.

Wardle, M. and Knapp, G.R., 1986, Astron. J., **91**, 23.
Weliachew, L., 1971, Astrophys. J., **167**, L47.
Whiteoak, J.B., Gardner, F.F. and Höglund, B., 1980, Mon. Not. Roy. Astron. Soc., **190**, 17p.
Whiteoak, J.B., Dahlem, M., Wielebinski, R. and Harnett, J.I., 1990, Astron. Astrophys., **231**, 25.
Wiklind, T., 1990, Ph.D. Thesis, Chalmers University of Technology. School of Electrical and Computer Engineering Technical Report No. 196.
Wiklind, T. and Henkel, C., 1989, Astron. Astrophys., **222**, 1.
Wolfendale, A.W., 1988, in *Molecular Clouds in the Milky Way and External Galaxies*, ed. R.L. Dickman, R.L. Snell, and J.S. Young, (Berlin:Springer), p. 76.
Young, J.S. and Scoville, N.Z., 1982, Astrophys. J., **258**, 467.
Young, J.S., Schloerb, F.P., Kenney, J. and Lord, S., 1986a, Astrophys. J., **304**, 443.
Young, J.S., Schloerb, F.P., Kenney, J. and Lord, S., 1986b, Astrophys. J., **311**, L17.
Young, J.S., 1990, in *The Interstellar Medium in Galaxies*, ed. H. Thronson and J.M. Shull, (Kluwer, Dordrecht, in press).

CO and magnetic fields in spiral galaxies

RAINER BECK[1], ELLY M. BERKHUIJSEN[1] and ESTEBAN BAJAJA[2]

Abstract A region with strong radio polarization in the SW quadrant of M31 was observed in the ^{12}CO $(1\rightarrow 0)$ line. Along the dust arm CO intensity is anticorrelated with the degree of polarized emission. On much larger scales this is observed in M81 and NGC 6946 which have higher star formation rates than M31. This indicates a close relation between the distribution and motion of CO clouds and the magnetic field structure. The influence of magnetic fields on the star formation rate may be very important and could be the clue to understand the correlation between far-infrared and radio continuum intensities within spiral galaxies.

1. The "magic circle" in the interstellar medium. The discovery of a tight correlation between the far-infrared and radio continuum luminosities of spiral galaxies (e.g. Wunderlich *et al.*, 1987) turns out to be essential to understand the physics of star formation in the interstellar medium. While the production of far-infrared emission is believed to be closely related to the star-formation rate, the connection of radio continuum emission, which involves both cosmic-ray electrons and magnetic fields, to star formation is far from being understood. Cosmic-ray acceleration is thought to occur in shock fronts of supernova remnants so that the total cosmic-ray energy density may also be related to the star-formation rate. However, the role of magnetic fields remains unclear. We call the sequence of processes in the interstellar medium leading from star formation to far-infrared emission on one side and to (closely related) radio continuum emission on the other side the "magic circle".

In a calorimeter model to explain the FIR-radio correlation Völk (1989) concludes that the energy density of the magnetic field has to be proportional to the energy density of the interstellar radiation field, whereas Helou and Bicay (1990) assume a relation between field strength and gas density. In both models the field is expected to be strongest in galaxies with the highest star-formation rates. High-resolution observations can show how strength and structure of the field changes in star-forming regions.

2. CO observations in M31. M31 is an almost ideal galaxy to study the relation between CO and magnetic fields. Its proximity allows high spatial resolution. Star-formation regions and radio continuum are concentrated in a "ring" at 10 kpc distance from the centre. The magnetic field structure is a torus with uniform direction (Beck, 1982), the basic configuration being generated by an α-ω-dynamo. This simple field structure could be a result of the ring-like gas configuration or, alternatively, the gas ring may be confined by the toroidal magnetic field.

A region with strong linearly polarized radio continuum emission in the SW quadrant of M31 was observed in the ^{12}CO $(1\rightarrow 0)$ line with the IRAM 30m

[1] MAX-PLANCK-INSTITUT FÜR RADIOASTRONOMIE, AUF DEM HÜGEL 69, D-5300 BONN 1, F.R.G
[2] INSTITUTO ARGENTINO DE RADIOASTRONOMIA, C.C. 5, VILLA ELISA (1894), ARGENTINA

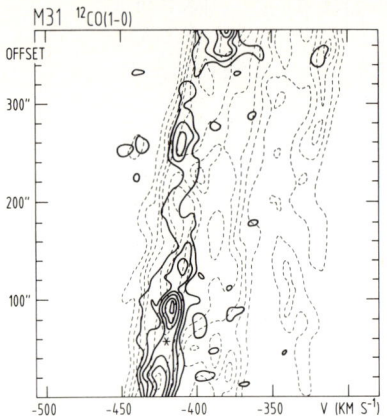
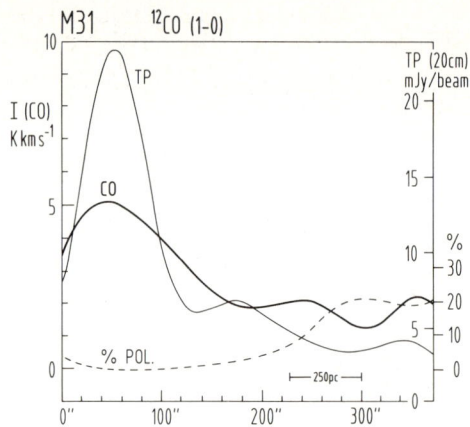

Figure 1: ^{12}CO $(1 \rightarrow 0)$ emission along a spiral arm in the SW quadrant of M31 observed with the IRAM 30m telescope at 23″ resolution. The asterisk marks the position of a strong radio continuum source. The dashed contours show the HI emission as observed by Brinks and Shane (1984).

Figure 2: Integrated CO and radio continuum emission and degree of linear polarization (all at 75″ resolution) along the same spiral arm as in Figure 1.

telescope on Pico Veleta (Bajaja et al., 1990). The CO spectra perpendicular to the spiral arm show strongest emission on the inner edge of the dust lane and HI arm. We find no significant velocity shift between HI and CO.

The same dust/HI arm contains one of the strongest radio continuum sources within M31. It has not been resolved by Walterbos et al. (1985) and hence must be smaller than 10″ (33 pc). Its radio flux density of 20 mJy at 20 cm corresponds to half that of Cas A. However, it is probably not a SNR as will be discussed in a subsequent paper (Bajaja et al., in preparation). It is also unlikely to be a background source since absorption in HI has not been detected. Alternatively, the strong radio emission could be due to an enhancement of the magnetic field strength in a giant molecular cloud complex. The CO emission peaks almost symmetrically at about 150 pc on either side of the radio continuum source (Figure 1). The CO minimum could be due to absorption by a cold foreground cloud, or to heating and de-population of the $J = 1$ level.

Both along and perpendicular to the dust lane CO intensity and degree of radio polarization are anticorrelated (Figure 2). Possible causes are Faraday depolarization, and/or beam depolarization due to a strong, unresolved ("turbulent") magnetic field in the spiral arm. On a larger scale, the molecular cloud complex may affect the structure of the uniform field: the field lines deviate from the direction of the torus (Beck et al., 1989). New observations in CO $(2 \rightarrow 1)$ and in radio polarization with higher resolution are required.

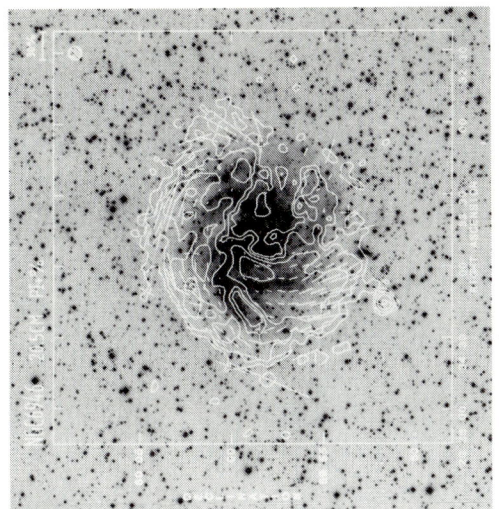
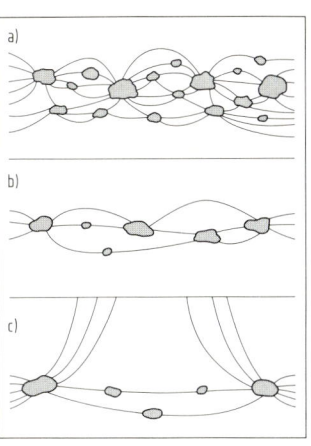

Figure 3: Linearly polarized radio continuum emission of NGC 6946 observed with the VLA in its D-configuration at 42″ resolution. The lengths of the E-vectors are proportional to the degrees of polarization.

Figure 4: Distribution of molecular clouds and magnetic field lines in spiral arms with a high (a), moderate (b) or low (c) number of molecular clouds.

3. Polarization observations in other spiral galaxies. High-resolution observations in CO and radio polarization are still rare in spiral galaxies. M81 and NGC 6946 possess prominent spiral arms with CO emission (Brouillet *et al.*, 1988, Casoli *et al.*, 1990). In radio continuum the degree of polarization was found to be highest in the *interarm* regions. Here the uniform magnetic field dominates, while the turbulent magnetic field is highest in the spiral arms (Krause *et al.*, 1989, Beck *et al.*, 1990, see Figure 3). Thus the anticorrelation between CO intensity and degree of polarization observed in a limited region of M31 appears to occur on a much larger scale in M81 and NGC 6946.

A large number of molecular clouds (Figure 4a) produces a high "turbulence" of the field because the superposition of field loops cannot be resolved in present radio maps. Enhanced turbulent motion of clouds by collisions or field turbulence induced by supernova remnants and stellar-wind bubbles can further tangle the field lines. The observed degree of polarization is low in the spiral arms (Figure 3). Galaxies with a moderate number of large molecular clouds like M31 show less disturbed field lines in spiral arms (Figure 4b) and hence higher degrees of polarization.

Regions with very few clouds (Figure 4c) should allow field lines to leave the galactic disk. Some regions in the SE quadrant of M31 show field lines with large angles to the plane (Berkhuijsen *et al.*, 1987). Star formation is low there but nothing is yet known about the distribution of molecular clouds. In the almost face-on galaxy NGC 6946 (Figure 3) field lines bending out of the plane may be responsible for high Faraday rotation and strong depolarization in the SW part. The striking similarity to coronal holes of the Sun suggests this phenomenon to be named a "galactic coronal hole". Detailed CO observations should be used to test this model.

4. Summary and discussion The correlation between far-infrared and radio continuum emission requires a connection between field strength and star formation. Radio polarization data show that also the field structure is related to star formation processes. These results are obvious if *magnetic fields control star formation*, and/or vice versa.

Magnetic fields are anchored in molecular clouds and certainly influence cloud motion and collisions between clouds (e.g. Clifford and Elmegreen, 1983). Stability of clouds essentially depends on magnetic fields. On the other hand, magnetic fields assist cloud collapse by transporting away angular momentum (e.g. Mouschovias, 1990). Putting all bits and pieces of evidence together, magnetic fields could well be the controlling force in the "magic circle" in the interstellar medium.

References

Bajaja, E., Berkhuijsen E.M. and Beck, R., 1990, in *Galactic and Intergalactic Magnetic Fields* ed. R. Beck, P.P. Kronberg and R. Wielebinski, (Kluwer Acad. Publ., Dordrecht), p. 203.
Beck, R., 1982, Astron. Astrophys., **106**, 121.
Beck, R., Loiseau, N., Hummel, E., Berkhuijsen, E.M., Graeve, R. and Wielebinski, R., 1989, Astron. Astrophys., **222**, 58.
Beck, R., Buczilowski, U.R. and Harnett, J.I., 1990, in *Galactic and Intergalactic Magnetic Fields*, ed. R. Beck, P.P. Kronberg and R. Wielebinski, (Kluwer Acad. Publ., Dordrecht), p. 213.
Berkhuijsen, E.M., Beck, R. and Graeve, R., 1987, in *Interstellar Magnetic Fields*, ed. R. Beck and R. Graeve, (Springer-Verlag, Heidelberg), p. 38.
Brinks, E. and Shane, W.W., 1984, Astron. Astrophys. Suppl., **55**, 179.
Brouillet, N., Baudry, A. and Combes, F., 1988, Astron. Astrophys., **196**, L17.
Casoli, F., Clausset, F., Viallefond, F., Combes, F. and Boulanger, F., 1990, Astron. Astrophys., **233**, 357.
Clifford, P. and Elmegreen, B.G., 1983, Mon. Not. Roy. Astron. Soc., **202**, 629.
Helou, G. and Bicay, M.D., 1990, in *Galactic and Intergalactic Magnetic Fields*, ed. R. Beck, P.P. Kronberg and R. Wielebinski, (Kluwer Acad. Publ., Dordrecht), p. 239.
Krause, M., Beck, R. and Hummel, E., 1989, Astron. Astrophys., **217**, 17.
Mouschovias, T.Ch., 1990, in *Galactic and Intergalactic Magnetic Fields*, ed. R. Beck, P.P. Kronberg and R. Wielebinski, (Kluwer Acad. Publ., Dordrecht), p.269.
Völk, H.J., 1989, Astron. Astrophys., 218, 67.
Walterbos, R.A.M., Brinks, E. and Shane, W.W., 1985, Astron. Astrophys. Suppl. Ser., **61**, 451.
Wunderlich, E., Klein, U. and Wielebinski, R., 1987, Astron. Astrophys. Suppl. Ser., **69**, 487.

Molecular absorption lines towards the nucleus of Centaurus A

A. ECKART[1], M. CAMERON[1], R. GENZEL[1], J.M. JACKSON[1],

H. ROTHERMEL[1], J. STUTZKI[1], G. RYDBECK[2] and T. WIKLIND[2]

Abstract We present measurements of absorption lines of ^{12}CO, ^{13}CO, HCO$^+$, HCN, CN, and CS towards the nuclear continuum source of Centaurus A obtained with the Swedish–ESO Submillimeter Telescope (SEST). The data indicate that the deepest absorption at the systemic velocity of Centaurus A originates in quiescent molecular clouds that are cold, dense, and clumped. Fractional molecular abundances are comparable to those found in Galactic molecular clouds and in other galaxies.

The study of the molecular interstellar medium in Centaurus A has begun only recently with the advent of large millimeter telescopes capable of accessing a declination of -43°. In a previous paper (Eckart et al., 1990a, hereafter Paper I) we presented a fully sampled map of the ^{12}CO(1 → 0) distribution in the disk which complements the ^{12}CO(2 → 1) cuts obtained by Phillips et al. (1987, 1989). From these data we determined that the ISM in the disk of Centaurus A is cold (T < 10 K), with densities of the order of the ^{12}CO critical density ($\sim 2 \times 10^4$ cm^{-3}). Absorption against the nuclear continuum source has been reported for HI (van Gorkom, 1987, van der Hulst et al., 1983, Gardner and Whiteoak, 1976) and a number of molecular species, such as OH, H$_2$CO, C$_3$H$_2$, ^{12}CO, and ^{13}CO (Gardner and Whiteoak, 1976, Bell and Seaquist, 1988, Phillips et al., 1987, Israel et al., 1990). Here we present measurements of absorption in the lines of ^{12}CO, ^{13}CO, HCO$^+$, HCN, CN and CS. A full discussion of the results is given in Eckart et al., 1991 (hereafter Paper II).

Line Features: In Figure 1 we show the HCO$^+$ spectrum in which the different absorption features are most easily discussed. This spectrum is free of hyper-fine structure line splitting and "contaminating" line emission, and all the features that can be identified in other spectra are present. Most prominent are three narrow features A,B, and C at v_{LSR}=550±1 km s^{-1}, 540±1 km s^{-1}, 544±1 km s^{-1}, respectively. The HCO$^+$(1 → 0) spectrum also reveals two broad troughs centered at about 580 km s^{-1} and 604 km s^{-1} probably corresponding to the HI absorptions at 576 and 595.5 km s^{-1} which have been attributed to infalling gas within 500 pc of the nucleus (van der Hulst et al., 1983).

Kinematics and Distribution of the Absorbing Gas: The spatially unresolved line features are most probably due to absorption of the continuum emission by molecular clouds in our line-of-sight to the compact, nuclear continuum source. From

[1] MAX-PLANCK INSTITUT FÜR EXTRATERRESTRISCHE PHYSIK, 8046 GARCHING, FRG
[2] ONSALA SPACE OBSERVATORY, CHALMERS UNIVERSITY OF TECHNOLOGY, S-43900 SWEDEN

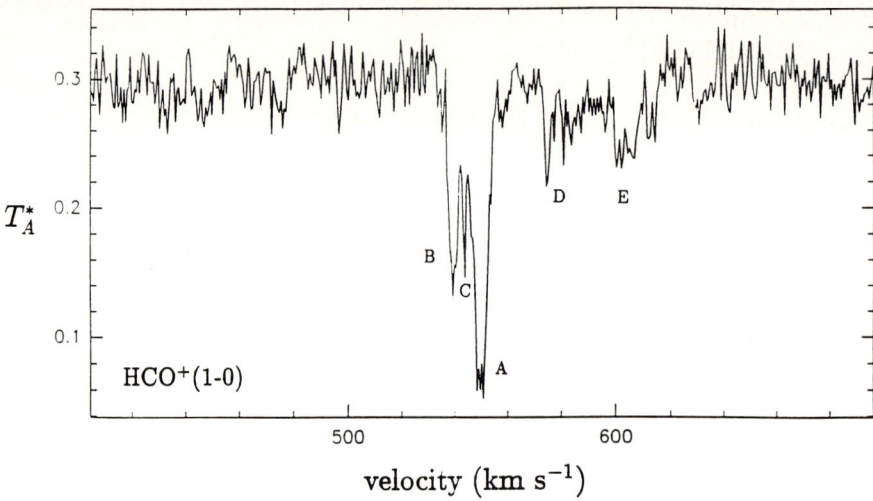

Figure 1: The HCO$^+$ absorption line spectrum observed towards the nucleus of Centaurus A. We indicated the individual absorption features A to E.

our data there is no indication that these clouds have to be particularly close to the nucleus. The observed velocity differences between individual absorption line features do not appear to be peculiar compared to the velocity dispersion of 60 km s^{-1} in the molecular disk (Paper I). The central HI absorption feature shows a frequency shift as a function of position against the jet components that is consistent with the rotation curve (van der Hulst *et al.*, 1983). This indicates that the absorption takes place in clouds *all along* the line of sight through the galaxy. Assuming that the absorbing atomic and molecular gas components are well "mixed" along this line we estimate a *mean distance* of the absorbing material from the nucleus of the order of 1 kpc (the diameter of the molecular disk is about 2.6 kpc; Paper I). The (FWHM) line widths of the different ^{12}CO(1 → 0) absorption features close to the systemic velocity of Centaurus A are well within the range of line widths measured for single molecular clouds in our Galaxy (Solomon *et al.*, 1987).

Molecular Excitation: For the ^{13}CO absorption lines we determine an optical depth ratio of $\tau_{2\to1}/\tau_{1\to0}$ = 0.57 ± 0.1 and a main beam brightness temperature ratio $T_{mb,2\to1}/T_{mb,1\to0}$ = 0.63 ± 0.1. Both values are consistent with an excitation temperature of T$_{ex}$ of 3 → 4 K *assuming LTE*. If the gas is subthermally excited the kinetic temperature will of course be larger. Off nucleus the ^{12}CO emission line ratios give a comparable result (Paper I). They range from 0.3 to 0.5 and indicate that the emission comes from very cold (T$_{ex}$ <10 K) or subthermally excited gas which is in agreement with the result obtained above. These results are supported by simple one component models of the absorbing gas as described by Genzel *et al.* (1990) and Eckart *et al.* (1990b), that take *non-LTE* conditions, clumpiness and opacity effects into account. The influence of the compact nucleus on excitation in the millimeter wavelength domain can be investigated using estimates of the Rayleigh-Jeans brightness temperature and the size of the nucleus (Paper II). At an assumed average distance of about 1 kpc it appears very unlikely that radiative excitation by the nuclear continuum emission is important for the molecular gas responsible for the absorption feature A at the systemic velocity of Centaurus A.

Hyperfine structure line ratios in HCN and CN: For the LTE case of optically thin absorbing clouds the intensity ratio of the hyperfine components are expected to be $R_{12} = I_{F=1\to1}/I_{F=2\to1} = 0.6$ and $R_{02} = I_{F=0\to1}/I_{F=2\to1} = 0.2$ (Walmsley et al., 1982). The HCN$(1\to0)$ spectrum clearly shows anomalous hyperfine structure line ratios. We measure $R_{12} = 0.8 \pm 0.1$ and $R_{02} = 0.5 - 1.0$, which is clearly not in agreement with the optically thin LTE ratios. Walmsley et al. (1982) have compared R_{12} and R_{02} for different Galactic clouds. Our data suggest that the absorption is probably due more to cold, dark clouds, rather than warm clouds associated with HII regions. The mere presence of hyperfine structure anomalies limits the density to less than the critical density of about 4×10^6 cm^{-3} for the HCN$(1\to0)$ line (Green and Thaddeus, 1974). Radiative transport model calculations (Hafener, 1985) indicate a kinetic temperature of $T_{kin} \leq 20$ K and hydrogen densities of $10^{6\pm0.5}$ cm^{-3}.

Molecular fractional abundances: For the ^{13}CO$(1\to0)$ line we obtain a column density of $N(^{13}CO) = 10^{16.2}$. If we assume that the atomic and molecular gas and the dust are well mixed along the line of sight through the galaxy a combination of CO, HI, and 9.7 μm silicate absorption data results in a ^{12}CO fractional abundance ranging between $(5 - 11) \times 10^{-5}$, which is in good agreement with the typical value of 8×10^{-5} for the dense molecular interstellar medium in the Galaxy. For an excitation temperature between 6 and 10 K the abundances agree reasonably well with measurements and model calculations of dense molecular clouds. Also the values are in the range of abundances covered by other well studied galaxies (paper II).

Conclusion: The main results on the properties of the absorbing molecular gas in Centaurus A can be summarized as follows:

1. The CO emission and absorption line ratios as well as a comparison to the HI data indicate that the absorption takes place against the compact nuclear continuum source in the line emitting clouds all along the line of sight through the molecular disk of Centaurus A. A mean distance from the nucleus for the clouds responsible for the central absorption feature at the systemic velocity of Centaurus A is therefore of the order of 1 kpc.
2. For the central absorption at the systemic velocity of Centaurus A the CO line ratios indicate that the bulk of the absorbing gas is cold with $T_{kin} < 10$ K.
3. Estimated ^{12}CO abundances of a few times $(5 - 11) \times 10^{-5}$ relative to H$_2$ as well as half power line widths of about 6 km s^{-1} for the absorption at the systemic velocity of Centaurus A, are comparable to typical values found for dense Galactic molecular clouds.
4. Abundances, excitation temperature, and density of the absorbing and

emitting molecular gas in the disk of Centaurus A are similar to the values obtained for the disk material in other galaxies. The number density of molecular hydrogen is of the order of the CO critical density of the $J=2$ rotational level (2×10^4 cm^{-3}).

5 The HCN and CN lines indicate that the absorbing molecular clouds are clumped. The hyperfine structure line ratios and the column density are consistent with a number density close to 10^6 cm^{-3} and a kinetic temperature of less than 20 K.

6 The fact that the absorbing molecular gas is cold and dense is in agreement with the finding, that at the mean distance of the absorbing gas of 1 kpc radiative excitation by the nuclear source is not important.

7 Molecular abundances relative to H$_2$ for CO, HCO$^+$, HCN, CN, and CS in Centaurus A are within the same range of values obtained for other well studied galaxies.

References

Bell, M.B. and Seaquist, E.R., 1988, Astrophys. J., **329**, L17.
Eckart, A., Downes, D., Genzel, R., Harris, A.I., Jaffe, D.T. and Wild, W., 1990b, Astrophys. J., **348**, 434.
Eckart, A., Cameron, M., Genzel, R., Jackson, J.M., Rothermel, H., Stutzki, J., Rydbeck, G. and Wiklind, T., 1991, Astrophys. J., (in press, paper I).
Eckart, A., Cameron, M., Rothermel, H., Wild, W., Zinnecker, H., Rydbeck, G., Olberg, M. and Wiklind, T., 1990a, Astrophys. J., **363**, 451 (paper II).
Gardner, F.F. and Whiteoak, J.B., 1976, Proc. Astr. Soc. Australia, **3**, 63.
Genzel, R., Stacey, G.J., Harris, A.I., Townes, C.H., Geis, N., Graf, U.U., Poglitsch, A. and Stutzki, J., 1990, (submitted to Astrophys. J.).
Green, S., and Thaddeus, P., 1974, Astrophys. J., **191**, 653.
Hafener, M., 1985, diploma thesis, University of Cologne, FRG.
Israel, F.P., van Dishoeck, E.F., Baas, F., Koornneef, J., Black, J.H. and de Graauw, Th., 1990, Astron. Astrophys., **227**, 342.
Phillips, T.G., Ellison, B.N., Keene, J.B., Leighton, R.B., Howard, R.J., Masson, C.R., Sanders, D.B., Veidt, B. and Young, K., 1987, Astrophys. J., **322**, L73
Phillips, T.G., Sanders, D.B. and Sargent, A.I., 1990, in *Submillimeter Astronomy*, ed. G.D. Watt and A.S. Webster (Kluwer, Dordrecht), p. 223.
Solomon, P.M., Rivolo, A.R., Barrett, J. and Yahil, A., 1987, Astrophys. J., **319**, 730.
van Gorkom, J.H., 1987, I.A.U. Symp. on *Structure and Dynamics of Elliptical Galaxies*, ed. T. de Zeeuw, p. 421.
van der Hulst, J.M., Golish, W.F. and Hashick, A.D., 1983, Astrophys. J., **264**, L37.
Walmsley, C.M., Churchwell, E., Nash, A. and Fitzpatrick, E., 1982, Astrophys. J., **258**, L75.

OB associations and HII regions in the SMC

PAOLO BATTINELLI

OSSERVATORIO ASTRONOMICO DI ROMA, V.LE DEL PARCO MELLINI, 84,

I-00136, ROMA, ITALY.

The study of stellar associations in galaxies can give important information on the sites of star formation and their influence on the surrounding interstellar matter. Moreover, the environmental effect on the processes of star formation in different galaxies can be investigated.

A new pattern recognition technique for OB associations in resolved galaxies (Battinelli, 1990), based only on the observed distribution of the OB stars, is used for the identification of OB associations in the SMC. Using a sample of 344 O/B2 stars selected from the recent catalogue of Azzopardi and Vigneau (1982), 31 OB associations have been identified in the region of SMC covered by the catalogue (i.e. $0^h 41^m \leq$ R.A.(1975) $\leq 1^h 22^m$). It is worth noting that the new technique used does not require any subjective evaluation by the researcher and is able to detect both compact and filamentary associations. The latter may be likely to be observed as a result of scenarios of star formation induced by shock wave compressions (by, for instance, ionization, wind, supernovae, *etc.*) in the interstellar medium which might produce such filamentary shaped regions of star formation. The mean size of the new identified associations is ≈ 90 pc, which is a value slightly larger than ≈ 77 pc found by Hodge (1985) in the only previous study of OB associations in SMC.

Since the ionization of the hydrogen in the HII regions is mainly due to ionizing UV–photons coming from nearby very hot stars (apart from the rarer case of supernovae), a strong correlation between OB stars and HII regions is expected. The spatial distribution of the OB associations is compared with the spatial distribution of HII regions in the SMC obtained from Davies *et al.* (1976) and most of the 31 associations are associated with one or more HII regions. More precisely: 17 associations (corresponding to 55% of the total number of OB associations) are found to be associated with bright (vb and b in the Davies *et al.* luminosity classification) HII regions; 10 (corresponding to 32%) are associated with fainter types of HII regions; only 4 (corresponding to 13%) are not associated or are doubtfully associated with HII regions.

In his identification of OB associations, Hodge (1985) reports only the correlation with bright HII regions which is weaker ($\approx 31\%$) than the one obtained in the present work. The percentage of bright HII regions not in associations is quite high in both Hodge's and the present identification, being 79% and 70% respectively. This may be due to difficulty of detection of compact groups of OB stars embedded in ionized gas nebulae, so that most of these stars are probably not included in Azzopardi and Vigneau (1982).

Incidentally, a comparison of the spatial distribution of the OB associations with the radio continuum map at 1.4 GHz of the SMC and with the HI distribution (both given by Loiseau *et al.*, 1987), shows that the HI distribution does not seem (as expected) to be tightly correlated with OB associations, whereas the radio continuum map is, in some ways, similar to the OB association distribution.

References

Azzopardi, M. and Vigneau, J., 1982, Astron. Astrophys. Suppl., **50**, 291.
Battinelli, P., 1990, Astron. Astrophys., (in press).
Davies, R.D., Elliott, K.H. and Meaburn, J., 1976, Mem. Roy. Astron. Soc., **81**, 89.
Hodge, P.W., 1985, Publ. Astron. Soc. Pacific, **97**, 530.
Loiseau, N., Klein, U., Greybe, A., Wielebinski, R. and Haynes, R.F., 1987, Astron. Astrophys. **178**, 62.

Interstellar masers

R. J. COHEN

UNIVERSITY OF MANCHESTER, NUFFIELD RADIO ASTRONOMY LABORATORIES, JODRELL BANK, MACCLESFIELD, CHESHIRE, SK11 9DL, U.K.

1 Introduction

Masers arise in the interstellar medium in a variety of settings, from comets through circumstellar envelopes and star-forming regions to active galactic nuclei. In each case a central source energises the surrounding molecular gas to produce powerful beams of microwave radiation by the process of stimulated emission. Maser action can be recognized by the intensity of the molecular line radiation and by the strongly non-equilibrium line ratios. In addition the lines are usually narrow and may be strongly polarized and time-variable. The theory and observation of cosmic maser sources are reviewed by Elitzur (1982), Cohen (1989), Moran (1990) and others.

Masers in star-forming regions are often termed interstellar masers because the maser molecules are interstellar in origin. These masers provide sophisticated but limited tools for studying the molecular gas close to young stars. In this paper I will review current knowledge of the maser physics and current observational developments. I hope to convince you that despite the difficulties in determining the physical conditions in maser regions the masers are nevertheless valuable probes of star-forming regions. They are important because they are the most readily detectable indicators of star-formation. They can be studied with milliarcsecond resolution using very-long-baseline-interferometry (VLBI), MERLIN and the VLA . These measurements provide direct evidence for outflow from young stars (through proper motions), for the existence of molecular discs around young stars on scales of 100–1000 AU, and evidence for the role of magnetic fields in star formation.

2 Physical Conditions

Maser action depends on the process of stimulated emission, which is very important at radio wavelengths. In fact the ratio of stimulated emission to spontaneous emission increases as the third power of the wavelength. This relation shows incidentally why only a desperate man would try to build an X-ray laser. Natural maser action in the interstellar medium relies critically on small departures from local thermodynamic equilibrium (LTE). The important parameter is the optical depth τ of the gas, given by

$$d\tau = \frac{h\nu}{4\pi\Delta\nu}B_{21}(g_2 n_1 - g_1 n_2)ds$$

where n_2 and n_1 are the number densities of the upper and lower states of the maser transition, g_i are the statistical weights, B_{21} is the Einstein coefficient for stimulated emission, $\Delta\nu$ is the bandwidth being considered and ds is an element along the path of the radiation. For a microwave transition the energy difference between the two states is typically 0.1K (dE/k), so the level populations should be very nearly equal in LTE. Only a small population shift is necessary to cause population inversion. For a uniform cloud and one-dimensional propagation the emergent radiation has a brightness temperature given by the usual expression

$$T_b = T_c e^{-\tau} + T_{ex}(1 - e^{-\tau})$$

where T_c is the brightness temperature of the continuum background and T_{ex} is the excitation temperature of the transition. However because of the population inversion both the optical depth and the excitation temperature are negative numbers. Hence the maser provides (exponentially large) gain, which may exceed e^{30} or 10^{13} for strong H_2O masers. It is easy to show how large gain can lead to line-narrowing and beaming of the radiation (reviews cited earlier). These factors together make it inherently difficult to infer the physical conditons from the observed maser radiation. Because of beaming we only see part of the gas responsible for the maser radiation; the intensity of the radiation is related to the molecular density only through the often poorly determined pumping processes and the beaming factor, and the linewidth is only loosely related to the temperature of the gas.

The maser transition itself may tell us something about the physical conditions. For example the strong 22 GHz water maser is a transition between two rotationally excited states each some 640 K above the ground state. Such masers can be excited only if they are close to a young star, or if they are exposed to an energetic stellar wind or outflow. This approach has proved very successful in understanding circumstellar masers, where the physical conditions change in a fairly regular way with distance from the evolved star, and different masers illuminate different zones of the circumstellar envelope (Chapman and Cohen, 1986, Cohen, 1989). However no such regularity has yet emerged in the case of interstellar masers. Perhaps the process of stellar birth is more turbulent and the physical conditions more inhomogeneous.

Table 1 lists the most important of the currently known interstellar masers. These are masers which are strong and for which at least ten sources are already known. The table shows explicitly that maser action is a symptom of a non-thermal distribution which affects many energy levels of a molecule. There can be several maser transitions from the same molecule in the same region of excited gas. Interferometric studies of the source W3(OH) bear this out for several transitions of OH and CH_3OH (Baudry et al., 1988, Menten et al., 1988a, b).

Table 1: *Important Interstellar Masers*

Molecule	Transition	Frequency (MHz)	E/k (K)	References
OH	$^2\Pi_{3/2}$, J=3/2, F=1 → 2	1612.231	0	McGee *et al.*, 1965
* OH	$^2\Pi_{3/2}$, J=3/2, F=1 → 1	1665.402	0	Weaver *et al.*, 1965
* OH	$^2\Pi_{3/2}$, J=3/2, F=2 → 2	1667.359	0	Weaver *et al.*, 1965
OH	$^2\Pi_{3/2}$, J=3/2, F=2 → 1	1720.530	0	Weinreb *et al.*, 1965
OH	$^2\Pi_{1/2}$, J=1/2, F=1 → 0	4765.562	182	Zuckerman *et al.*, 1968
OH	$^2\Pi_{3/2}$, J=5/2, F=3 → 3	6035.092	121	Yen *et al.*, 1969
CH$_3$OH	$2_0 \to 3_{-1}$ E	12178.60	19	Batrla *et al.*, 1987
* H$_2$O	$6_{16} \to 5_{23}$	22235.08	644	Cheung *et al.*, 1969
CH$_3$OH	$4_{-1} \to 3_0$ E	36169.24	21	Morimoto *et al.*, 1985
CH$_3$OH	$7_0 \to 6_1$ A^+	44069.43	65	Morimoto *et al.*, 1985
H$_2$O	$3_{13} \to 2_{20}$	183310.09	205	Cernicharo *et al.*, 1990
H$_2$O	$10_{29} \to 9_{36}$	321225.64	1861	Menten *et al.*, 1990

* More than 100 sources known.

Field and Gray (1988) have considered the many-level problem theoretically with a semi-classical approach. The radiative transfer shows many interesting effects whereby the growth and saturation of one maser transition is strongly coupled to the growth and saturation of other maser transitons. The emergent maser intensities could provide useful diagnostics of the physical conditions in principle. At present however the unknown beaming factors (which can be different for different transitions) limit the accuracy with which theoretical and observed line ratios may be compared.

3 Signposts of Star-Formation

Masers are the most readily detectable tracers of star-formation with ground-based techniques (Cohen, 1989). For example the 22 GHz water masers have luminosities of up to 10^{26} W, which enable them to be detected throughout the

Galaxy and even in external galaxies (Churchwell et al., 1977). It has been estimated that the Galaxy contains thousands of detectable water masers, but to find them all would be an heroic task. With a typical beamwidth of 1 arcmin it would take over one million measurements just to survey a 1 degree strip of the Galactic plane. At present only a few hundreds of 22 GHz maser sources are known. However the situation is improving rapidly. The IRAS all-sky infrared survey has provided the starting point for targetted maser searches of likely star- forming regions selected by their infrared colours. These targetted searches are proving very successful at finding not only water masers (Wouterloot et al., 1988, Braz et al., 1989 and others), but also OH masers (Cohen et al., 1988) and methanol masers (Kemball et al., 1988).

Figure 1 shows results from a recent Jodrell Bank survey for methanol 12 GHz masers associated with IRAS sources (Lim, 1989). The sources were searched previously for OH, and so we can begin to investigate the statistics of occurrence. A clear heirarchy emerges. The sample contains 17 methanol masers, 49 OH masers, and 66 catalogued water masers. All of the 17 methanol masers are associated with OH masers, and at least 40 of the OH masers are associated with catalogued water masers. It is highly likely that the remaining OH masers have simply not been searched for water masers. This heirarchy can be partly understood as follows. OH masers are found almost exclusively near compact HII regions (e.g. Turner, 1982), and so trace the most massive young stars ($M \geq 8\ M_\odot$). HII regions can help establish conditions suitable for OH masers through photodissociation chemistry and through shocks. 22 GHz H_2O masers on the other hand are associated with a wider range of young stars down to about one solar mass (e.g. Rodriguez et al., 1987, Wilking and Claussen, 1987). They are also intrinsically brighter than OH masers, and hence more easily detected. The physical conditions necessary for methanol masers are not yet established.

Methanol masers are more interesting than these results might suggest however. They appear to fall into two distinct spectral classes. The sources we have been discussing are Class B sources, of which the prototype is W3(OH) (Batrla et al., 1987). They radiate strongly at 12 GHz and are always associated with compact HII regions, OH and H_2O masers. The 12 GHz emission covers the same velocity range as the 1.6 GHz OH emission. Class A sources on the other hand radiate strongly at 25 GHz ($J_{k=2} \rightarrow J_{k=1}$ E series of transitions) and at 44 GHz (Haschick et al., 1990), and can be found offset by as much as 1 pc from compact HII regions, OH and H_2O masers. The nature of these class A sources is not yet certain, but it has been suggested that they trace an earlier phase of stellar evolution (Menten et al., 1986). Suggestions of this kind are always difficult to confirm because the masers are so much brighter than any of the other signposts of star-formation!

4   Kinematics

Radio interferometers enable us to map the distribution and motions of maser sources on angular scales down to one milliarcsecond or less (the sizes of the

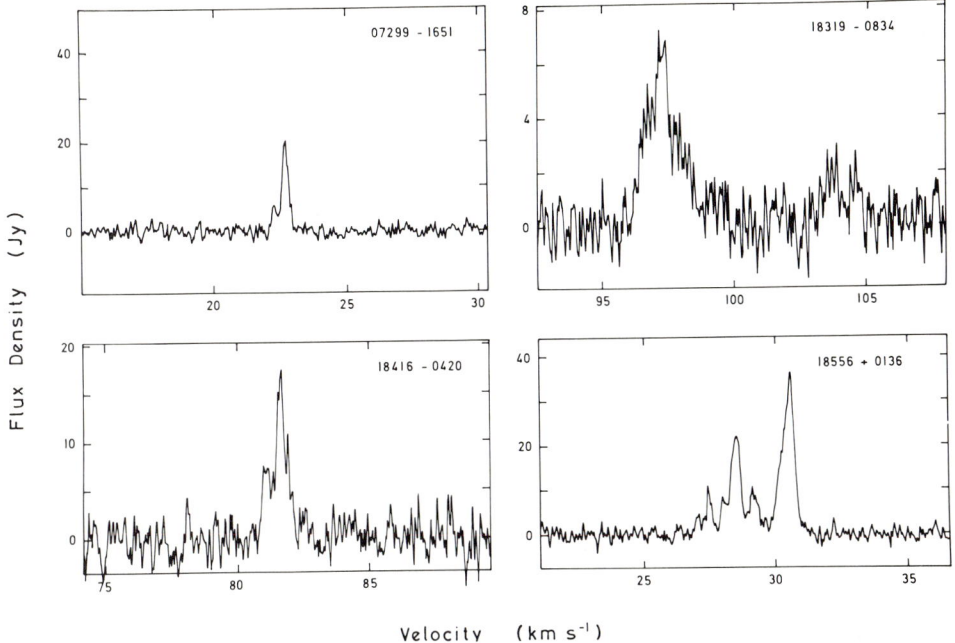

Figure 1: Methanol 12 GHz masers discovered in a survey of IRAS point sources, using the Jodrell Bank Mk2 telescope. The spectra have a velocity resolution of 0.4 km/s. Adapted from Lim (1989).

maser spots). Only the 22 GHz H_2O and 1.6 GHz OH masers have been studied extensively to date. The picture which emerges is one of turbulent outflow. Individual maser spots cluster around young stars on angular scales of one arcsecond (Gaume and Mutel, 1987, Forster and Caswell, 1989 and refs therein). Within a maser cluster the velocity range is usually so great that we can immediately deduce that the cluster is gravitationally unbound. There is seldom any obvious regularity in the velocity pattern. However repeated observations of H_2O masers have revealed systematic outflow in some sources (Genzel et al., 1981, Reid et al., 1988). The proper motions imply transverse velocities comparable with the radial velocities. Under suitable assumptions we can then deduce the distance to the source. The technique is extrememly labour intensive on account of the complexities of VLBI measurements, but has already proved useful in determining the galactic distance scale, as it is independent of the calibration difficulties of optical methods.

There has been debate over the nature of interstellar OH masers, and whether they too are part of the outflow phenomenon, or whether they belong to an (earlier) accretion phase. The kinematic argument indicates that many OH maser clusters are not gravitationally bound (Baart and Cohen, 1985). However in some cases the OH masers appear red-shifted with respect to the central HII region, suggesting infall (Garay et al., 1985). The first results from OH proper motion studies are just becoming available, and they indicate outflow. Bloemhof et al. (1989) have measured outflow of OH masers in the prototypical source W3(OH) using VLBI, and at Jodrell Bank we have detected outflow of the OH masers in Cepheus A using MERLIN (unpublished). The apparent red-shifts of OH masers in W3(OH) may after all reflect a blue-shift of the compact HII region, due to expansion of the HII region and to optical depth effects (Welch and Marr, 1987). Proper motion studies need to be carefully designed to avoid being compromised by maser variability. It appears from our work at Jodrell Bank that the natural timescale dl/dV, where l is the linear scale size of the source and V the radial velocity range, is a good empirical guide to the maser variability. Sources with small dl/dV do indeed vary on smaller timescales.

5 Maser Discs

One of the exciting recent developments has been the detection of protoplanetary sized maser discs surrounding young stars. This is not by any means the first time that radio astronomers have claimed to see disc structure in maser sources, but the recently observed discs are certain to stand the test of time! A good example is the massive molecular disc in the source G35.2-0.7N (Brebner et al., 1987), results for which are reproduced in Figure 2. The VLA NH_3 observations show an unambiguous signature of a rotating disc orthogonal to a well-collimated bipolar outflow (Little et al., 1985). The size of the NH_3 emission region decreases systematically with increasing excitation, indicating that the disc is heated from within. The density also increases towards the centre of the NH_3 disc. The OH masers are coincident with an ultracompact HII region at the centre of the disc and the outflow. The OH masers trace the innermost parts of the molecular disc, on a scale of 1000 AU. Their motion is chaotic, with little evidence for the rotation seen on larger angular scales. Brebner (1988) has found similar OH maser discs in several other molecular outflow sources.

On an even smaller scale is the remarkable SiO maser disc in Orion-KL. It has been known for some time that the SiO masers in the source closely coincide with the infrared source IRc2. The SiO masers are rotational transitions of vibrationally excited SiO, with excitations of several thousands of Kelvins. The emission has a twin-peaked line profile with a strikingly regular pattern of linear polarization, which can be explained by a disc structure (Barvainis, 1984). New observations of the $J=2 \rightarrow 1$ masers at 86 GHz with the Hat Creek millimetre array have revealed the disc: an elliptical pattern of masers only 80 AU in radius. Regular gradients of velocity with position suggest that the motion of the disc is a combination of rotation plus expansion (Plambeck et al., 1990). The disc is orthogonal to the large scale outflow, and inclined to the line-of-sight by some 45 degrees.

6 Magnetic Fields

The polarization of interstellar masers provides a unique probe of the magnetic

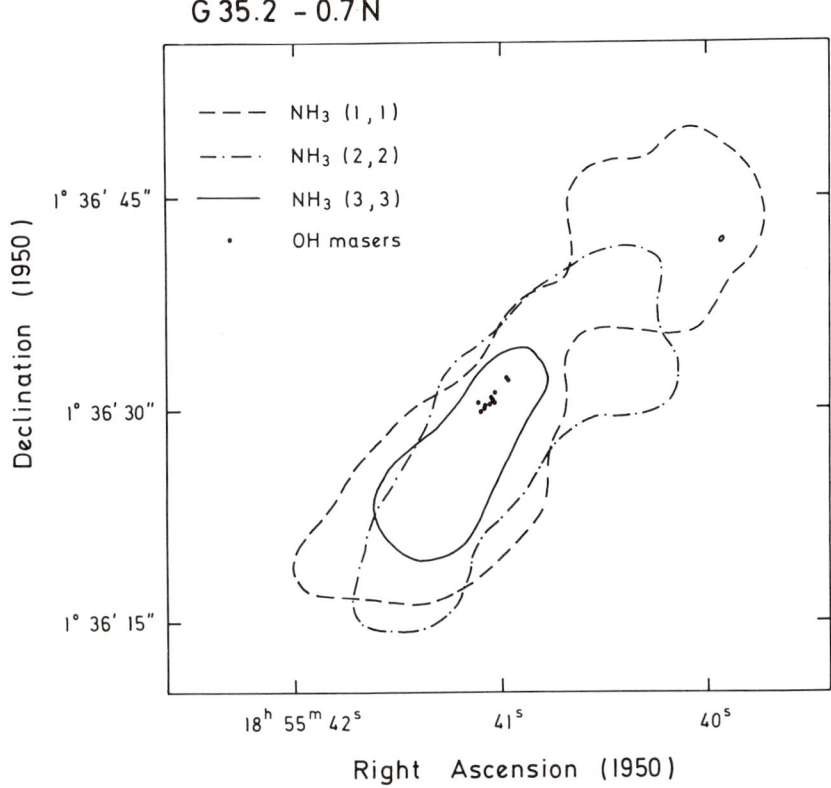

Figure 2: Schematic view of the inner molecular disc of G35.2-0.7N, showing the half-power contours of the NH_3 (1,1), (2,2) and (3,3) emission, and the 1665 MHz OH masers (Brebner et al., 1987). The OH masers coincide with an ultracompact HII region near the geometrical centre of the disc and bipolar outflow.

field in star-forming regions. Circular polarization is produced by the Zeeman effect, and can reach almost 100%. The theoretical treatment of the radiative transfer is very difficult, but the results can be simply summarized. In the case where the Zeeman splitting exceeds the maser linewidth then the magnetic field becomes as important as the radial velocity component in determining whether an element of molecular gas contributes to a particular maser beam or not. In effect the right and left hands of circular polarization are decoupled and the emergent beams can be up to 100% circularly polarized (Deguchi and Watson, 1986). This is the case applicable to the 1.6 GHz OH masers. The detection of strongly circularly polarized emission itself indicates a magnetic field strength of a milligauss or more. In some cases complete Zeeman patterns are seen and the magnetic field strength can be measured directly. One of the best examples is in Orion-KL, where 1665 and 1612 MHz OH Zeeman groups have been observed by Hansen (1982) and Hansen and Johnston (1983), indicating a magnetic field of 4 mG directed towards us. Another good example is Cepheus A, where 1665 and 1667 MHz Zeeman groups indicate a magnetic field of 3 mG directed away from us (Wouterloot et al., 1980, Cohen et al., 1990). These values show that the

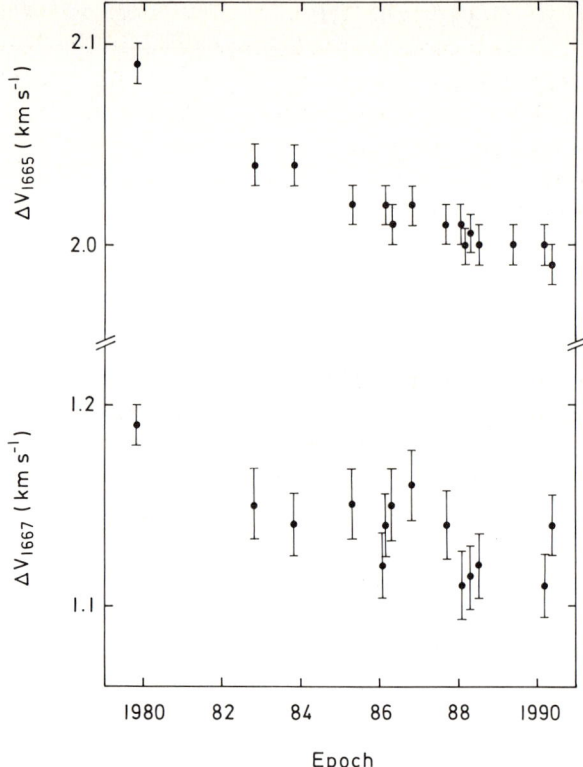

Figure 3: Zeeman splittings of the OH 1665 and 1667 MHz doublets in Cepheus A plotted as functions of time. The systematic decrease corresponds to a magnetic field decay of 0.4% per annum (Cohen et al., 1990).

magnetic field is strong enough to be important for the dynamics of the region (Myers and Goodman, 1988, Myers, this volume, p. 133). In Cepheus A we have recently detected a systematic decrease in the Zeeman splitting with time, corresponding to a magnetic field decay of 0.4% per annum (Cohen et al., 1990). This can be interpreted in terms of an expansion of the maser region, with a timescale of 500 years which is consistent with the proper motions mentioned earlier. The observed effect is shown in Figure 3.

Clearly we would like to know more about the field configuration than just its direction towards or away from the observer. For sources showing linear as well as circular polarization then more can be deduced. Garcia-Baretto et al. (1988) have conducted a VLBI experiment and measured full Stokes parameters of the 1665 MHz OH masers in W3(OH). Assuming that several elliptically polarized features are σ-components they were able to derive three-dimensional magnetic field vectors at several points in the maser region. They found the magnetic field to be some 6 mG, with a surprisingly uniform magnitude and direction across the source.

For maser transitions where the Zeeman splitting is much smaller than the linewidth then little polarization is produced. The emergent maser lines should have only a small amount of polarization in their wings, a case familiar from the optical polarization of sunspots, or the Zeeman effect in the 21 cm hydrogen line. This effect has recently been observed in the 22 GHz water masers by Fiebig and Güsten (1989). It is a result I never expected to see, for two reasons. First it implies very high magnetic fields (some 50 mG, the highest values ever measured in the interstellar medium), and secondly it is an extremely difficult measurement to bring off technically, with many systematic errors to be overcome. The high magnetic field strengths follow the increase with gas density which is already apparent in the case of 1.6 GHz OH masers. The increase follows approximately a power law with exponent one half, which is consistent with flux freezing of the Galactic magnetic field during gravitational collapse.

Finally I want to mention two old results which are still remarkable. Davies (1974) pointed out many years ago that the magnetic fields deduced from Zeeman studies of OH masers vary in a systematic way around the sky, following the sense of the large-scale Galactic magnetic field (at least in its line-of-sight direction). Secondly Cohen et al. (1984) noted that bipolar molecular outflows also show a preferential alignment with the Galactic magnetic field (as traced by optical polarization of starlight). In both cases we see evidence of a magnetic link between very large and very small structures in the interstellar medium. This suggests that the role of the magnetic field in star-formation may be a crucial one. Interstellar masers offer unique opportunities to investigate it.

References

Baart, E. E. and Cohen, R. J., 1985, Mon. Not. Roy. Astron. Soc., **213**, 641.
Barvainis, R., 1984, Astrophys. J., **279**, 358.
Batrla, W., Matthews, H. E., Menten, K. M. and Walmsley, C. M., 1987. Nature, **326**, 49.
Baudry, A., Diamond, P. J., Booth, R. S., Graham, D. and Walmsley, C. M., 1988, Astron. Astrophys., **201**, 105.
Bloemhof, E. E., Reid, M. J. and Moran, J. M., 1989, *Physics and Chemistry of Interstellar Molecular Clouds*, ed. G. Winnewisser and J. T. Armstrong (Springer-Verlag, New York), p. 228.
Braz, M., Scalise Jr., E., Gregorio Hetem, J. C., Monteiro do Vale, J. L. and Gaylard, M., 1989, Astron. Astrophys. Suppl., **77**, 465.
Brebner, G. C., 1988, PhD thesis, University of Manchester.
Brebner, G. C., Heaton, B., Cohen, R. J. and Davies, S. R., 1987, Mon. Not. Roy. Astron. Soc., **229**, 679.
Cernicharo, J., Thum, C., Hein, H., John, D., Garcia, P. and Mattioco, F., 1990, Astron. Astrophys., **231**, L15.
Chapman, J. M. and Cohen, R. J., 1986, Mon. Not. Roy. Astron. Soc., **220**, 513.
Cheung, A. C., Rank, D. M., Townes, C. H., Thornton, D. D. and Welch, W. J., 1969, Nature, **221**, 626.
Churchwell, E., Witzel., A., Huchtmeier, W., Pauliny-Toth, I., Roland, J. and Seiber, W., 1977, Astron. Astrophys., **54**, 969.
Cohen, R. J., 1989, Rep. Prog. Phys., **52**, 881.
Cohen, R. J., Baart, E. E. and Jonas, J. L., 1988, Mon. Not. Roy. Astron. Soc., **231** 205.
Cohen, R. J., Brebner, G. C. and Potter, M. M., 1990, Mon. Not. Roy. Astron. Soc., **246**, 3P.
Cohen, R. J., Rowland, P. R. and Blair, M. M., 1984, Mon. Not. Roy. Astron. Soc., **210**, 425.

Davies, R. D., 1974, IAU Symp. No. 60, ed. F. J. Kerr and S. C. Simonson III (Reidel, Dordrecht, Holland) p. 275.
Deguchi, S. and Watson, W. D., 1986, Astrophys. J., **300**, L15.
Elitzur, M., 1982, Rev. Mod. Phys., **54**, 1225.
Fiebig, D. and Güsten, R., 1989, Astron. Astrophys., **214**, 333.
Field, D. and Gray. M. D., 1988, Mon. Not. Roy. Astron. Soc., **234**, 353.
Forster, J. R. and Caswell, J. L., 1989, Astron. Astrophys., **213**, 339.
Garay, G., Reid, M. J. and Moran, J. M., 1985, Astrophys. J., **289**, 681.
Garcia-Barreto, J. A., Burke, B. F., Reid, M. J., Haschick, A. D. and Schilizzi, R. T., 1988, Astrophys. J., **326**, 954.
Gaume, R. A. and Mutel, R. L., 1987, Astrophys. J. Suppl., **65**, 193.
Genzel, R., Reid, M. J., Moran, J. M. and Downes, D., 1981, Astrophys. J., **244**, 884.
Hansen, S. S., 1982, Astrophys. J., **260**, 599.
Hansen, S. S. and Johnston, K. J., 1983, Astrophys. J., **267**, 625.
Haschick, A. D., Menten, K. M. and Baan, W. A., 1990, Astrophys. J., **354**, 556.
Kemball, A. J., Gaylard, M. J. and Nicolson, G. D., 1988, Astrophys. J., **331**, L37.
Lim, T. L., 1989, MSc thesis, University of Manchester.
Little, L. T., Dent, W. R. F., Heaton, B. H., Davies, S. R. and White, G. J., 1985, Mon. Not. Roy. Astron. Soc., **217**, 227.
McGee, R. X., Robinson, B. J., Gardner, F. F. and Bolton, J. G., 1965, Nature, **208**, 1193.
Menten, K. M., Walmsley, C. M., Henkel, C. and Wilson. T. L., 1986, Astron. Astrophys., **157**, 318.
Menten, K. M., Johnston, K. J., Wadiak, E. J., Walmsley, C. M. and Wilson, T. L., 1988a, Astrophys. J., **331**, L41.
Menten, K. M., Reid, M. J., Moran., J. M., Wilson, T. L., Johnston, K. J. and Batrla, W., 1988b, Astrophys. J., **333**, L83.
Menten, K. M., Melnick, G. J. and Phillips, T. G., 1990, Astrophys. J., **350**, L41.
Moran, J. M., 1990, *Molecular Astrophysics, A Volume Honouring Alexander Dalgarno*, ed. T. Hartquist (Cambridge University Press) p. 397.
Morimoto, M., Ohishi, M. and Kanzawa, T., 1985, Astrophys. J., **288**, L11.
Myers, P. C. and Goodman, A. A., 1988, Astrophys. J., **326**, L27.
Plambeck, R. L., Wright, M. C. H. and Carlstrom, J. E., 1990, Astrophys. J., **348**, L65.
Reid, M. J., Schneps, M. H., Moran, J. M., Gwinn, C. R., Genzel, R., Downes, D. and Ronnang, B., 1988, Astrophys. J., **330**, 809.
Rodriguez, L. F., Haschick, A. D., Torrelles, J. M. and Myers, P. C., 1987, Astron. Astrophys., **186**, 319.
Turner, B. E., 1982, *Regions of Recent Star Formation*, ed. R. S. Roger and P. E. Dewdney (Reidel, Dordrecht, Holland) p. 425.
Weaver, H., Williams, D. R. W., Dieter, N. H. and Lum, W. T., 1965, Nature, **208**, 29.
Weinreb, S., Meeks, M. L., Carter, J. C., Barrett, A. H. and Rogers, A. E. E., 1965, Nature, **208**, 440.
Welch, W. J. and Marr, J., 1987, Astrophys. J., **317**, L21.
Wilking, B. A. and Claussen, M. J., 1987, Astrophys. J., **320**, L133.
Wouterloot, J. G. A., Brand, J. and Henkel, C., 1988, Astron. Astrophys., **191**, 323.
Wouterloot, J. G. A., Habing, H. J. and Herman, J., 1980, Astron. Astrophys., **81**, L11.
Yen, J. L., Zuckerman, B., Palmer, P. and Penfield, H., 1969, Astrophys. J., **156**, L27.
Zuckerman, B., Palmer, P., Penfield, H. and Lilley, A. E., 1968, Astrophys. J., **153**, L69.

$^{15}NH_3$ masers toward NGC 7538–IRS 1

PETER SCHILKE

MAX-PLANCK-INSTITUT FÜR RADIOASTRONOMIE, AUF DEM HÜGEL 69,

D-5300 BONN 1, F.R.G.

Introduction: Ammonia towards NGC 7538 has a complex history (see e.g. Wilson et al., 1983, Mauersberger et al., 1986a, b). There is an extremely hot component observed in absorption in many transitions of $^{14}NH_3$. This absorption takes place in a clump of diameter 0.01 pc and must be very close to the ultracompact HII region associated with NGC 7538–IRS 1. This source is the only known source in the Galaxy where the 141 GHz line (1450 K above ground) of vibrationally excited ammonia is seen in *absorption* (Schilke et al., 1990).

An additional unique feature is the $^{15}NH_3(3,3)$ maser observed by Mauersberger et al. (1986b) in Effelsberg and by Johnston et al. (1989) with the VLA. There are theoretical reasons to expect such a maser (see e.g. Flower, Offer and Schilke, 1990) in low density regions where ortho-H_2 is underabundant. With this model in mind, we have attempted to observe a complete set of $^{15}NH_3$ towards this source in order to understand the excitation.

Observations: We succeeded in observing the (1,1) and (2,2) lines of $^{15}NH_3$ in absorption and, surprisingly, the (4,3) line in emission (see Figure 1). We conclude that the (4,3) line is a weak maser amplifying the continuum background of NGC 7538-IRS 1. This is based on the fact that all high excitation lines of $^{14}NH_3$ (apart from the (9,6) maser towards the source) show absorption. Observation of emission in the same velocity range as the $^{15}NH_3(3,3)$ maser and the $^{14}NH_3$ absorption is a strong hint that maser amplification of the continuum background is occuring in $^{15}NH_3(4,3)$.

In a Boltzmann–plot (Figure 2), the $^{14}NH_3$ absorption lines give a rotation temperature of 150 K. The linear shape of the plot suggests thermalization and therefore high H_2–densities ($> 10^6\,cm^{-3}$).

Theory: It is possible to extend the model of Flower, Offer and Schilke (1990) to explain the (4,3) emission. However, one is then forced to place the two masers in spatially distinct regions with different densities and temperatures.

Hence, we decided to look for other pumping models. The detection of the vibrationally excited line (Schilke et al., 1990) gave us a hint that the first vibrationally excited state may be involved in the pumping process. And indeed, a simple estimate shows that transitions via the vibrationally excited state lead to inversion (see Figure 3).

The question remains: under which conditions do transitions via the vibrationally excited level play a role? A simple estimate based on an approach of Carroll and Goldsmith (1981) for diatomic molecules is to compare the probabilities for rotational transitions (radiative and collisional transitions) with the probability for vibrational transitions (only absorption of IR photons emitted by hot, optically thick local dust is taken into account).

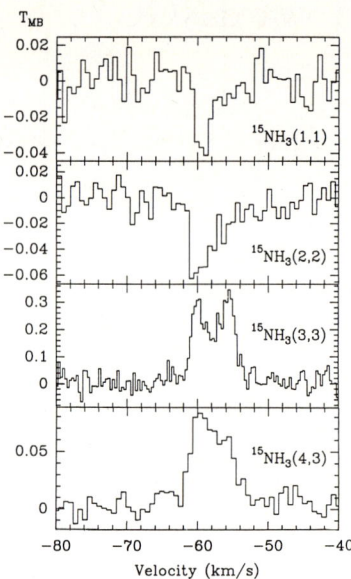

Figure 1: Effelsberg 100m spectra of ^{15}NH$_3$ towards NGC 7538–IRS 1. The intensity scale is main beam brightness temperature.

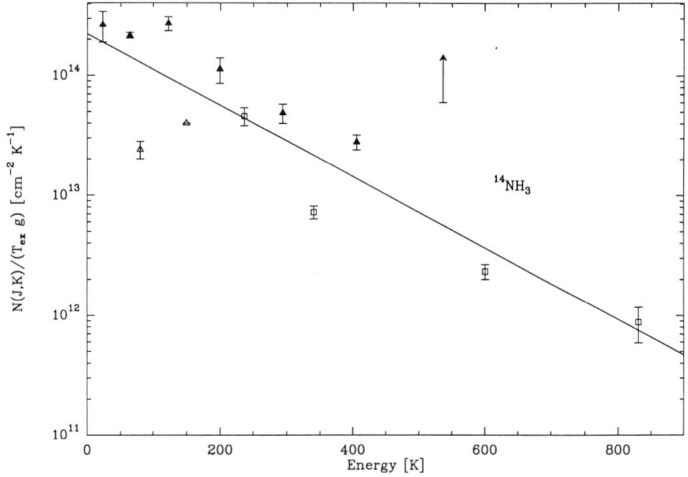

Figure 2: Boltzmann plot of ^{14}NH$_3$. The logarithm of the column density N(J,K) divided by the statistical weight g is plotted versus the excitation energy of the level (J,K). A linear shape of the plot implies thermalization, the slope of the line is $-1/T_{\rm rot}$

Here we have to use the effective Einstein A-coefficient $A^{\rm eff} = A\beta$, (the escape probability is given by $\beta = (1 - e^{-\tau})/\tau$), because the lines usually are optically thick. For the vibrational transitions, the probability for a IR photon to be absorbed by dust is much greater than the probability for a reabsorption by an

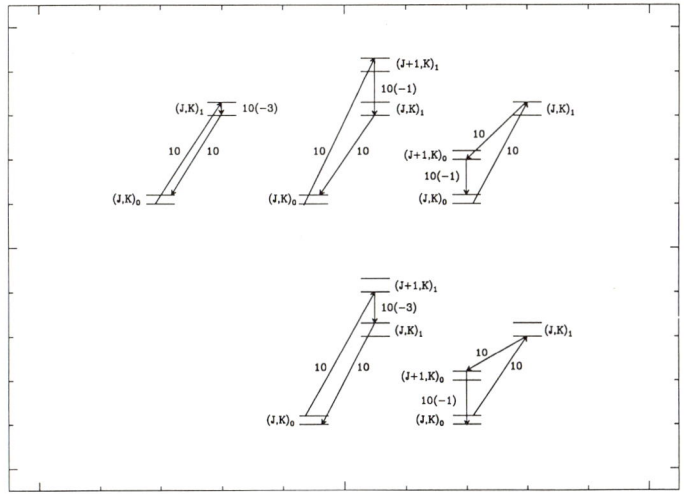

Figure 3: In this figure the subscripts on the (J,K) labels denotes the vibrational quantum number v_2. The numbers are the Einstein A-coefficients. At the top there are three processes which change the state of a molecule from the lower level of an inversion doublet to an upper one involving vibrational transitions. The three processes are: top left, absorption of 10 μm photon, inversion transition in $v_2 = 1$ state and decay to same (J,K) level of ground state; top center, same process, but with rotational transition in vibrationally excited state; top right, absorption of 10 μm photon, immediate decay into (J+1,K) state and rotational transition in ground state. The first process has no counterpart which reverses the process, the second one has a counterpart (bottom center), but less effective due to the larger Einstein A–coefficient (caused by the large frequency difference) and the third one has a counterpart with equal strength (bottom right). The second process is presumably dominant because of the large difference between the Einstein A–coefficients of the process and its counterpart, the first process is less effective and the third cancels with its counterpart.

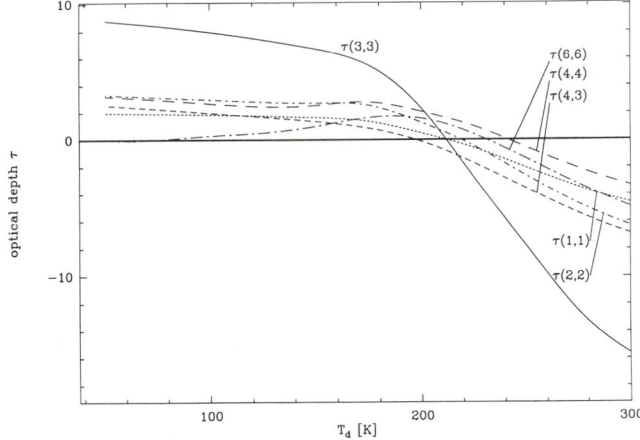

Figure 4: Optical depth as a function of dust temperatures. The gas temperature is held constant at 180 K. The molecular hydrogen density is 10^7 cm^{-3} and the ammonia concentration is taken to be 10^{-9}.

ammonia molecule, and we can set the escape probability equal to 1. If we set T_d^{cr} as the critical temperature where $p_{vib} > p_{rot}$, we get

$$T_d^{cr} = \frac{h\nu/k}{\log\left[1 + A_{vib}/(A_{rot}^{eff} + n(H_2)C_{rot})\right]}$$

If we insert typical values ($h\nu/k = 1400/$,K, $A_{vib} = 10\,s^{-1}$, $A_{rot} = 0.01\,s^{-1}$, $\tau_{rot} = 100$, $n(H_2) = 10^7\,cm^{-3}$, $C_{rot} = 10^{-11}\,cm^3\,s^{-1}$), we find $T_d^{cr} = 130\,K$. This is a lower limit for the dust temperature required to perturb the ground state populations.

Calculations: Detailed calculations were performed using a 260 level LVG statistical equilibrium program. The dust was taken into account following the approach of Deguchi (1981). The results are seen in Figure 4. One sees essentially the behaviour expected: at a critical temperature higher than 130 K the optical depths (and therefore the excitation temperatures) are heavily perturbed by the presence of the vibrationally excited state. At high dust temperatures the levels become inverted. The same calculation for $^{14}NH_3$, (i.e. 300 times higher NH_3–abundance) yields no inversion, presumably due to the effects of trapping in the highly optically thick rotational lines. Hence, the absence of the $^{14}NH_3$ counterpart of the $^{15}NH_3$ masers can be explained in the frame of the model.

Conclusions:

1. We observed several lines of $^{14}NH_3$ towards the ultracompact HII region NGC 7538–IRS 1. We derived a rotation temperature of 150 K and a $^{14}NH_3$ column density of $5 \times 10^{18}\,cm^{-2}$. We have indications that the lines are thermalized and that therefore the H_2–density is greater than $10^6\,cm^{-2}$.
2. We observed the (4,3) line of $^{15}NH_3$ in emission, indicating weak maser action.
3. We present a model to explain the maser process by 10 μm IR pumping via the vibrationally excited state 1450 K above ground.

References

Carroll T.J. and Goldsmith P.F., 1981, Astrophys. J., **245**, 891.
Deguchi S. 1981, Astrophys. J., **249**, 145.
Flower D.R., Offer A. and Schilke P., 1990, Mon. Not. Roy. Astron. Soc., **244**, 4P.
Johnston K.J., Stolovy S.R., Wilson T.L., Henkel C. and Mauersberger R., 1989, Astrophys. J., **343**, L41.
Mauersberger R., Henkel C., Wilson T.L. and Walmsley C.M., 1986a, Astron. Astrophys., **162**, 199.
Mauersberger R., Wilson T.L. and Henkel C., 1986b, Astron. Astrophys., **160**, L13.
Schilke P., Mauersberger R., Walmsley C.M. and Wilson T.L., 1990, Astron. Astrophys., **227**, 220.
Wilson T.L., Mauersberger R., Walmsley C.M. and Batrla W. 1983, Astron. Astrophys., **127**, L19.

Properties of H$_2$O masers from the Arcetri atlas

G. COMORETTO[1], J. BRAND[1], M. CATARZI[1], R. CESARONI[2],

C. GIOVANARDI[1], M. FELLI[1], M. MASSI[1], F. PALAGI[3], F. PALLA[1] and

G. TOFANI[1]

1 Introduction

We have re-observed with a single instrument all the known galactic H$_2$O masers north of $\delta = -30°$. We present the preliminary results of the statistical study based on this catalogue. Other projects carried out include the search for new maser sources in selected lists of IRAS sources or in star forming regions, and variability studies for a small sample of objects.

The observations have been carried with the 32m Medicina radiotelescope, operated by the Istituto di Radioastronomia of the Italian Research Council (CNR), and with a 512 channel digital autocorrelator built by the Radio Astronomy group of the Arcetri Observatory. The detection limit, with the adopted resolution and integration time, is $\simeq 11$ Jy. Absolute calibration was done using continuum sources, and the absolute flux scale is accurate to $\simeq 20\%$.

The basis for the observations was a catalogue of all the detected H$_2$O maser sources north of declination $-30°$, compiled by Cesaroni et al. (1988), further referred to as Paper I, and including a total of 526 sources. Most of these sources were associated with a star-forming region, or with a late-type star.

All the sources in the reference catalogue were re-observed. Some sources are not resolved by the 1.5' beam of this telescope, and therefore a total of 509 distinct positions has been observed. A total of 203 sources were detected. For each one we measured the peak and integrated flux, the interval in velocity over which emission was detected, the LSR velocity of the peak, and its linewidth (Comoretto et al., 1990, further referenced as Paper II).

2 Statistical analysis

Comparing the peak flux densities of Paper I (referred to as S_R) with those of Paper II (S_A) we note a systematic decrease. This effect can be explained by considering that usually in the literature the maximum flux densities are reported, while our observations represent a "random snapshot" of the maser activity.

In a histogram of $\log(S_A/S_R)$ the distribution can be fitted with a gaussian of

---

[1] OSSERVATORIO ASTROFISICO DI ARCETRI, FLORENCE, ITALY.
[2] DEPARTMENT OF ASTRONOMY, UNIVERSITY OF FLORENCE, ITALY, AND MAX-PLANK-INSTITUT FÜR RADIOASTRONOMIE, BONN, F.R.G.
[3] GRUPPO NAZIONALE DI ASTRONOMIA–CNR, FLORENCE, ITALY.

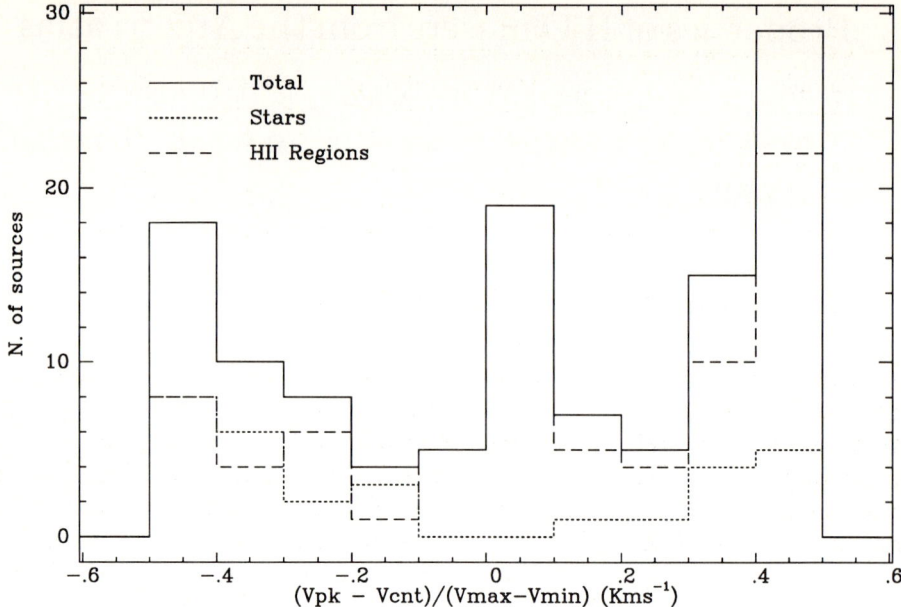

Figure 1: Position of the strongest line in the emission velocity range

mean -0.08 and $\sigma = 0.67$, with an excess of about 20 sources with $S_A = S_R$. The average decrease indicates that the flux densities given in literature are statistically greater than their mean value by about 15–25%. Most sources have varied in intensity by almost an order of magnitude.

A similar comparison between the peak velocities shows that this is rather constant. 37% of the detected sources show a velocity variation of less than $1\,\mathrm{km\,s^{-1}}$, and only 17% show a variation greater than $10\,\mathrm{km\,s^{-1}}$. These larger variations are to be interpreted as onset of new components at slightly different velocities.

A measure of the spectrum compactness is given by the fraction h of the integrated flux contained in the stronger line. The distribution for h is markedly different for star forming regions (almost uniform distribution), and for late type stars (distribution peaked at $h = 1$). This is consistent with the observation that 71% of the masers associated with star forming regions have multiple line spectra, while this is true for only 45% of the detected stellar masers.

For 120 out of the 203 detected sources more than one line has been measured. The position of the strongest peak in the emission interval (the interval over which maser emission has been observed) is reported in figure 1. The quantity in the ascissa is the normalized peak position, $(v_{peak} - v_{mean})/(v_{max} - v_{min})$, with $v_{mean} = (v_{max} + v_{min})/2$. We observe that the peak has a strong tendency to be located either at one extreme of the spectrum, particularly for stellar masers, or in the centre of the emission range, particularly for star forming regions.

Table 1. *New detected masers*

Sample	Sources	Observed	Detected	(by us)
CO Outflow	144	142	54	(11)
IRAS HCO$^+$	261	261	43	(30)
Selfabsorbed and Wings	70	69	12	(12)

3 Correlation with the IRAS catalogue

To study the relation of the maser emission with the infrared spectrum, we correlated the sources in our catalogue with the IRAS Point Source Catalogue.

We analyzed separately the sources associated with late type stars and with star forming regions (HII). We found an IRAS counterpart for 74% of the HII sources, and for 97% of the late type stars. The lack of infrared emission for a significant fraction of the HII sources seems to preclude a radiative pumping of the maser in these sources. The two classes of sources show also very different FIR spectra. HII sources show typically red spectra, while stellar sources are typically blue.

Wood and Churchwell (1989) identified a region occupied by ultracompact HII regions in the (f_{25}/f_{12})-(f_{60}/f_{12}) color-color plane, i.e. $\log f_{25}/f_{12} \geq 0.57$ and $\log f_{60}/f_{12} \geq 1.30$. Of our 151 sources, 60% satisfy the above criteria. By comparison, only 45% of a sample of compact (< 40 arcsec) Sharpless HII regions with infrared counterparts fall in the box quoted by Wood and Churchwell.

A clear correlation of maser emission with FIR colors exists only for the $25\,\mu$m/$12\,\mu$m and the $60\,\mu$m/$12\,\mu$m colors. This indicates that sources with very high reddening in the $12\,\mu$m band have a higher chance to show maser emission.

4 New detections

Detection of new masers were obtained in selected classes of objects (Table 1). The selection criteria for the various samples were: (1) Star forming regions with known CO outflow, (2) IRAS sources in colour intervals characteristic of compact molecular clouds (often associated with HCO$^+$ emission) (Palla et al., this volume, p. 207), (3) Compact molecular clouds having CO line profiles with broad wings or self-absorbed.

We are studying now the correlation of the maser energy with the mechanical energy in the CO outflow for the first sample. We found a preliminary relation between the maser luminosity $L_{\mathrm{H_2O}}$ and the mechanical luminosity L_{HMF}:

$$\log \left[\frac{L_{\mathrm{H_2O}}}{L_\odot}\right] = -6.6 + 0.73 \log \left[\frac{L_{\mathrm{HMF}}}{L_\odot}\right] \qquad (2)$$

which nicely fits the theoretical predictions.

In all the samples the detected masers tend to satisfy the same f_{25}/f_{12} and f_{60}/f_{12} criteria of ultracompact HII regions.

References

Cesaroni, R., Palagi, F., Felli, M., Catarzi, M., Comoretto, G., Di Franco, S., Giovanardi C. and Palla F., 1988, Astron. Astrophys. Suppl., **76**, 445.

Comoretto, G., Palagi, F., Cesaroni, R., Felli, M., Bettarini, A., Catarzi, M., Curioni, G.P., Curioni, P., Di Franco, S., Giovanardi, C., Massi, M., Palla, F., Panella, D., Rossi, E., Speroni, N. and Tofani, G., 1990, Astron. Astrophys. Suppl., **84**, 179.

Wood, D.O.S. and Churchwell, E., 1989, Astrophys. J., **340**, 265.

Survey of H₂O masers associated with compact molecular clouds and ultracompact HII regions

F. PALLA, J. BRAND, R. CESARONI, G. COMORETTO and M. FELLI

OSSERVATORIO ASTROFISICO DI ARCETRI, L.GO ENRICO FERMI 5,

50125 FIRENZE, ITALY.

1 Introduction

The search for and the detection of new maser sources, both in H$_2$O and OH, has been improved tremendously by using the IRAS–Point Source Catalogue (PSC) as database (*cf.* Wouterloot and Walmsley, 1986, Braz *et al.*, 1989). The main outcome of the studies carried out in the post–IRAS years has been that of establishing reliable selection criteria to select *well defined* samples from the large PSC database. To test the goodness of selection criteria, and to place stronger constraints on them, we have set up an experiment to search for H$_2$O masers associated with compact molecular clouds (CMC), due to their HCO$^+$ emission, and ultracompact HII regions (UCHII), as suggested by Wood and Churchwell (1989). The aim of this project is to study the maser occurence in the earliest evolutionary phases of massive stars that are still deeply embedded in dense molecular clouds. The observations have been carried out with the Istituto di Radioastronomia–CNR 32m. radiotelescope located in Medicina (Bologna).

2 Selection Criteria

For the CMCs we follow Richards *et al.* (1987): (1) sources with galactic latitude limited to $|b| \leq 10°$; (2) FIR colours: $0.61 \leq [F_{60}/F_{25}] \leq 1.74$ and $0.087 \leq [F_{100}/F_{60}] \leq 0.52$; (3) no upper limits at 25, 60, 100 μm; (4) $F_{60\,\mu m} \geq 100\,\mathrm{Jy}$; (5) declination $\delta \geq -30°$. A total of 453 sources was thus selected. Taking out those with a positional coincidence with known HII regions reduces the sample to a total of 261 unassociated sources. Wood and Churchwell (1989) characterize UCHII regions by (6) FIR colours: $[F_{25}/F_{12}] \geq 0.57$ and $[F_{60}/F_{12}] \geq 1.30$; (7) no upper limits at 25, 60 μm. Cross-correlation of the UCHII sample with the 261 CMC candidates identifies 126 sources that satisfy both sets of requirements (defined as "*high*", while the remaining 135 are defined "*low*").

3 Results

The survey resulted in the detection of 43 (16%) new maser sources, 38 of which represent first time detection. The "*high*" and "*low*" samples have quite different

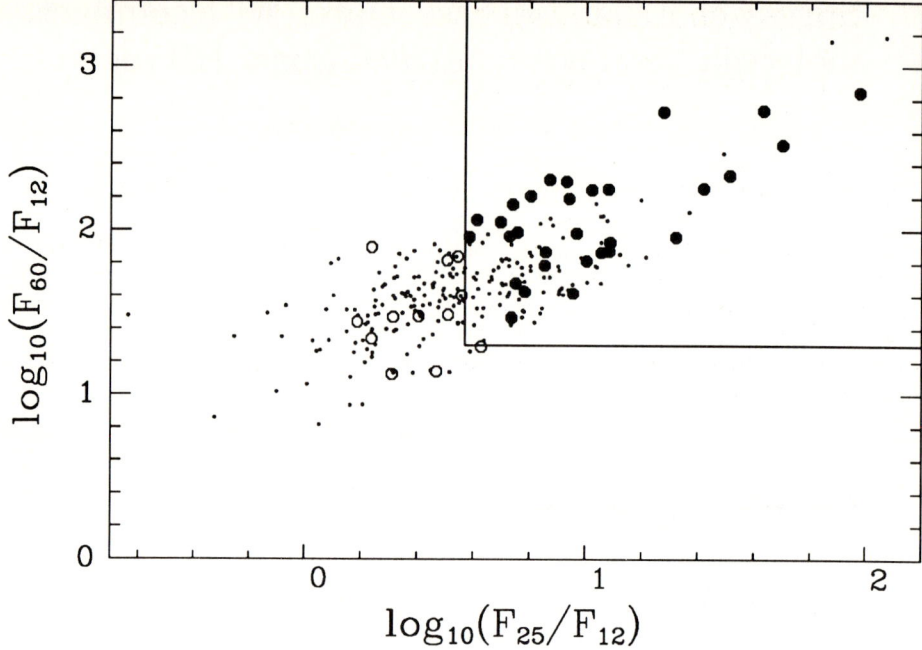

Figure 1. Colour-colour diagram of the 261 sources of our sample. The 126 sources within the box satisfy colour criteria for UCHII regions from Wood and Churchwell. Detected sources are showed as filled (high) and empty (low) circles. Undetected sources are shown as small dots.

statistics of maser occurrence: for the former we find $31/126 = 24\%$ while for the latter $12/135 = 9\%$. The occurrence of maser emission increases for sources with high flux at $60\,\mu$m. The stronger constraint set by $[60–12] \geq 1.9$ does not depend on the flux density: masers are equally found in the $F_{60\,\mu m}$ interval considered here. The [60–25] colour is a poor diagnostic. Detected and undetected sources are rather well mixed in the various FIR colour-colour diagrams. There is a clear indication that for $[60–12] \geq 1.9$ and $[25–12] \geq 0.9$ the detection rate is higher than 50% (cf. figure 1). No segregation is found in the [100–60]/[60–25] plane.

References

Braz, M.A., Scalise, E. Jnr., Hetem, J.C.G., Monteiro do Vale, J.L. and Gaylard, M., 1989, Astron. Astrophys. Suppl., **77**, 465.

Richards, P.J., Little, L.T., Toriseva, M. and Heaton, B.D., 1987, Mon. Not. Roy. Astron. Soc., **228**, 43.

Wood, D.O.S. and Churchwell, E., 1989, Astrophys J., **340**, 265.

Wouterloot, J. and Walmsley, C.M., 1986, Astron. Astrophys., **168**, 237.

Interstellar chemistry

ERIC HERBST[1] and T.J. MILLAR[2]

Abstract In this review, we discuss some aspects of the chemistry of molecular clouds. After a consideration of the chemistry occurring both on the surfaces of dust particles and in the gas phase, we discuss how gas phase reactions can produce the bulk of the gaseous molecules detected outside star-forming regions. Detailed gas phase models of non-star-forming regions are considered and their successes and failures illuminated. A very recent approach to the gas phase formation of very large molecules is discussed. The chemistry of star-forming regions is then considered and recent work involving the desorption of molecules from grain mantles emphasized. A special model of the so-called "Compact Ridge" source in Orion is also discussed.

1 Chemical Processes

1.1 Surface processes

For a typical gas-to-dust ratio in interstellar clouds, the time-scale for the collision of any gas phase particle with a grain is $\sim 3 \times 10^9/n$ years, where n is the total particle density in cm^{-3}. For $n > 100$ cm^{-3}, gas phase particles collide at least once with a dust grain within a typical lifetime of an interstellar cloud. In dense clouds, $n \geq 10^4$ cm^{-3}, collisions are frequent unless some process, such as grain coagulation, acts to reduce the effective surface area of the dust. The interaction of the gas and dust is therefore potentially important in almost all environments in which interstellar molecules are detected. At low temperatures, collisions between the neutral component of the gas and dust are likely to lead to retention of the gas phase particles on the surface because their thermal energy is much less than typical adsorption energies and because excess kinetic energy can be transferred rapidly to the surface. Detailed calculations show that the fraction of collisions which lead to retention is $\sim 0.3 - 1.0$, depending on the species and the nature of the surface (Leitch-Devlin and Williams, 1985). However, depending on the grain charge, this fraction may be zero for ions.

A species which is bound with energy E to a grain with temperature T will remain on the surface for a time t given by $t = \nu_n^{-1} \exp(E/kT)$, where ν_n is the vibrational frequency normal to the surface, after which it will be desorbed thermally. At low temperatures and for reasonable binding energies, this time may be longer than the cloud lifetime. Once on the surface, absorbed species can diffuse thermally to neighbouring sites on time scales $\sim 10^{-2}$ s, or, particularly for light particles such as H and D atoms, quantum tunnel through the diffusion barrier on a time-scale of $\simeq 10^{-12}$ s. Tielens and Allamandola (1987) have shown that adsorbed species can migrate over all the 10^6 or so surface sites before

---

[1] DEPARTMENT OF PHYSICS, DUKE UNIVERSITY, DURHAM, NORTH CAROLINA 27706 U.S.A.
[2] MATHEMATICS DEPARTMENT, UMIST, P.O. BOX 88, MANCHESTER M60 1QD, U.K.

thermal desorption. The important point, however, is that H-atoms can scan the surface so rapidly by tunnelling that the heavy species remain essentially fixed. Thus, provided that there is a sufficient supply of H atoms from the gas phase, one expects gas phase species to be hydrogenated rather efficiently once they arrive on the grain surface. Detailed numerical calculations (*e.g.* Brown, Charnley and Millar, 1988) show that simple saturated species such as H_2O, NH_3 and CH_4 will be the dominant constituents of the grain mantle. The surface chemistry is likely to be much more complex than this for a number of reasons. Firstly, it is possible that surface reactions which possess activation energies can be driven by H-atoms tunnelling through the energy barriers (Tielens, 1983). Secondly, in regions in which the H-atom abundance is small, radical-radical surface reactions can become important. Because of the low mobility of heavy radicals, such reactions occur over a long time-scale, but this can be shortened appreciably if some of the excess energy released in a surface reaction goes into kinetic energy of the products (Brown, 1990). Finally, the influence of ultraviolet photons in generating radical species on the surface can be important, particularly if molecular clouds are very clumpy, in which case each clump of gas may experience an essentially unattenuated radiation field. We shall return to grain surface processes in section 5.2, where we shall argue that complex molecules can be formed efficiently on grains at low temperatures, but are probably not released into the gas phase until some catastrophic event liberates the molecular mantle.

There is, however, one molecule, namely H_2, which we know forms by surface reactions in interstellar clouds. For this molecule there are no simple and efficient gas-phase formation processes. The recombination of H atoms on grains to form H_2 is essential in understanding the *Copernicus* satellite observations of H_2 in diffuse clouds (Jura, 1974a, b). It is interesting to note that the next most simple molecule, HD, needs to be formed in gas-phase ion-neutral reactions in order to account for its observed abundance in diffuse clouds (Watson, 1973). It was this realisation, together with observations of molecular ions and isotopic fractionation, which focused much attention on gas-phase processes and led to the relative neglect of surface processes for many years.

1.2 Gas phase processes

Most exothermic chemical reactions in the gas phase are dominated by the phenomenon of activation energy. Activation energy stems from short-range barriers or maxima in the potential energy surfaces governing reactions; the barriers arise from the breaking of old chemical bonds before new ones are formed. For chemical reactions that possess activation energy barriers, the rate coefficient k is given by the simple Arrhenius equation

$$k = A(T)\exp[-E_a/k_B T] \qquad (1)$$

where E_a is the activation energy, k_B is the Boltzmann constant, T is the absolute temperature and $A(T)$, the so-called pre-exponential factor, is at most a weakly temperature-dependent term. Pre-exponential factors vary widely in their absolute values and temperature dependence; a large number of theories ranging from simple hard sphere collision models to the standard statistical approach (the "activated complex" theory) have been used to rationalize experimental values of $A(T)$ (Weston and Schwarz, 1972). However, the temperature dependence of most rate coefficients is determined mainly by the activation energy term. For neutral-neutral reactions among ordinary molecules in singlet electronic states, activation energies are $\geq 1\,\mathrm{eV}$ so that reactions at interstellar cloud temperatures, e.g. under 100 K, are very slow.

Although there are neutral reactants (both atoms and so-called radicals in non-singlet electronic states) that appear to react with small or zero activation energy, the major class of reactions in which activation energy does not play a major role are ion-molecule reactions. Studied by a variety of investigators at temperatures from under 2 K to well above room temperature (Anicich and Huntress, 1986, Rowe, 1988, Hawley et al. 1990), ion-molecule reactions ordinarily exhibit two important characteristics: (1), there is no hint of activation energy and therefore no exponential dependence of the rate coefficient on temperature and (2), the pre-exponential factor in equation (1) – which is equal to the rate coefficient if $E_a = 0$ – is either independent of temperature or inversely dependent on it.

The reason that ion-molecule reactions ordinarily do not possess activation energy barriers is still somewhat obscure in general. What appears to happen is that strong long-range attractive forces (e.g. charge – dipole) correlate with short-range potential minima rather than maxima; the minima are called reaction complexes. The reaction rate coefficients are governed by the nature of the long-range forces in the sense that the maximum impact parameter for reaction is determined by the centrifugal barrier at long range (Levine and Bernstein, 1974). The strong long-range forces lead to very large rate coefficients at room temperature ($k \approx 10^{-9}\,\mathrm{cm^3\,s^{-1}}$) and to two different types of temperature dependence.

If the neutral molecule does not possess a permanent dipole, the long-range force depends on its polarizability and the pre-exponential factor turns out to contain no temperature dependence whatsoever (Levine and Berstein, 1974). The expression for the rate coefficient in this instance, first derived by Langevin, is

$$k = 2\pi e (\alpha/\mu)^{0.5} \qquad (2)$$

in e.s.u-c.g.s units, where e is the electronic charge, α is the polarizability and μ is the reduced mass of reactants. This expression for k pertains if one potential surface of reaction correlates with reactants; if the reactants are in degenerate states and more than one potential surface for reaction occurs, then one must consider what percentage of these are attractive in nature and whether or not surface hopping occurs (Clary, Dateo and Smith, 1990, Herbst and Knudson, 1981). The result is that the Langevin expression may be somewhat too large.

If the neutral molecule has a permanent dipole, no simple expression can be derived for the rate coefficient but a variety of theoretical studies have shown that the rate coefficient increases as temperature decreases, typically as the inverse square root of temperature (Sakimoto, 1981, Bates, 1982, Clary, 1985), if the reaction is strongly exothermic (Herbst, 1986). Experimental studies are in the main in agreement with the theories (Rowe, 1988). The result is that rate coefficients as large as $10^{-7}\,\mathrm{cm^3\,s^{-1}}$ can exist at temperatures of 10 K.

Not all ion-molecule reactions are as simple as the two classes discussed above. Some possess activation energy barriers and, if the barriers are low enough, the reactions may still be important at low interstellar temperatures because of the phenomenon of tunnelling. In particular, the two ion-molecule reactions

$$NH_3^+ + H_2 \longrightarrow NH_4^+ + H \tag{3}$$

$$C_2H_2^+ + H_2 \longrightarrow C_2H_3^+ + H \tag{4}$$

are slow at room temperature due to small potential energy barriers, become slower as temperature decreases and then eventually become more rapid as the temperature gets significantly under 100 K (Barlow and Dunn, 1987, Hawley and Smith, 1989). As shown in great detail by Herbst *et al.* (1990) for reaction (3), the reason for this behaviour is that the complex formed in the reaction is so long-lived at very low temperature that the tunnelling probability to form products becomes appreciable. Both of these reactions are important in interstellar models. Although the temperature dependence of (3) from 10 – 500 K is well studied experimentally, the temperature dependence of (4) is not yet known at important temperatures between 2 and 80 K. Since there are many slow ion-molecule reactions between ions and molecular hydrogen at room temperature, these reactions may be surprisingly rapid at very low temperature due to the tunnelling phenomenon.

Although the vast majority of ion-molecule reactions that are rapid at low temperatures are exothermic, endothermic reactions may also occur if the endothermicities are quite small and if the reactants are not quite thermal. The only well-studied example of such a reaction is

$$N^+ + H_2 \longrightarrow NH^+ + H \tag{5}$$

which is a primary reaction in the interstellar synthesis of ammonia (Galloway and Herbst, 1989, Marquette *et al.*, 1988, Brown, Cragg and Bettens, 1990). Under interstellar conditions, reaction (5) is powered by an excess of N^+ translational energy and/or an excess of H_2 rotational energy.

The experimental study of neutral-neutral reactions at low temperatures has lagged behind that of ion-molecule systems (Smith, 1988). It is customarily thought that only reactions involving atoms or radicals (molecules in non-singlet electronic states) are devoid of activation energy and hence rapid at low temperatures. However, there is little experimental evidence that even these reactions are rapid at temperatures under 100 K. Long-range theories akin to those used to treat ion-molecule systems have been utilized (Graff, 1989). The long-range forces involved in neutral-neutral reactions are weaker than in ion-molecule systems and are complicated at low temperature by fine structure effects. In particular, atoms such as C and Si, which possess spherically symmetric ground fine structure states, may be much less reactive than atoms such as O and S which, because of their greater than half-filled outer electronic shells, have non-spherical ground fine structure states (Graff, 1989).

1.2.1 Radiative association reactions

When the collision of two species can be described by an attractive potential energy surface leading to a potential minimum (collision complex), there is a possibility that the supermolecule formed by the two reactants can be stabilized as the complex structure. Although such stabilization, normally referred to as association, can occur for both neutral-neutral and ion-molecule systems, the lack of activation energy in ion-molecule reactions renders this process a salient feature in such systems. Stabilization in association reactions can occur via two mechanisms – by inelastic collisions with a third body and by spontaneous emission of radiation. The former has been studied extensively in the laboratory (Bates and Herbst, 1988a) but the latter, referred to as radiative association, is more important for interstellar chemistry since it is dominant at low gas densities. For many years, the determination of rate coefficients for ion-molecule radiative association reactions was a theoretical affair only (Bates and Herbst, 1988a) due to difficulties in studying such inefficient processes at very low laboratory densities. The theoretical treatments showed that association reactions between reactants with small numbers of atoms are normally quite inefficient, but that the efficiency improves as the size of the reactants increases, as the temperature is lowered and as the potential energy surface becomes more attractive (leading to larger well depths). These conclusions are testable by comparison with experimental measurements of three-body association reactions. In addition to the above factors, the mechanism of radiative stabilization is of importance (Herbst and Bates, 1988) in radiative association reactions. For most reactive systems, the complex is formed in its ground electronic state and spontaneous emission occurs from vibrational states lying above the dissociation limit to such states lying below the dissociation limit, which are therefore stable.

On the other hand, far larger stabilization rates can be achieved for complexes formed in excited electronic states that possess dipole-allowed transitions to stable vibrational levels of the ground electronic state or for complexes formed in the ground electronic state that possess such transitions to lower-lying vibrational levels of excited electronic states.

Much effort has gone into the calculation of the rate coefficient for the radiative association reaction

$$C^+ + H_2 \longrightarrow CH_2^+ + h\nu \tag{6}$$

because of its importance in diffuse interstellar clouds (Black and Dalgarno, 1977). The latest result (Smith, 1989), in good agreement with an earlier result (Herbst, 1982), is that roughly one out of every one million collisions

($k \approx 10^{-15}\,\text{cm}^3\,\text{s}^{-1}$) leads to sticking at a temperature of 10 K. The efficiency would be even less were it not for the fact that the CH_2^+ complex is stabilized by an efficient mechanism involving electronic de-excitation. Interestingly, the theoretical value for the low temperature rate coefficient is in good agreement with that derived from diffuse cloud chemical models (Black and Dalgarno, 1977).

Another well-studied radiative association reaction is

$$CH_3^+ + H_2 \longrightarrow CH_5^+ + h\nu \qquad (7)$$

which leads to methane synthesis (see below) in dense interstellar clouds. Although the mechanism for complex stabilization in this sytem has been debated for years (Herbst and Bates, 1988), recent quantum chemical calculations (Talbi, 1990) show that there is no low-lying complex electronic state so that stabilization must occur via pure vibrational emission. With this stabilization mechanism, the latest theoretical value for the rate coefficient of this reaction reaches $\approx 10^{-14} - 10^{-13}\,\text{cm}^3\,\text{s}^{-1}$ at 10 K (Smith, 1989). This system has recently been studied in the laboratory by Gerlich and Kaefer (1989), whose results are in tolerable agreement with the theoretical predictions of Smith (1989). An older experimental determination by Barlow, Dunn and Schauer (1984), for which $k > 10^{-13}\,\text{cm}^3\,\text{s}^{-1}$ at 13 K, when properly interpreted (Bates, 1986) is in some conflict with the work of Gerlich and Kaefer (Bates and Herbst, 1988a), a conflict which is unresolved even when detailed theory is invoked (Bates, 1990).

Although ion-molecule radiative association reactions involving smaller reactants are normally only critical in interstellar chemistry if the reactants cannot react more rapidly to produce exothermic products, it may be the case that as the reactants get larger, radiative association reactions become so efficient that they dominate the chemistry. Recent calculations by Herbst and McEwan (1990) show that at low temperatures, radiative association reactions for systems with roughly ten or more atoms begin to approach unit efficiency, even in the presence of competitive exothermic channels. Some experimental evidence for this assertion is also available (Herbst and McEwan, 1990, Sen, Anicich and McEwan, 1990).

1.2.2 Dissociative recombination reactions

Ion-molecule syntheses in interstellar clouds produce positive molecular ions which are then converted into neutral species via recombination reactions with the dominant carriers of negative charge. There has been some discussion recently concerning the possibility that negatively charged polycyclic aromatic hydrocarbon species (PAH's) are this dominant carrier (Omont, 1986, Lepp and Dalgarno, 1988) although the hypothesis is disputed for all but the largest PAH's in a more recent work (Allamandola, Tielens, and Barker, 1989). Assuming the dominant carrier of negative charge in interstellar clouds to be electrons, the neutralization of positive molecular ions takes place predominantly in dissociative recombination reactions, in which the parent neutral species formed in the collision dissociates into smaller fragments. Such fragmentation can occur via a direct process in which a repulsive potential surface of the parent neutral species crosses the bound ionic surface near its equilibrium point. In addition, there is an indirect process in which the ion-electron system first crosses into excited rovibrational states of a high-lying bound Rydberg electronic state of the neutral species before a second crossing occurs to a repulsive state of the parent neutral (Mitchell and Guberman, 1989).

Dissociative recombination reactions have been studied in the laboratory and typically occur with large rate coefficients ($k = 10^{-7}-10^{-6}\,\mathrm{cm^3\,s^{-1}}$) at room temperature and with a weak inverse dependence on temperature (Adams and Smith, 1988). There is however controversy concerning the rate of the reaction between H_3^+ and electrons, with widely varying results obtained by three different experimental groups (Adams and Smith, 1988, Mitchell, 1990, Amano, 1988, 1990). From the point of view of interstellar chemistry, however, a more important deficiency in our knowledge of dissociative recombination reactions concerns the neutral products of these processes. Until recently, next to no knowledge existed concerning these products and interstellar modellers were forced to use simple theories for the neutral product branching ratios (Bates and Herbst, 1988b). In particular, the idea of Green and Herbst (1979) that single or double hydrogen atom removal is favoured over the breaking of bonds between heavy atoms has been utilized extensively.

Very recently, some progress has been made in the laboratory in studying the products of dissociative recombination reactions although for no system have all the possible exothermic channels been probed. The laboratory techniques involve both laser induced fluorescence (Herd, Adams and Smith, 1990) and vacuum ultraviolet absorption (Rowe, 1990). The former technique has been used to monitor OH product whereas the latter technique has been utilized to measure atomic hydrogen product. Both techniques have been utilized for the reaction between H_3O^+ and e, the interstellar importance of which is discussed below. At room temperature, OH is produced on 65 (± 10) % of reactive collisions while the number of H atoms produced per reactive collision is in the range 1.0 – 1.2. This information is still insufficient to determine the branching ratio for production of $H_2O + H$, although it does show that the product channels $OH + H_2$ and $OH + H + H$ are roughly equal. The observation that OH is the dominant product put several theories to rest but is reasonably consistent with a statistical approach of Herbst (1978), especially as modified by Galloway and Herbst (1990) and a recent treatment of Bates (1989).

1.2.3 Photodestruction

It has long been known that in unshielded regions of the interstellar medium photodestruction of atomic and molecular species occurs on a time-scale of 100 – 1000 years. Even in diffuse clouds such as ζ Oph, the photodestruction time-scale is $\sim 10^4$ years. However the two most important molecular species H_2 and CO are destroyed not by continuum photons, but by line photons which enables these species to self-shield against the UV radiation field – some mutual shielding of CO by H_2 also occurs. There is some laboratory information available on the photodissociation and photoionisation cross-sections of some simple interstellar species, but a lack of detailed knowledge of the intensity of the interstellar UV radiation field and of the absorption and scattering properties of dust grains means that calculated photorates are highly uncertain and can lead to large uncertainties in calculated molecular abundances. Kirby (1990) and van Dishoeck (1988) have discussed the detailed approach to calculations of photodestruction rates and van Dishoeck (1988) has listed photodissociation rates for many important species. Photoionisation is also potentially important, many interstellar molecules having ionisation energies less than the Lyman cut-off, 13.6 eV. Van Dishoeck (1988) lists some relevant rates.

It has long been thought that photodestruction by the interstellar radiation field is unimportant in dust clouds or in regions far from star-formation in giant molecular clouds. This may no longer be a valid approach if molecular clouds turn out to be highly clumped, and may help to resolve the discrepancy between observed CI/CII/CO distributions and those calculated using standard, non-clumped, cloud models (Boissé, 1990). In addition, the cosmic-ray ionisation of H_2 in dense clouds leads to an internally generated flux of ultraviolet photons caused by fluorescent emission from H_2 which is excited by electron collisions. Sternberg, Dalgarno and Lepp (1987) and Gredel et al. (1989) have calculated photorates due to these photons. Although somewhat dependent on physical conditions, the rates vary from 5–5000 times the cosmic-ray ionisation rate, with most species having photorates $\sim 10^{-15}\,\mathrm{s^{-1}}$ for a standard cosmic-ray ionisation rate. For stable, neutral molecules, this rate is about 10–50 % of that due to loss in ion reactions in dense clouds, but for radicals, which are lost most rapidly by reaction with atoms, the photorate is $10^{-3}-10^{-4}$ times smaller than the chemical rate. Thus except for certain species, one does not expect very profound effects to occur as a result of these ultraviolet photons.

2 Gas phase syntheses

2.1 The synthesis of selected smaller species

Ion-molecule chemistry in dense clouds commences with the ionisation of neutral species by cosmic ray bombardment. The dominant neutral species, formed on the surfaces of dust particles, is molecular hydrogen and the dominant cosmic ray ionisation is the reaction

$$H_2 + \text{Cosmic Ray} \longrightarrow H_2^+ + e + \text{Cosmic Ray} \qquad (8)$$

which is normally assumed to occur with a (first-order) rate coefficient of $\sim 10^{-17}\,\mathrm{s^{-1}}$. This rate is sufficiently small that dense clouds remain overwhelmingly neutral although the small abundances of ions produced are critical for the gas phase formation of polyatomic neutral molecules. In diffuse clouds, which are not discussed in this article, the dominant form of ionisation is via ultraviolet irradiation and the fractional ionisation is orders of magnitude greater (van Dishoeck and Black, 1988).

Upon formation, the H_2^+ ion "immediately" (within a day or so at the standard gas density of $10^4 \, cm^{-3}$) reacts to form the polyatomic ion H_3^+ via the well-studied ion-molecule reaction

$$H_2^+ + H_2 \longrightarrow H_3^+ + H \quad . \tag{9}$$

The H_3^+ ion can then react with many of the atomic species assumed to be present in the initial chemical stages of the dense cloud gas via proton transfer processes. An important example is the reaction with atomic oxygen

$$H_3^+ + O \longrightarrow OH^+ + H_2 \tag{10}$$

to produce the OH^+ ion which, as many other ions, reacts rapidly with molecular hydrogen via hydrogen atom transfer:

$$OH^+ + H_2 \longrightarrow H_2O^+ + H \quad . \tag{11}$$

The water ion produced in the reaction is itself reactive with molecular hydrogen:

$$H_2O^+ + H_2 \longrightarrow H_3O^+ + H \tag{12}$$

and the hydronium ion is produced. This ion does not react with molecular hydrogen. An important destruction reaction for hydronium is dissociative recombination with electrons, in which, as discussed above, the dominant product is now known to be OH although H_2O may occur as a product in up to approximately 35% of collisions.

The above ion-molecule syntheses of OH and H_2O are representative in the sense that they illustrate the processes of protonation, hydrogenation and recombination which can produce many simple polyatomic ionic and neutral species. However, the syntheses of the analogous species ammonia (NH_3) and methane (CH_4) and their radical counterparts (NH_2, NH, CH_3, CH_2, CH) are both somewhat different. The reaction between H_3^+ and N to produce $NH^+ + H_2$ does not occur because it is quite endothermic and the reaction to produce the alternative products $NH_2^+ + H$ has a large activation energy barrier (Herbst, DeFrees and McLean, 1987). Ammonia must be synthesized via a process that commences with molecular nitrogen as a reactant. Molecular nitrogen itself is formed by a series of atom-radical reactions (Herbst and Klemperer, 1973, Pineau des Forêts, Roueff and Flower, 1990) and leads to the production of N^+ via the reaction

$$He^+ + N_2 \longrightarrow N^+ + N + He \tag{13}$$

where He^+ is produced via direct cosmic ray bombardment. The atomic nitrogen ion generated in this reaction is sufficiently energetic to react with H_2 despite the slight endothermicity and a sequence of H-atom transfer reactions from NH^+ leads eventually to the ion NH_4^+, from which ammonia (NH_3) and less saturated species (poorer in hydrogen) are produced via dissociative recombination reactions.

For the case of methane, the primary reaction

$$H_3^+ + C \longrightarrow CH^+ + H_2 \tag{14}$$

has not been studied in the laboratory but quantum chemical calculations show that there is no barrier to reaction. Another possible primary reaction involves ionised atomic carbon:

$$C^+ + H_2 \longrightarrow CH^+ + H \tag{15}$$

but turns out to be endothermic by 0.4 eV. This reaction is thought to be quite important in regions of the diffuse interstellar medium subjected to shock waves (Flower et al., 1988); the interesting chemistry of shocked regions is beyond the limited scope of this review paper. In low temperature regions, the radiative association to form CH_2^+, discussed above, is a competitive channel. Ionized atomic carbon derives from precursors such as carbon monoxide.

Once CH^+ is produced, normal H-atom transfer reactions with molecular hydrogen lead to the methyl (CH_3^+) ion. This ion does not undergo a normal reaction with H_2, but the rather slow radiative association reaction to produce CH_5^+ is known to occur and has been previously discussed. Dissociative recombination of CH_5^+ leads to methane and less saturated single-carbon-atom hydrocarbons.

The analogous chemistries of the second-row elements sulphur (Oppenheimer and Dalgarno, 1974, Millar and Herbst, 1990a) and phosphorus (Thorne et al., 1984, Millar, Bennett and Herbst, 1987, Millar, 1990a) also merit discussion. The reaction between H_3^+ and S

$$H_3^+ + S \longrightarrow HS^+ + H_2 \tag{16}$$

is thought to occur in analogy with the O atom reaction, which has been studied in the laboratory. The HS^+ ion does not react exothermically with H_2 but the radiative association to produce H_3S^+ has been calculated to occur (Herbst, DeFrees and Koch, 1989). The H_3S^+ ion then undergoes dissociative recombination to produce HS and H_2S, although preliminary laboratory results do not indicate the latter to be a major channel (Rowe, 1990). As for phosphorus, the reaction between P and H_3^+ to produce $PH^+ + H_2$ is exothermic and should occur; it has not been studied in the laboratory. The H-atom transfer reactions between the ions P^+, PH^+ and PH_3^+ with H_2 are endothermic. Thus if PH^+ is to react further with H_2 it would have to be by radiative association. Adams, McIntosh and Smith (1990) have measured the analogous three-body association at both 80 K and 300 K, and deduced a temperature dependence of $T^{-1.4}$. The reaction is slow, however, $\sim 3 \times 10^{-15}$ cm^3 s^{-1} at 10 K, and does not lead to significant formation of P-H bonds. Interstellar phosphous chemistry is currently thought to proceed from the reactions (Millar, 1990a)

$$P^+ + H_2O \longrightarrow HPO^+ + H \qquad (17)$$

$$PH^+ + H_2O \longrightarrow HPO^+ + H_2 \qquad (18)$$

$$P + O_2 \longrightarrow PO + O \qquad (19)$$

although reaction (19) is slow. P-O bonds thus form readily (Thorne et al., 1984) but any PO produced is destroyed rapidly to produce large amounts of PN (Millar, Bennett and Herbst, 1987, Millar, 1990a).

The chemistry of silicon has been investigated in some detail (Herbst et al., 1989, Langer and Glassgold, 1990). Like phosphorus, silicon is unreactive with H_2 but very reactive with oxygen bearing molecules such as H_2O and O_2 and the calculations predict that, because of its bond strength, SiO should be the most abundant silicon-bearing molecule, taking up nearly all of the available silicon. The failure to detect SiO in cold, dark clouds is probably due to a severe depletion of silicon, and possibly water, onto grain surfaces.

The reactions mentioned above are only those which are important in initiating chemistry and are a small subset of a large number of ion-neutral reactions involving smaller species that have been written down over the years to explain the formation and destruction of molecules in dense clouds. The latest release of the UMIST ratefile for astrochemistry contains almost 3000 gas phase reactions among 313 species involving 12 elements (Millar et al., 1990); over 70% of these are ion-neutral reactions. For an up-to-date list of reactions, the reader is advised to contact one of the authors.

2.2 The synthesis of hydrocarbons and other complex species

The gas phase synthesis of hydrocarbons proceeds via several pathways. Foremost are so-called insertion reactions involving C^+ and C reacting with smaller hydrocarbon neutrals and ions, respectively. The ionic products of such reactions will subsequently react with H_2 if possible; otherwise they are depleted by dissociative recombination with electrons and by reaction with heavy neutral species. The following syntheses of the well-known interstellar molecules C_2H, C_2H_2, C_3H, cyclo-C_3H, cyclo-C_3H_2 and C_4H from methane are illustrative:

$$C^+ + CH_4 \longrightarrow C_2H_2^+ + H_2;\ C_2H_3^+ + H \qquad (20)$$

$$C_2H_2^+ + H_2 \longrightarrow C_2H_3^+ + H;\ C_2H_4^+ + h\nu \qquad (21)$$

$$C_2H_3^+ + H_2 \longrightarrow \text{no reaction} \qquad (22)$$

$$C_2H_3^+ + e \longrightarrow C_2H_2 + H; \ C_2H + H_2(2H) \tag{23}$$

$$C^+ + C_2H_2 \longrightarrow C_3H^+ + H \tag{24}$$

$$C_3H^+ + H_2 \longrightarrow cyclo-C_3H_3^+/C_3H_3^+ + h\nu \tag{25}$$

$$cyclo-C_3H_3^+ + e \longrightarrow cyclo-C_3H_2 + H; \ cyclo-C_3H + H_2 \tag{26}$$

$$C_3H_3^+ + e \longrightarrow C_3H_2 + H; \ C_3H + H_2 \tag{27}$$

$$C^+ + C_3H_2 \longrightarrow C_4H^+ + H \tag{28}$$

$$C_4H^+ + H_2 \longrightarrow C_4H_2^+ + H \tag{29}$$

$$C_4H_2^+ + e \longrightarrow C_4H + H \tag{30}$$

and based on a large number of laboratory studies (Anicich and Huntress, 1986, Millar et al., 1987, Herbst and Leung, 1989, Hawley and Smith, 1989). Reactions involving hydrocarbon ions and neutral C have not been shown since, although competitive in interstellar cloud models, they have not yet been studied in the laboratory.

Since hydrocarbon ions with more than one carbon atom tend not to react rapidly with H_2 unless they are very unsaturated (hydrogen-poor), the above synthetic pathway cannot produce the more saturated neutral hydrocarbons. These can be produced by so-called condensation reactions between smaller hydrocarbon ions and neutrals followed by dissociative recombination reactions. Some well studied condensation reactions are:

$$CH_3^+ + CH_4 \longrightarrow C_2H_5^+ + H_2 \tag{31}$$

$$C_2H_2^+ + C_2H_2 \longrightarrow C_4H_3^+ + H; \ C_4H_2^+ + H_2. \tag{32}$$

As the reacting species get larger, radiative association reactions gain in importance; one such reaction studied at room temperature is (Sen, Anicich, and McEwan, 1990)

$$C_4H_3^+ + C_2H_2 \longrightarrow C_6H_5^+ + h\nu \tag{33}$$

which has been measured to occur in one out of every ten collisions.

The above three ion-molecule synthetic pathways have been utilized by Herbst and Leung (1989) to produce hydrocarbons through nine carbon atoms in size.

In general, the larger the hydrocarbon, the less the amount of laboratory information and the more guessing required. For these mainly unsaturated hydrocarbons, the C^+ and C insertion pathways are the most rapid. However, the situation is expected to change for larger hydrocarbons since radiative association is expected to dominate (see below).

Hydrocarbons are the simplest large polyatomic molecules and other classes of species are thought to be produced from them. Probably the most important non-hydrocarbon class of complex interstellar molecules are the cyanoacetylenes or cyanpolyynes, which are linear organo-nitrogen species with structure $H(CC)_n - CN$. Although a variety of ion-molecule syntheses have been deduced for these molecules (Winnewisser and Herbst, 1987), the dominant ion-molecule pathway appears to involve insertion reactions between atomic nitrogen and hydrocarbons ions. For example, the synthesis of HC_3N can occur via

$$N + C_3H_3^+ \longrightarrow H_2C_3N^+ + H \tag{34}$$

which has been studied in the laboratory, followed by dissociative recombination. A similar synthesis starting from the hydrocarbon ion $C_2H_4^+$ has been used to explain the observation of the interstellar radical CH_2CN (Irvine et al., 1988, Herbst and Leung, 1990). Recently, Herbst and Leung (1990) have shown that if neutral-neutral reactions between the CN radical and neutral hydrocarbons occur rapidly at low temperature, they constitute a more rapid synthesis for cyanopolyynes than the ion-molecule route.

The synthesis of oxygen-containing organic molecules is perhaps the most difficult to explain via gas phase theories. A variety of very unsaturated species in this class can be explained by radiative association reactions involving hydrocarbon ions and CO (Adams et al., 1989). Also, there is a general consensus that methanol is produced via the radiative association reaction

$$CH_3^+ + H_2O \longrightarrow CH_3OH_2^+ + h\nu \tag{35}$$

followed by dissociative recombination. However, there is great uncertainty involving complex species such as dimethyl ether, ethanol and methyl formate (Herbst, 1987). It is fairly clear from observational studies that the complex oxygen-containing molecules are localized in star-forming regions. Gas phase reactions starting from methanol and protonated methanol have been suggested by Blake et al. (1987) as responsible for the oxygen-containing species seen in the Compact Ridge source of the Orion Nebula; according to Blake et al. (1987) these are operative if a large amount of water from the plateau source interacts with the Compact Ridge. Some recent calculations by us (Millar, Herbst and Charnley, 1990; see section 5.2) are not in accord with this view however. Surface processes (Brown, 1990) have also been suggested. A totally unexplored gas phase route involves reactions between neutral oxygen atoms and hydrocarbon ions. Interestingly, the formation of organo-sulphur molecules such as CCS, CCCS, and H_2CS appears much easier to explain because of the large abundance of ionised atomic sulphur – S^+ – and its ability to react with neutral hydrocarbons. A detailed discussion is to be found in Millar and Herbst (1990a).

Like S^+ and indeed Si^+, the P^+ ion also reacts readily with neutral hydrocarbon molecules (Smith, McIntosh and Adams, 1989), for example,

$$P^+ + C_2H_2 \longrightarrow PC_2H^+ + H \tag{36}$$

$$PH^+ + C_2H_2 \longrightarrow PC_2H_2^+ + H \tag{37}$$

The PH^+ abundance is calculated to be larger than that of P^+ in dense clouds so that reaction (37), which produces a cyclic ion, dominates. This ion, which does not react with H_2, can recombine to form CCP or HCCP, the abundances of which can be significant, particularly if these organo-phosphorus compounds do not react with atomic oxygen at low temperatures (Millar, 1990a).

2.3 Loss processes

Neutral molecules formed via sequences of gas phase reactions are then destroyed by similar processes. A stable neutral molecule such as water will be depleted only by ion-molecule reactions. The dominant ions in dense clouds that can react with neutral molecules such as water are H_3^+, HCO^+, H_3O^+, HN_2^+, H^+, He^+ and C^+. Although there is no question that reactions with the latter three ions result in depletion of the neutral at the full rate of reaction, reactions with the first four (protonating) ions may be less efficient depending on subsequent reactions. Consider, for example, the reaction between water and H_3^+

$$H_3^+ + H_2O \longrightarrow H_3O^+ + H_2. \tag{38}$$

If the H_3O^+ ion product is then depleted principally by dissociative recombination with electrons and if the products of this reaction are simply $H_2O + H$, then the reaction between H_3^+ and H_2O cannot be said to deplete water. Of course, as is now known from laboratory studies, the radical OH is formed in 65($\pm$10)% of dissociative recombination reactions between H_3O^+ and electrons so that reaction does lead to depletion of H_2O but not with 100% efficiency. Although the neutral products of dissociative recombination reactions are in general not known, it is in our view important to avoid the assumption that only one set of products can occur and that this set consists of H atoms plus the remainder of the neutral parent molecules. Otherwise reaction with protonating ions will not deplete molecules and overall depletion rates will be lowered greatly. This effective recycling of products can lead to much larger calculated abundances of neutral species than is the case if several sets of products are obtained in dissociative recombination reactions (Millar et al., 1987).

Unlike their stable counterparts, radical neutrals probably react rapidly with abundant atoms at low temperature and are therefore depleted more rapidly in molecular clouds. Although chemists normally think of radicals as species with unpaired electrons (*e.g.* OH, C_2H), unusual molecules with paired electrons may also be quite reactive. For example, the species C_2 has a ground electronic state of $^1\Sigma$ symmetry but very few chemists would call this species unreactive. At low temperatures however, even small activation energy barriers can dramatically curtail reaction efficiencies in the absence of efficient tunnelling. It has been found (Millar *et al.*, 1987) that the rate of depletion reactions between atomic oxygen and radical and other abnormal hydrocarbons is critical to whether or not complex species can be built up with observable abundances. In selected models (Millar *et al.*, 1987, Herbst and Leung, 1989, 1990) the assumption has been made that for hydrocarbons with more than two carbon atoms, only the bare carbon clusters C_n react with O rapidly. Herbst (1990) has found that if truly giant hydrocarbons of sizes in the range discussed for PAH's are to be synthesized via gas phase reactions in dense interstellar clouds, reactions between such species, even if bare, and oxygen atoms must be slow.

2.4 Isotopic fractionation

Ion-molecule reactions can lead to rather strong fractionation effects at low temperatures due to differences in zero-point energies between reactants and products. Although the ion-molecule chemistry of fractionation among carbon and oxygen isotopes has been studied (Langer *et al.*, 1984), the strongest effects are seen between deuterium and hydrogen. The first reaction considered in such fractionation was (Guélin *et al.*, 1977, Snyder *et al.*, 1977)

$$H_3^+ + HD \longrightarrow H_2D^+ + H_2 \tag{39}$$

because much of the deuterium in dense clouds is tied up in the form of HD and the reaction is known to occur rapidly, a far from universal feature for exothermic ion-molecule exchange reactions (Henchman and Paulson, 1989). In any event, zero-point energy differences make the reaction as shown exothermic, whereas the reverse reaction is sufficiently endothermic to be quite slow in low temperature interstellar clouds. The result is that the H_2D^+/H_3^+ abundance ratio can become orders of magnitude greater than the HD/H_2 abundance ratio of a few x 10^{-5}. The large H_2D^+/H_3^+ abundance ratio produced via the above reaction can lead to large abundance ratios between the deuterated and normal forms of other trace species via ion-molecule reactions such as

$$H_2D^+ + CO \longrightarrow DCO^+ + H_2. \tag{40}$$

It is now appreciated that reaction (39) is not the only important exchange reaction in deuterium fractionation. The reaction between CH_3^+ and HD also plays a role (Wootten, 1987) as do reactions with atomic deuterium, formed principally via the dissociative recombination of DCO^+ at low temperature (Dalgarno and Lepp, 1984). Indeed, a large number of intertwined reactions are involved (Brown and Rice, 1986, Millar, Bennett and Herbst, 1989). Currently it appears that ion-molecule reactions can account for the large amount of fractionation seen in abundance ratios between deuterated and normal forms of trace species in non-star-forming regions of interstellar clouds (Millar, Bennett and Herbst, 1989). Gas phase reactions cannot account however for the unexpectedly large deuterium fractionation seen recently in high temperature star-forming regions but must be supplemented by deuterium atom addition reactions on the surfaces of dust particles. These are discussed in Section 5 below.

3 Gas phase (pseudo-time-dependent) models

3.1 Homogeneous cloud models

Many of the most detailed chemical kinetic models for interstellar clouds have been of the pseudo-time-dependent variety in which chemical abundances vary as a function of time under fixed physical conditions (Leung, Herbst and Huebner, 1984, Langer et al., 1984, Millar and Nejad, 1985, Brown and Rice, 1986a, b, Millar, Leung and Herbst, 1987, Millar et al., 1988, Langer and Graedel, 1989, Herbst and Leung, 1989, Herbst et al., 1989, Millar and Herbst, 1990a, b, Brown, Cragg and Bettens, 1990, Millar, 1990a). Calculations in which physical conditions are allowed to change are discussed in sections 3.2 and 3.3 below.

While the models differ somewhat in their choice of rate coefficients, species considered, physical conditions adopted and so on, they do show a number of similarities. In particular, the abundances reached at steady-state, which occurs in general for times greater than about 10^7 yr, are often orders of magnitude different from abundances at earlier times, $\sim 10^5$ yr, independent of the initial conditions chosen. Molecules such as the hydrocarbons and related species have large abundances at early time before decreasing to their steady-state values. This peak at early times occurs because there is a significant abundance of C atoms which drive the hydrocarbon chemistry as described in Section 2.2. On a longer time-scale, carbon is chemically processed into the extremely stable CO molecule which takes up essentially all of the available carbon. However certain molecules, particularly those which do not contain carbon and which are formed in neutral-neutral reactions show the opposite behaviour. That is, their abundances increase from early-time to steady-state. Examples include species such as N_2, O_2, NH_3, SO and SO_2. As an example of how abundances vary with time we give in Table 1 a compilation of results from Herbst and Leung (1989, 1990) and Millar and Herbst (1990a) for both early-time and steady-state and compare these to those derived from observations of the cold, dark cloud TMC-1. Note that all of these calculations adopted ion-dipolar rate coefficients and low metal abundances. While it can be noticed that for most species, the early-time results, that is $\sim 10^5$ yr, give best agreement with the observations, this is not always the case. In particular, N_2H^+, NH_3, SO and SO_2 have a better agreement for later times. No best-fit solution has been sought, but recent work by Brown, Cragg and Bettens (1990) suggests that, from the smaller species at

Table 1. *Unless otherwise noted, all results come from Herbst and Leung (1989). Note that early-time results are for 10^5 years. The SO and SO_2 abundances possess a severe time-dependence. At 5×10^5 yr, they are 3.4×10^{-9} and 3.0×10^{-10} respectively. The large rise in these abundances occurs at times $\sim 10^5$ yr, the exact values of which vary by a factor of about two depending on the choice of certain neutral-neutral rate coefficients involved in the formation and destruction of SO.*

Species	Fractional abundance[a]			Species	Fractional abundance[a]		
	e.t.	s.s.	obs.		e.t.	s.s.	obs.
CO	8.1(-5)	1.5(-4)	8(-5)	C_2	2.6(-8)	2.3(-10)	5(-8)
OH	9.5(-8)	2.4(-7)	3(-7)	CH	4.2(-8)	1.9(-10)	2(-8)
C_2H	7.2(-8)	2.8(-10)	8(-8)	C_3H	2.7(-8)	1.0(-10)	5(-10)
C_4H	8.2(-8)	5.7(-12)	2(-8)	C_5H	5.3(-9)	1.8(-13)	4(-10)
C_6H	4.9(-9)	4.0(-14)	1(-9)	C_3H_2	2.3(-8)	1.3(-10)	2(-8)
CH_3CCH	4.0(-10)	3.1(-12)	6(-9)	CH_3C_4H	2.6(-9)	4.3(-14)	2(-9)
CN[b]	7.0(-8)	4.2(-9)	3(-8)	HCN	1.5(-7)	1.6(-9)	2(-8)
HNC	9.3(-8)	1.8(-9)	2(-8)	CH_2CN[b]	3.9(-8)	9.2(-11)	5(-9)
CH_2CHCN[b]	7.8(-11)	1.7(-14)	2(-10)	CH_3CN[b]	5.7(-9)	5.7(-13)	1(-9)
HC_3N[b]	3.2(-8)	4.4(-11)	6(-9)	HC_5N[b]	4.7(-9)	1.2(-13)	3(-9)
HC_7N	6.8(-10)	1.2(-15)	1(-9)	HC_9N	1.1(-10)	3.9(-18)	3(-10)
CH_3C_3N[b]	7.8(-11)	1.7(-14)	2(-10)	C_3N[b]	1.0(-8)	3.4(-11)	1(-9)
NH_3	2.9(-9)	5.8(-9)	2(-8)	N_2H^+	1.7(-12)	2.8(-10)	5(-10)
HCO^+	4.5(-9)	9.5(-9)	8(-9)	C_3O	3.8(-11)	5.8(-12)	1(-10)
H_2CO	3.3(-7)	3.3(-9)	2(-8)	CH_2CO	7.3(-8)	5.7(-10)	1(-10)
CH_3OH	2.3(-9)	5.4(-12)	4(-9)	CH_3CHO	9.5(-9)	1.1(-10)	6(-10)
CS[c]	1.9(-8)	3.4(-8)	1(-8)	HCS^+ [c]	3.4(-10)	5.2(-10)	6(-10)
OCS[c]	6.4(-9)	1.3(-9)	2(-9)	SO[c]	1.7(-11)	1.1(-7)	5(-9)
NO	2.6(-8)	2.2(-7)	<3(-8)	SO_2[c]	3.4(-12)	9.2(-9)	<1(-9)
H_2S[c]	4.7(-11)	9.9(-11)	7(-10)	H_2CS[c]	2.1(-9)	2.6(-10)	3(-9)
C_2S[c]	1.6(-9)	2.0(-11)	8(-9)	C_3S[c]	4.0(-10)	4.2(-12)	2(-9)

a) $\alpha(-\beta)$ means $\alpha \times 10^{-\beta}$.
b) Results from Herbst and Leung (1990).
c) Unpublished results from Millar and Herbst (1990a).

least, an age of $\sim 10^6$ yr might be appropriate for TMC-1. What is clear is that the abundances of up to about 35 molecules, including the cyanopolyynes up to HC_9N, can be explained by the models.

In addition to uncertainties attached to individual rate coefficients, the results of these calculations are affected by more global considerations such as the influence of ultraviolet photons generated by cosmic rays (Gredel *et al.*, 1989), the choice of branching ratios in the dissociative recombination of molecular ions with electrons (Millar *et al.*, 1988), the efficiency of oxygen atom reactions with hydrocarbons (Millar, Leung and Herbst, 1987) and variations in elemental abundances (Langer and Graedel, 1989, Millar and Herbst, 1990b). Millar (1990b) has presented the results of a model of dark cloud chemistry in which some of these effects have been examined in a systematic and self-consistent manner.

In general the effects of cosmic-ray-induced photons are minimal. Gredel et al. (1989) present photodestruction rates for several species with values ranging from $100\zeta - 10^4\zeta\,\mathrm{s}^{-1}$, where ζ is the total cosmic-ray ionisation rate and is typically $10^{-17}\,\mathrm{s}^{-1}$. Since most molecules have values $< 500\zeta$, we shall take this to be a representative value for comparison with other loss processes. Photodissociation by the ambient interstellar ultraviolet radiation field occurs at a rate of $\sim 10^{-10}\exp(-\beta A_v)\,\mathrm{s}^{-1}$ with $\beta \sim 2-3$. Thus cosmic-ray photons are more important than interstellar photons once $A_v \gtrsim 2-3$ magnitudes. In dark clouds, radicals are destroyed by reactions with neutral atoms, particularly O and C atoms. For a rate coefficient of $10^{-11}\,\mathrm{cm}^3\,\mathrm{s}^{-1}$ and an atom abundance of $10^{-4}n\,\mathrm{cm}^{-3}$ (or $1\,\mathrm{cm}^{-3}$ for $n = 10^4\,\mathrm{cm}^{-3}$), one sees that the time-scale for loss of radicals by cosmic-ray photons is much longer than for loss by neutral atom reactions. For stable neutrals, loss is via ion-molecule reactions with rate coefficients $\sim 10^{-8} - 10^{-9}\,\mathrm{cm}^{-3}\,\mathrm{s}^{-1}$ and ion abundances of $\sim 10^{-8}n$. For these species, loss by ion reactions is faster than, though comparable to, loss by cosmic-ray photons. One therefore expects little difference in abundances when such photons are included, a result substantiated by detailed numerical calculations (Gredel et al., 1989, Millar, 1990b).

Reactions involving atoms can be important both in the formation and destruction of molecules. For example insertion reactions involving C atoms synthesise large hydrocarbon species, while O atom reactions may destroy these same species. Although the O atom abundance is larger than that of C atoms at all times abundances of hydrocarbons are affected appreciably at early time only if $O + C_nH_m$ ($m \geq 0$) reactions are fast, although some of these reactions may involve a spin-flip and possess an activation energy (Millar, Leung and Herbst, 1987). In certain instances, hydrocarbon abundances can be reduced by several orders of magnitude because, as shown above, fast neutral-neutral reactions involving atoms occur at rates $\sim 10 - 100$ times faster than ion-neutral reactions. In fact the "raw rate" i.e. $kn(X^+)$, at which an ion X^+ destroys a neutral is not always the relevant rate to use, since it is often the case that recycling of the neutral can occur. Thus the choice of the branching ratios in dissociative recombination reactions is crucial to estimates of molecular abundances. As an example, consider the loss of the C_3H radical in ion-neutral reactions. This species can undergo proton-transfer reactions with ions such as H_3^+, HCO^+ and H_3O^+ to form $C_3H_2^+$, which is unreactive with H_2 at low temperatures. Millar and Nejad (1985) assumed that dissociative recombination of this ion led back to C_3H, so that destruction of C_3H occurred in reactions with less abundant ions such as C^+ and He^+. However, if one assumes that one-half of the recombinations lead to C_3 and that this species reacts rapidly with O-atoms, as has been done by Herbst and Leung (1986), then the effective destruction rate of C_3H is much faster since now one-half of all protonation reactions lead to its destruction.

Millar et al. (1988) have investigated the sensitivity of model results to the branching ratios chosen in dissociative recombination reactions and discuss effects such as that above in some detail. They find that the branching ratios adopted for the $H_3O^+ + e$ reaction have the most widespread effect on molecular abundances because the OH radical, which is a product of this recombination, is a highly reactive intermediate in the formation of many species, particularly oxides such as CO, NO, SO, SiO, PO as well as O_2 and CO_2.

Finally, the choice of elemental abundances can affect calculated abundances, in some instances in dramatic fashion. For trace elements, such as S, Si, P and Cl, it is known that as long as fractional abundances $< 10^{-6}$ are adopted, molecular abundances scale directly with the amount of available material (Herbst *et al.*, 1989, Millar and Herbst, 1990a, Millar 1990a). For the major elements, C, N and O, the picture is more complex. For example, most calculations of dark cloud chemistries adopt an O/C abundance ratio greater than 1, which results in smaller CI abundances than observed – though only in one dark cloud - if the cloud age is more than a few times 10^5 yr. In order to overcome this problem, some calculations have been carried out with an O/C ratio less than one (Herbst and Leung, 1986, Langer *et al.*, 1984). The excess of carbon also allows for very efficient hydrocarbon formation. Millar and Herbst (1990b) have discussed cases in which the C, N and O abundances all vary and applied their calculations to models of dark clouds in the Large and Small Magellanic Clouds. They find that hydrocarbon abundances do not depend only on the O/C ratio, but also on the abundance difference, O−C. One of their dark cloud models for the SMC has an O/C ratio of 10.5 – the Galactic value is 2.4 – and a total carbon abundance only 4% of the Galactic value, but has abundances of hydrocarbon radicals such as C_2H and C_4, close to or larger than the Galactic values. This occurs because the abundance of atomic oxygen, which is important in destroying these radicals, is dependent on O−C, since CO takes up essentially all the available oxygen, leaving only a small abundance of atomic oxygen.

3.2 Toward more physical diversity

The models discussed above, whether or not they include cosmic-ray induced photodestruction processes, contain a gross oversimplification of the physical conditions present in dense interstellar clouds. These clouds exhibit clumpiness on the order of any distance scale probed, contain outer regions where the density is lower and the photodestruction due to external photons more important, and often contain regions of active star formation. We defer a discussion of the chemistry of active star forming regions until section 5. In this section, the effects of large scale clumpiness, photodestruction and dynamics on the chemistry will be discussed.

3.2.1 Clumpiness

Imagine a low density medium pervaded by external ultra-violet radiation. In this medium are high density clumps in which cosmic ray-induced ionisation initiates the ion-molecule chemistry discussed previously. If an especially strong ultra-violet source is nearby (*e.g.* an HII region) the edges of the dense clumps may contain sizeable "photodissociation zones" (Tielens and Hollenbach, 1985a, b, Genzel et al., 1988), in which ultra-violet radiation strongly affects the chemistry, as it would the outer cloud layers of relatively homogeneous dense clouds. Large abundances of C and even C^+ are found in "photodissociation zones" in addition to warm carbon monoxide. But inside the photodissociation zone layers, if they exist, the ion-molecule chemistry of dense clumps would be relatively unaffected by the clumpiness of the interstellar material were it not for the possibility of mixing between diffuse and dense regions. Chièze and Pineau des Forêts (1989) have recently published a small chemical model in which mixing occurs so as to remove inhomogeneities in the abundances of individual species between diffuse and dense cloud portions in the absence of intermediate "photodissociation zones". The mixing occurs with an empirical time scale that can be varied. Most interestingly, the mixing puts large amounts of C and C^+, which are present in abundance in the diffuse gas which is pervaded by normal ultra-violet radiation, into the dense clumps. These atomic species are precursors for the gas phase formation of organic molecules and, in homogeneous models, are present in abundance only at times well before steady-state conditions are achieved. The consequence is that large abundances of organic molecules are calculated to exist only at early time and not at steady state. This extreme time dependence in the calculated abundances of organic molecules is bothersome. If C and C^+ are continually injected into the dense clumps it is possible that the steady-state abundances of complex molecules will rise to the point that the distinction between early-time and steady state concentrations will vanish.

The possibility of producing significant amounts of organic molecules at steady state was mentioned by Chièze and Pineau des Forêts (1989) although their model did not include large molecules. This deficiency has now been remedied by Chièze, Pineau des Forêts and Herbst (1990) who have utilized the reaction set of Herbst and Leung (1989, 1990) to represent the dense clump chemistry. These authors have found that mixing of C and C^+ is not a panacea for producing large abundances of organic molecules at steady state. Let us consider their specific results for one organic molecule – HC_9N. In the homogeneous dark cloud calculation of Herbst and Leung (1990) the fractional abundance of HC_9N at early-time (occurring approximately 10^5 yr after the onset of the chemistry) is 1.1×10^{-10} whereas it is 3.9×10^{-18} at steady state ($> 10^7$ yr). The observed fractional abundance in TMC-1 is 3×10^{-10}. With a very short mixing time scale of 10^5 yr, the steady-state abundance of HC_9N rises only to 1.7×10^{-14} if the material mixing into the dense clumps is mainly C^+. However, if the material is mainly neutral atomic carbon, the steady-state abundance of HC_9N reaches 5.3×10^{-11} which is essentially indistinguishable from the early-time abundance. This effect occurs for most large molecules in the model and leads to the conclusion that mixing is effective in producing large abundances of organic molecules at steady state if and only if the gas streaming in from the diffuse layer is mainly neutral in its atomic carbon content. The reason for this unexpected dichotomy between C^+ and C is that since C^+ is an ion, it is also effective in depleting neutral species.

How physically reasonable is this picture? One objection is that it is not clear how diffuse cloud gas can be rich in C and not C^+. It is possible however that the processing of C^+ into C occurs quickly during the transit from diffuse cloud gas into the inner regions of dense clumps. Even if the mechanism is operative, one can argue against the use of calculated abundances at times much greater than $10^5 - 10^6$ yr because the clumps probably do not live long enough. And even if they do, the problem of why the gaseous material is not all adsorbed onto the surfaces of dust particles on a time scale of 10^6 yr still exists. It may be necessary to invoke periodic shock waves to drive the material off the dust grains (Williams, 1988). Finally, both the physical mechanism driving the mixing and whether or not rapid mixing is consistent with observed line widths of radioastronomical lines must be considered in more detail (de Boisanger and Chièze, 1990).

3.2.2 Photodestruction

Unlike the case of diffuse clouds, in which the effects of external ultra-violet radiation are pervasive (van Dishoeck and Black, 1988), these effects in dense clouds were at first unrecognized. However, whether one considers the outer edges of otherwise homogeneous dense clouds, or one considers photodissociation zones surrounding dense clumps in a more diffuse medium, it is necessary to consider the photodestructive effects of ultra-violet radiation (van Dishoeck, 1988), especially if the normal radiation field is enhanced. Such enhancement will occur near HII regions for example. In the model of Tielens and Hollenbach (1985a, b), the effect of intense radiation is followed up to a visual extinction of approximately 10. Three zones can be identified. In the zone nearest the ultra-violet source ($A_v < 3$), temperatures in excess of 100 K are reached and the gaseous material becomes mainly atomic. In addition, carbon is mainly ionised. In the next zone ($A_v = 3-5$), the temperature drops to 100 K, the atomic carbon is mainly neutral, and some H_2 and CO exist. Finally, in the third and last zone, the temperature drops below 100 K and H_2 and CO dominate, although plenty of atomic O is still around. Tielens and Hollenbach (1985a, b) used their model to show that large abundances of interstellar C seen originally by Phillips and co-workers (Phillips and Huggins, 1981) could be accounted for by photodissociation along the outer edges of interstellar clouds. However, the more modern interpretation, due to Genzel and co-workers (Genzel *et al.*, 1988, Stutzki *et al.*, 1988) and based on C^+ as well as C observations, is to place the zones along the outer edges of cloud clumps spread throughout the cloud. Thus the original idea of Phillips and Huggins (1981) that the neutral C exists within clouds and not on their outer layers has survived but in a different form. It is also important to remember that large amounts of neutral C exist inside dense clumps as well if they are represented by the early-time predictions of pseudo-time-dependent models.

Another aspect of photodestruction processes in the chemistry of dense interstellar clouds is provided by studies of external galaxies such as the LMC and SMC where the metallicity is considerably lower than in our galaxy. In this case, the paucity of ultra-violet-shielding dust particles renders ultra-violet radiation more penetrating and "edge" effects in dense clouds more pervasive (Maloney and Black, 1988).

3.2.3 Dynamics

Nearly all model calculations have ignored the effects of molecular accretion onto dust grains because, as discussed in section 1.1, the accretion time-scale is much shorter than cloud lifetimes. Brown and Charnley (1990) have, however, developed a model which incorporates accretion and some simple surface hydrogenation reactions and applied it to TMC-1. They searched for a best-fit solution which occurred at a time of $\sim 10^6$ yr, by which time the grains had accreted substantial "ice-type" mantles. An alternative approach to the incorporation of accretion has been discussed by Tarafdar *et al.* (1985) who followed the hydrodynamic collapse of an initially diffuse cloud into a dense phase and incorporated a substantial chemistry which coupled to the hydrodynamics via an energy balance equation. These authors found that substantial chemical evolution could occur while the cloud was at low density, and accretion relatively unimportant. Because the free-fall time is proportional to $n^{-0.5}$, a cloud spends most of its lifetime at low densities. In more recent work (Prasad *et al.*, 1987), they have shown that in certain cases, collapse may be followed by an expansion. It is thus possible that two clouds which appear physically similar – in terms of density and temperature – may be chemically different because the one undergoing an expansion will have evolved chemically for a longer period than the one undergoing a contraction.

Charnley *et al.* (1988) have also included some dynamic effects in a global model of cloud chemistry. They, too, include a collapse phase and accretion, with some limited surface chemistry. The collapse of a particular clump is assumed to be halted by a wind from a newly-formed star which shocks the clump, releasing mantle material from the grains and initiating a high-temperature shock chemistry. Eventually this material cools, forming another dense clump of gas which begins to collapse, beginning the cycle once more. These authors find that a chemical limit cycle develops and that the chemistry never reaches a steady-state but rather appears to be young. Their initial model has now been extended to include nitrogen chemistry (Nejad, Williams and Charnley, 1990) and calculations which relax some of the original assumptions, such as the release of the entire mantle in the shock phase, are underway.

4 Towards complex molecules

The current gas phase model of Herbst and Leung (1989, 1990) contains hydrocarbons through nine carbon atoms in complexity. The question of whether or not gas phase reactions can produce much larger species in observable quantities is an intriguing one. However it is a question that cannot currently be answered unequivocally via pseudo-time-dependent models of dense interstellar clouds due to the lack of knowledge of rate coefficients for all processes germane to the synthesis of larger species and to the sheer number of rate processes that must be included in such detailed models. Herbst (1990) has recently attempted to avoid the latter difficulty with a very simplified treatment. Considering hydrocarbons only, he does not treat individual molecules but rather groups of hydrocarbons designated $A(m)$ and $B(m)$ where m is the number of carbon atoms, A represents neutral hydrocarbons, and B represents positively charged ones. In other words, the numbers of hydrogen atoms in the hydrocarbon molecules are not followed.

The synthesis of groups of hydrocarbons consisting of $m + n$ carbon atoms occurs via the processes

$$A(m) + B(n) \longrightarrow B(m+n) \tag{41}$$

$$A(n) + B(m) \longrightarrow B(m+n) \tag{42}$$

which can be thought of as radiative association reactions, as condensation reactions, or as C^+ and C insertion reactions, since the loss of hydrogen atoms is not followed. It is expected, though, that as the number of carbon atoms significantly exceeds 10–15 atoms, radiative association will dominate (Herbst and McEwan, 1990), and the rate coefficient for processes (41) and (42) has been designated k_{ra}. In the initial version of the model, it has been assumed that k_{ra} has no dependence on hydrocarbon size.

Once hydrocarbon ions with $m + n$ carbon atoms are produced, they are neutralized via dissociative reactions with electrons to produce neutrals with $m + n$ or fewer carbon atoms:

$$B(m+n) + e \longrightarrow A(m+n) + \text{H atoms} \tag{43}$$

$$B(m+n) + e \longrightarrow A(r) + A(p) + \text{H atoms} \tag{44}$$

where $r + p = m + n$. All dissociative recombination rate coefficients in the model are designated k_e and assumed to be equal; the fraction of such reactions leading to neutrals without breakage of carbon-carbon bonds (43) is designated f and is assumed to be independent of $m + n$.

After hydrocarbon neutrals of a given size are produced, they are depleted by ion-molecule reactions with a generic rate coefficient k_i. Other processes occurring in the model are depletion of both hydrocarbon ions and neutrals via reaction with neutral atomic oxygen atoms with generic rate coefficients k_{bo} and k_{ao} respectively.

The time-dependent abundances of groups of hydrocarbons $A(10 \leq m \leq 30)$ and $B(10 \leq m \leq 30)$ are found to be strong functions of the generic rate coefficients and branching ratios utilized. Because these parameters are not known, it is difficult at this stage to assess whether the more or the less efficient models calculated by Herbst (1990) are correct. In a particularly efficient model defined principally by $k_{ra} = 3 \times 10^{-9}\,\text{cm}^3\,\text{s}^{-1}$, $f = 0.90$, and $k_{ao} = 3 \times 10^{-14}\,\text{cm}^3\,\text{s}^{-1}$; that is, by rapid radiative association, non-disruptive dissociative recombination, and slow reactions between oxygen atoms and hydrocarbon neutrals on the average, the total fractional abundance of all neutral hydrocarbons with $m \geq 20$ reaches 3×10^{-8} within 10^6 yr. This total fractional abundance is close to the value of the 10^{-7} figure discussed for PAH's in dense clouds (Lepp and Dalgarno, 1988). Calculations are planned in which the assumption of fixed abundances for species in the model other than the large hydrocarbons will be removed, to determine whether the results are sensitive to this simplifying assumption. The ultimate question of whether or not gas phase reactions produce large abundances of very complex species within dense interstellar clouds will not be answered unambiguously however until much more information is known about the relevant rate coefficients.

5 The chemistries of star-forming regions

Molecular line observations of giant molecular clouds undergoing massive star-formation show that these objects are extremely complex and may contain outflows, photodissociation regions, shocked gas, and be very fragmented (Wilson and Walmsley, 1989, Genzel, this volume, p. 75). The Orion Molecular Cloud is the nearest such object and shows many of these characteristics. In particular, there appear to be three distinct regions in the vicinity of IRc2 – near the core of the cloud – in which the chemical compositions and abundances may not be determined by conventional gas-phase ion-neutral processes. These regions are the Hot Core, the Compact Ridge and the Plateau. We shall discuss each of these in turn and describe the chemistries thought to occur in each.

5.1 The hot core

A simple picture of this suggests that it is a small, dense ($n \sim 10^7\,\text{cm}^{-3}$, $N \sim 10^{22.5}\,\text{cm}^{-2}$) clump of gas heated radiatively to $T \sim 100 - 200\,\text{K}$ by IRc2. The gas is rich in small saturated molecules such as NH_3 and H_2S and contains much larger deuterium fractionation in species such as HDO and NH_2D than expected at such high temperatures (Blake et al., 1987, Minh et al., 1990, Jacq et al., 1988, 1990). It had previously been suggested (Sweitzer, 1978, Pauls et al., 1983) that the abundance anomaly in NH_3 was due to the evaporation of grain mantles rich in molecular ices, the mantles having been accreted during a previous cold phase of the gas. Brown, Charnley and Millar (1988) developed a numerical model incorporating gas phase chemistry and accretion with surface hydrogenation of accreted atoms and radicals. Subsequently Brown and Millar (1989a, b) extended these models to include the formation of mono- and multi-deuterated species. These calculations give a satisfactory description of the observed abundances in the Hot Core, although they calculate larger abundances of H_2O and NH_3 than observed. This discrepancy may be due to their adoption of a free-fall collapse in the cold clump of gas. If the collapse to a density of $\sim 10^7\,\mathrm{cm^{-3}}$ is retarded, then more O and N atoms can be processed into stable molecules such as CO, O_2 and N_2 before accretion dominates. As a result the amount of accreted O and N atoms decreases and less H_2O and NH_3 are formed by surface hydrogenation reactions. This hydrogenation on grain surfaces can also account for the H_2S observations (Minh et al., 1990), for which the hot core fractional abundance, $\sim 10^{-6}$, is 10^3 times larger than that in the quiescent ridge clouds. This can only occur, however, if the initial elemental sulphur abundance is $\sim 10^{-6}$, which is much larger, by ~ 50, than sulphur abundances adopted in dark cloud models (Millar and Herbst, 1990a). Such low abundances are adopted because current chemical models predict that essentially all sulphur is processed into CS and SO. However, there are a number of uncertainties in the neutral-neutral rate coefficients which link CS and SO. It is possible that, in the dark cloud models, a large elemental sulphur abundance could account for the CS and SO abundances while at the same time raising the calculated abundance of H_2S which is an order of magnitude less than that observed in dark clouds (Millar and Herbst, 1990a).

5.2 The compact ridge

The Compact Ridge is also small, hot, dense and situated close to IRc2. Like the Hot Core, it shows substantial abundance anomalies, particularly in molecules containing oxygen, and large deuterium fractionations. In addition, certain molecules such as $HCOOCH_3$, $(CH_3)_2O$ and CH_3CH_2OH are extremely abundant (Turner, 1990), while CH_3COOH is not detected. Blake et al. (1987) discussed a model of this source in which the injection of water from the IRc2 outflow enhanced the formation rate of methanol and, in turn, helped synthesise larger molecules and particular isomers of these molecules. Their calculations adopted a steady-state approach and neglected the wide range of chemical reactions which the injected H_2O can undergo. Subsequently, Millar, Herbst and Charnley (1990) performed a full pseudo-time-dependent calculation for this scenario. An important point to note is that conventional ion-neutral chemistry cannot account for the abundances of the complex oxygen-bearing molecules, especially now that the abundance estimates may need to be revised upwards by large factors due to optical depth effects (Turner, 1990). Millar et al. (1990) found that the injection of water did not lead to abundance enhancements in methanol and other species, since the water reacts primarily with H_3^+ and not CH_3^+. Following proton transfer and dissociative recombination with electrons, OH is formed. Millar et al. considered an alternative model in which methanol was injected and found reasonably successful agreement between their theoretical results and observation for several complex species. However, Turner (1990) has argued that several weak ethanol transitions are anomalously strong in Orion-KL, indicating that ethanol is very optically thick and that previous estimates of its abundance have been too small by a factor of 300 or so. His new abundance estimate of $\sim 10^{-7} - 10^{-6}$ is much larger than can be accounted for in the model of Millar *et al.* (1990). These optical depth effects should be checked through measurements of ^{13}C-ethanol.

If the methanol injection model is valid, then the source of the methanol is unlikely to be the wind from IRc2, since a fractional abundance $\sim 10^{-4}$ is required, but it may result from the evaporation of grain mantles. Although the chemistry by which methanol forms on grains is uncertain, methanol has been identified as a major constituent of grain mantles from infrared spectroscopy (Baas *et al.*, 1988). Although the conclusion must be tentative, it appears possible that the evaporation of mantles - albeit of differing composition - may determine the gas phase chemical abundances in both hot cores in Orion. Indeed, if Turner's abundance estimates are correct, it is likely that the most complex molecules in these sources are formed on grains (Brown, 1990).

5.3   The plateau

"Plateau" emission in Orion has both a high velocity ($|\Delta v| > 30\,\mathrm{km\,s^{-1}}$) and a low-velocity ($|\Delta v| \sim 18\,\mathrm{km\,s^{-1}}$) component, the former corresponding to a bipolar outflow centred to the north of IRc2 and the latter outflow centred on IRc2. Blake *et al.* (1987) have noted the presence of several molecules including SO, SO_2, SiO and H_2CO. It appears that "fragile" molecules such as H_2CO may be confined to small, dense clumps of gas entrained in the flow, while molecules such as SO and SO_2 exist in a thin shell of dense ($\sim 10^6\,\mathrm{cm^{-3}}$) gas where the outflow impacts on the molecular cloud. The molecular abundances can possibly be explained by shock-induced chemistry (Hartquist, Oppenheimer and Dalgarno, 1980, Leen and Graff, 1988); the abundances of H_2S and SiO are very difficult to explain in conventional ion-neutral models, particularly since sulphur and silicon depletions are likely to be large in dense clouds (see Blake *et al.*, 1987, for a more complete discussion of this issue).

Acknowledgements

Eric Herbst wishes to acknowledge the support of the National Science Foundation (US) for his programme in astrochemistry. T. J. Millar acknowledges the support of the Science and Engineering Research Council.

References

Adams, N.G., McIntosh, B.J. and Smith, D., 1990, Astron. Astrophys, **232**, 443.
Adams, N.G. and Smith, D., 1988 in *Rate Coefficients in Astrochemistry* ed. T.J. Millar and D.A. Williams (Kluwer:Dordrecht), p.173.
Adams, N.G., Smith, D., Giles, K. and Herbst, E., 1989, Astron. Astrophys. **220**, 269.
Allamandola, L.J., Tielens, A.G.G.M. and Barker, J.R., 1989, Astrophys. J. Suppl. **71**, 733.
Amano, T., 1988, Astrophys. J., **329**, L121.

Anicich, V.G. and Huntress, W.T. Jr., 1986, Astrophys. J. Suppl., **62**, 553.
Baas, F, Grim, R.J.A., Geballe, T.R., Schutte, W. and Greenberg, J.M., 1988, in *Dust in the Universe* ed. M.E. Bailey and D.A. Williams (Cambridge University Press), p. 55.
Barlow, S.E. and Dunn, G.H., 1987, Int. J. Mass Spectrom. Ion Proc., **80**, 227.
Barlow, S.E., Dunn, G.H. and Schauer, K., 1984, Phys. Rev. Letts., **52**, 902.
Bates, D.R., 1982, Proc. Roy. Soc. London, **A384**, 289.
Bates, D.R., 1986, Phys. Rev., **A34**, 1878.
Bates, D.R., 1989, Astrophys. J., **344**, 531.
Bates, D.R., 1990, private communication.
Bates, D.R. and Herbst, E., 1988a, in *Rate Coefficients in Astrochemistry* ed. T.J. Millar and D.A. Williams (Dordrecht:Kluwer), p. 17.
Bates, D.R. and Herbst, E., 1988b, in *Rate Coefficients in Astrochemistry* ed. T.J. Millar and D.A. Williams (Dordrecht:Kluwer), p. 41.
Black, J.H. and Dalgarno, A., 1977, Astrophys. J. Suppl., **34**, 405.
Blake, G.A., Sutton, E.C., Masson, C.R. and Phillips, T.G., 1987, Astrophys. J., **315**, 621.
Boissé, P., 1990, Astron.Astrophys., **228**, 483.
Brown, P.D., 1990, Mon. Not. Roy. Astron. Soc., **243**, 65.
Brown, P.D. and Charnley, S.B., 1990, Mon. Not. Roy. Astron. Soc., **244**, 432.
Brown, P.D., Charnley, S.B. and Millar, T.J., 1988, Mon. Not. Roy. Astron. Soc., **231**, 409.
Brown, P.D. and Millar, T.J., 1989a, Mon. Not. Roy. Astron. Soc., **237**, 661.
Brown, P.D. and Millar, T.J., 1989b, Mon. Not. Roy. Astron. Soc., **240**, 25P.
Brown, R.D., Cragg, D.M. and Bettens, R.P.A., 1990, Mon. Not. Roy. Astron. Soc., **245**, 623.
Brown, R.D. and Rice, E.H.N., 1986a, Mon. Not. Roy. Astron. Soc., **223**, 405.
Brown, R.D. and Rice, E.H.N., 1986b, Mon. Not. Roy. Astron. Soc., **223**, 429.
Charnley, S.B., Dyson, J.E., Hartquist, T.W. and Williams, D.A., 1988, Mon. Not. Roy. Astron. Soc., **231**, 269.
Chièze, J.P. and Pineau des Forêts, G., 1989, Astron. Astrophys., **221**, 89.
Chièze, J.P., Pineau des Forêts, G. and Herbst, E., 1990, Astrophys. J. (submitted).
Clary, D.C., 1985, Mol. Phys. **54**, 605.
Clary, D.C., Dateo, C.E. and Smith, D., 1990, Chem. Phys. Letts., **167**, 1.
Dalgarno, A. and Lepp, S., 1984, Astrophys. J., **287**, L47.
de Boisanger, C. and Chièze, J.P., 1990, Astron. Astrophys. (in press).
Flower, D.R., Monteiro, T.S., Pineau des Forêts, G. and Roueff, E., 1988, in *Rate Coefficients in Astrochemistry* ed. T.J. Millar and D.A. Williams (Kluwer:Dordrecht), p. 271.
Galloway, E.T. and Herbst, E., 1989, Astron. Astrophys, **211**, 413.
Galloway, E.T. and Herbst, E., 1990 (in preparation).
Genzel, R., Harris, A.I., Jaffe, D.T. and Stutzki, J., 1988, Astrophys. J., **332**, 1049.
Gerlich, D. and Kaefer, G., 1989, Astrophys. J., **347**, 389.
Graff, M.M., 1989, Astrophys. J., **339**, 239.
Gredel, R., Lepp, S., Dalgarno, A. and Herbst, E., 1989, Astrophys. J., **347**, 289.
Green, S. and Herbst, E., 1979, Astrophys. J., **229**, 121.
Guélin, M., Langer, W.D., Snell, R.L. and Wootten, H.A., 1977, Astrophys. J., **217**, L165.
Hartquist, T.W., Oppenheimer, M.A. and Dalgarno, A., 1980, Astrophys. J., **236**, 182.
Hawley, M., Mazely, T.L., Randeniya, L.K., Smith, R.S., Zeng, X.K. and Smith, M.A., 1990, Int. J. Mass Spectrom. Ion. Proc., **97**, 55.
Hawley, M. and Smith, M.A., 1989, J. Amer. Chem. Soc., **111**, 8293.
Henchman, M. and Paulson, J.F., 1989, J. Chem. Soc. Far. Trans., **85**, 1673.
Herbst, E., 1978, Astrophys. J., **222**, 508.
Herbst, E., 1982, Astrophys. J., **252**, 810.
Herbst, E., 1986, Astrophys. J., **306**, 667.
Herbst, E., 1987, Astrophys. J., **313**, 867.
Herbst, E., 1990, Astrophys. J., (in press).
Herbst, E. and Bates, D.R., 1988, Astrophys. J., **329**, 410.
Herbst, E., DeFrees, D.J. and Koch, W., 1989, Mon. Not. Roy. Astron. Soc., **237**, 1057.

Herbst, E., DeFrees, D.J. and McLean, A.D., 1987, Astrophys. J., **321**, 898.
Herbst, E., DeFrees, D.J., Talbi, D., Pauzat, F. Koch, W. and McLean, A.D., 1990, J. Chem. Phys. (submitted).
Herbst, E. and Klemperer, W., 1973, Astrophys. J., **185**, 505.
Herbst, E. and Knudson, S.K., 1981, Chem. Phys. **55**, 293.
Herbst, E. and Leung, C.M., 1986, Mon. Not. Roy. Astron. Soc., **222**, 689.
Herbst, E. and Leung, C.M., 1989, Astrophys. J. Suppl., **69**, 271.
Herbst, E. and Leung, C.M., 1990, Astron. Astrophys., **233**, 177.
Herbst, E. and McEwan, M.J., 1990, Astron. Astrophys. **229**, 201.
Herbst, E., Millar, T.J., Wlodek, S. and Bohme, D.K., 1989, Astron. Astrophys., **222**, 205.
Herd, C.R., Adams, N.G. and Smith, D., 1990, Astrophys. J., **349**, 388.
Irvine, W.M., Friberg, P., Hjalmarson, A., Ishikawa, S., Kaifu, N., Kawaguchi, K., Madden, S.C., Matthews, H.E., Ohishi, M., Saito, S., Suzuki, H., Thaddeus, P., Turner,B.E., Yamamoto, S and Ziurys, L.M., 1988, Astrophys. J., **334**, L107.
Jacq, T. Jewell, P.R., Henkel, C., Walmsley, C.M. and Baudry, A., 1988, Astron. Astrophys., **199**, L5.
Jacq. T., Walmsley, C.M. Henkel, C., Baudry, A. Mauersberger, R. and Jewell, P.R., 1990, Astron. Astrophys., **228**, 447.
Jura, M., 1974a, Astrophys. J., **197**, 575.
Jura, M., 1974b, Astrophys. J., **197**, 581.
Kirby, K.P., 1990, in *Molecular Astrophysics - A Volume Honouring Alexander Dalgarno* ed. T.W. Hartquist (Cambridge University Press), p. 159.
Langer, W.D. and Glassgold, A.E., 1990, Astrophys. J., **351**, 123.
Langer, W.D. and Graedel, T.E., 1989, Astrophys. J. Suppl., **69**, 241.
Langer, W.D., Graedel, T.E., Frerking, M.A. and Armentrout, P.B., 1984, Astrophys. J., **277**, 581.
Leen, T.M. and Graff, M.M., 1988, Astrophys. J., **325**, 411.
Leitch-Devlin, M.A. and Williams, D.A., 1985, Mon. Not. Roy. Astron. Soc., **213**, 295.
Lepp, S. and Dalgarno, A., 1988, Astrophys. J., **324**, 553.
Leung, C.M., Herbst, E. and Huebner, W.F., 1984, Astrophys. J. Suppl., **56**, 231.
Levine, R.D. and Bernstein, R.B., 1974, *Molecular Reaction Dynamics* (Oxford University Press).
Marquette, J.B., Rebrion, C. and Rowe, B.R., 1988, J. Chem. Phys., **89**, 2041.
Maloney, P. and Black, J.H., 1988, Astrophys. J., **325**, 389.
Millar, T.J., 1990a, Astron. Astrophys. (in press).
Millar, T.J., 1990b, in *Molecular Astrophysics - A Volume Honouring Alexander Dalgarno* ed. T.W. Hartquist (Cambridge University Press), p. 115.
Millar, T.J., Bennett, A. and Herbst, E., 1987, Mon. Not. Roy. Astron. Soc., **229**, 41P.
Millar, T.J., Bennett, A. and Herbst, E., 1989, Astrophys. J., **340**, 906.
Millar, T.J., DeFrees, D.J., McLean, A.D. and Herbst, E., 1988, Astron. Astrophys., **194**, 250.
Millar, T.J. and Herbst, E., 1990a, Astron. Astrophys., **231**, 466.
Millar, T.J. and Herbst, E., 1990b, Mon. Not. Roy. Astron. Soc., **242**, 92.
Millar, T.J., Herbst, E. and Charnley, S.B., 1990, Astrophys. J. (in press).
Millar, T.J., Leung, C.M. and Herbst, E., 1987, Astron. Astrophys., **183**, 109.
Millar, T.J. and Nejad, L.A.M., 1985, Mon. Not. Roy. Astron. Soc., **217**, 507.
Millar, T.J., Rawlings, J.M.C., Bennett, A., Brown, P.D. and Charnley, S.B., 1990, Astron. Astrophys. Suppl. (in press).
Minh, Y., Ziurys, L.M., Irvine, W.M. and McGonagle, D., 1990, Astrophys. J., **360**, 136.
Mitchell, J.B.A., 1990, Phys. Rep., **186**, 215.
Mitchell, J.B.A. and Guberman, S.L., 1989 *Dissociative Recombination: Theory, Experiment and Applications* (World Scientific Press: Singapore).
Nejad, L.A.M., Williams, D.A. and Charnley, S.B., 1990, Mon. Not. Roy. Astron. Soc., **246**, 183.
Omont, A., 1986, Astron. Astrophys., **164**, 159.
Oppenheimer, M. and Dalgarno, A., 1974, Astrophys. J., **187**, 231.
Pauls, T.A., Wilson, T.L., Bieging, J.H. and Martin, R.N., 1983, Astron. Astrophys., **124**, 123.
Phillips, T.G. and Huggins, P.J., 1981, Astrophys. J., **251**, 533.
Pineau des Forêts, G., Roueff, E. and Flower, D.R., 1990, Mon. Not. Roy. Astron. Soc., **244**, 668.

Prasad, S.S., Tarafdar, S.P., Villere, K.R. and Huntress, W.T. Jr. 1987, in *Interstellar Processes* ed. D.J. Hollenbach and H.A. Thronson (Reidel:Dordrecht), p. 631.
Rowe, B.R., 1988, in *Rate Coefficients in Astrochemistry* ed. T.J. Millar and D.A. Williams (Kluwer:Dordrecht), p. 135.
Rowe, B.R., 1990 in *Chemistry and Spectroscopy of Interstellar Molecules* ed. N. Kaifu, (Tokyo University Press, in press).
Sakimoto, K., 1981, Chem. Phys., **63**, 419.
Sen, A.D., Anicich, V.G. and McEwan, M.J., 1990, in *Chemistry and Spectroscopy of Interstellar Molecules* ed. N. Kaifu, (Tokyo University Press, in press).
Smith, D., McIntosh, B.J. and Adams, N.G., 1989, J. Chem. Phys., **90**, 6213.
Smith, I.W.M., 1988, in *Rate Coefficients in Astrochemistry* ed. T.J. Millar and D.A. Williams (Kluwer:Dordrecht), p. 103.
Smith, I.W.M., 1989, Astrophys. J., **347**, 282.
Snyder, L.E., Hollis, J.M., Buhl, D. and Watson, W.D., 1977, Astrophys. J., **218**, L61.
Sternberg, A., Dalgarno, A. and Lepp, S., 1987, Astrophys. J., **320**, 676.
Stutzki, J., Stacey, G.J., Genzel, R., Harris, A.I., Jaffe, D.T. and Lugten, J.B., 1988, Astrophys. J., **332**, 379.
Sweitzer, J.S., 1978, Astrophys. J., **225**, 116.
Talbi, D., 1990, private communication.
Tarafdar, S.P., Prasad, S.S., Huntress, W.T. Jr., Villere, K.R. and Black, D.C., 1985, Astrophys. J., **289**, 220.
Thorne, L.R., Anicich, V.G., Prasad, S.S. and Huntress, W.T. Jr., 1984, Astrophys. J., **280**, 139.
Tielens, A.G.G.M., 1983, Astron. Astrophys., **119**, 117.
Tielens, A.G.G.M. and Allamandola, L.J., 1987, in *Interstellar Processes* ed. D.J. Hollenbach and H.A. Thronson (Reidel:Dordrecht), p. 397.
Tielens, A.G.G.M. and Hollenbach, D.J., 1985a, Astrophys. J., **291**, 722.
Tielens, A.G.G.M. and Hollenbach, D.J., 1985b, Astrophys. J., **291**, 747.
Turner, B.E., 1990, Astrophys. J. Suppl. (submitted).
van Dishoeck, E.F., 1988, in *Rate Coefficients in Astrochemistry* ed. T.J. Millar and D.A. Williams (Kluwer:Dordrecht), p. 49.
van Dishoeck, E.G. and Black, J.H., 1988, in *Rate Coefficients in Astrochemistry* ed. T.J. Millar and D.A. Williams (Kluwer:Dordrecht), p. 209.
Watson, W.D., 1973, Astrophys. J., **182**, L73.
Weston, R.E. Jr. and Schwarz, H.A., 1972, *Chemical Kinetics* (Englewood Cliffs, N.J. : Prentice-Hall).
Williams, D.A., 1988, in *Rate Coefficients in Astrochemistry* ed. T.J. Millar and D.A. Williams (Kluwer:Dordrecht), p. 281.
Wilson, T.L. and Walmsley, C.M., 1989, Astron. Astrophys. Rev., **1**, 141.
Winnewisser, G. and Herbst, E., 1987, Topics in Current Chemistry, **139**, 119

CRESUS studies of ion-molecule reactions

J.-B. MARQUETTE

LABORATOIRE D'AÉROTHERMIQUE DU CNRS, 4TER ROUTE DES GARDES,
92190 MEUDON, FRANCE

1 Introduction

Since the early and extremely successful developments of the flowing afterglow, FA, and the subsequent selected ion flow tube, SIFT, respectively by Ferguson and his collaborators (1969), and by Smith and Adams (1987), various experiments were designed to study gas-phase ion-molecule reactions at kinetic temperatures or energies comparable to those pertaining in dense interstellar clouds (a few tens of Kelvins at most). For example, the first direct determination of the rate coefficient of a radiative association reaction, namely

$$CH_3^+ + H_2 \rightarrow CH_5^+ + h\nu \tag{1}$$

was obtained at 13 K by Dunn and his group by using a cooled RF ion trap (Barlow *et al.*, 1984a, b). More recently, a guided beam apparatus was developed by Gerlich and his collaborators (1987) to obtain reaction cross sections at kinetic energies in the centre-of-mass frame in the *meV* domain.

Among flow reactor experiments it is worth citing here the pulsed supersonic free jet apparatus developed by Mark Smith and his group in Tucson (Hawley *et al.*, 1990). Although this apparatus is not working under truly thermal energy conditions, it is possible to derive from experimental data bulk rate coefficients down to the astoundingly low translational temperature of 0.3 K. Indeed, the cooling principle of this technique is similar to that of the CRESUS experiment developed in Meudon (Rowe *et al.*, 1989). This acronym is a French translation for "Reaction Kinetics in Uniform Supersonic Flow with Selection". The apparatus is briefly described below and some results and future directions of investigation are exposed afterwards.

2 Experiment

The main originality of the CRESUS experiment is to yield very low temperatures by the isentropic expansion of a buffer gas (helium in the present case) through a contoured nozzle. If the nozzle profile is suitably calculated the flow is uniform, so that the kinetic analysis of the reaction under study is almost identical to that used with FA and SIFT flow tube techniques. The first version of the experiment (CRESU configuration) was indeed a kind of "supersonic flowing afterglow" since primary ions were produced by direct ionisation of the buffer

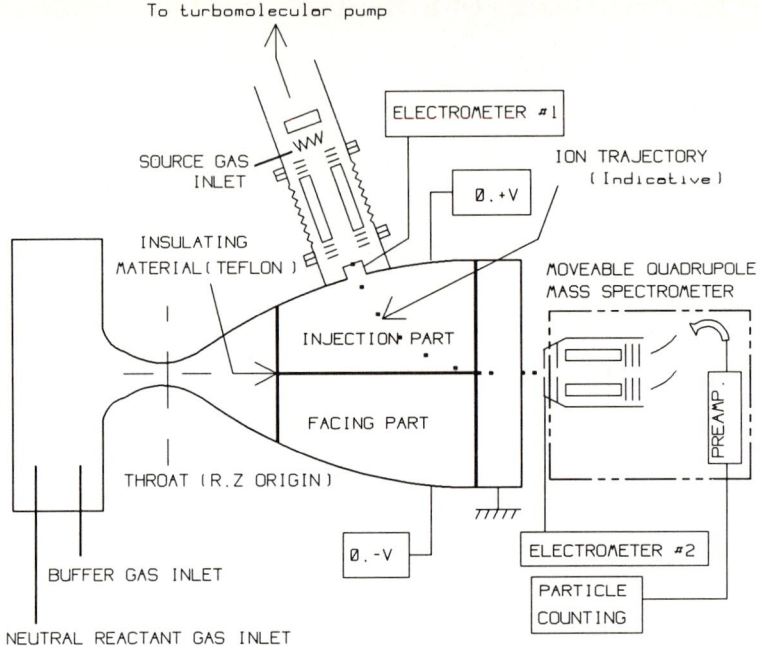

Figure 1: Schematic view of the CRESUS apparatus.

gas, close downstream from the nozzle exit, and addition of a convenient parent gas. Some results were obtained in that way in the 8–163 K range (Rowe and Marquette, 1987). However, to gain in flexibility, it appeared highly desirable to replace the 20 keV electron beam used as the ion source by a more convenient device. Thus, the same kind of ion injector as in the SIFT technique was incorporated. The CRESUS apparatus being described in detail elsewhere (Rowe et al., 1989), only basic features are briefly explained below.

Figure 1 gives a schematic sketch of the experiment. Owing to the lack of diffusion in uniform supersonic flows the ions need to be drifted nearly radially from the exit hole of the injector toward the flow axis.

To properly combine the drift and flow velocities the potential applied to the nozzle wall (named "injection part" in Figure 1) is adjusted to the mobility properties of the ion under study. As in previous experiments (Rowe and Marquette, 1987) data are obtained by a moveable quadrupole mass spectrometer associated to a channeltron multiplier and a particle counting device.

3 Results

Various ions have already been used to measure rate constants of both binary and ternary processes (Rowe et al., 1989). In the present study first attempts have been made to determine branching ratios at low temperature (70 K). Figure 2 presents typical results obtained on the reaction $He^+ + O_2$ which is known

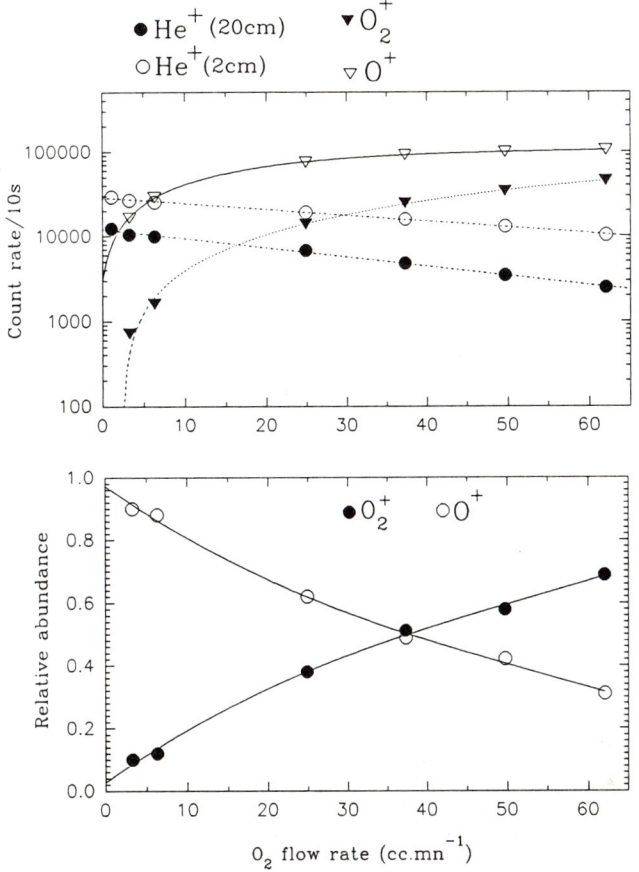

Figure 2 CRESUS data for the reaction of He$^+$ with O$_2$ at 70 K and product distribution plot.

to have two ionic products at 300 K : O$^+$(97%), and O$_2^+$ (3%) (Adams and Smith, 1976). The present data yield a rate coefficient of 8.0×10^{-10} cm^3 s^{-1}, a value in good agreement with previous determinations within the uncertainty range ($\pm$ 30%). Also important is the fact that the distribution of ionic products at 70 K is exactly the same as at room temperature. This gives great support to the present experimental technique. Further studies are in progress on reactions having two exit channels (He$^+$ + N$_2$, C$^+$ + O$_2$) or more (C$^+$ + NH$_3$). Also planned are studies of some processes involving isotope exchanges and/or polyatomics ions.

References

Adams, N.G. and Smith, D., 1976, J. Phys. B, **9**, 1439.
Barlow, S.E., Dunn, G.H. and Schauer, M., 1984a, Phys. Rev. Letters, **52**, 902.
Barlow, S.E., Dunn, G.H. and Schauer, M., 1984b, Phys. Rev. Letters, **53**, 1610.
Ferguson, E.E., Fehsenfeld, F.C. and Schmeltekopf, A.L., 1969, Adv. Atom. Mol. Phys., **5**, 1.

Gerlich, D., Disch, R. and Scherbarth, S., 1987, J. Chem. Phys., **87**, 350.
Hawley, M., Mazely, T.M., Randeniya, L.K., Smith, R.S., Zeng, X.K. and Smith, M.A., 1990, Int. J. Mass Spectrom. Ion Proc., **97**, 55.
Rowe, B.R. and Marquette, J.B., 1987, Int. J. Mass Spectrom. Ion Proc., **80**, 239.
Rowe, B.R., Marquette, J.B. and Rebrion, C., 1989, J. Chem. Soc. Faraday Trans. 2, **85**, 1631.
Smith, D. and Adams, N.G., 1987, Adv. Atom. Mol. Phys., **24**, 1.

Molecular hydrogen line emission from C-shocks

M. D. SMITH[1,2], P. W. J. L. BRAND[3] and A. MOORHOUSE[4]

Abstract   Emission lines of shocked H_2 are observed from within bipolar outflows and are often associated with bow-like arcs and jets. We have developed a model to test the hypothesis that this emission arises from C-shocks. The model is tuned to predict the emission from both planar and curved shocks. We approach one of the outstanding problems in this field: the extremely wide H_2 lines observed within powerful bipolar sources. Line widths reach at least 140 km s^{-1} in the OMC-1 outflow (e.g. Brand et al., 1989). Numerous line ratios and smooth profile shapes, accurately measured in OMC-1, must also be explained. We outline a model – the Shock Absorber – which contains all the observed features. We consider bow shocks moving through dense gas of low ionization and high magnetic field. In fact, the field must be sufficiently strong to cushion the C-shock impact. The possibilities for a high magnetic field in active star-forming environments are discussed.

Fast shocks destroy molecules. How fast can H_2 be accelerated without wholesale dissociation? It has been shown that H_2 is dissociated for shock speeds exceeding a breakdown value v_b of about 25 to 50 km s^{-1} for hydrodynamic J-shocks (Kwan, 1977) and MHD C-shocks (Draine et al., 1983, Smith and Brand, 1990a) respectively. In C-shocks, dissociation occurs either directly at high temperature or via self-ionization at high ion-neutral streaming speeds. The higher breakdown velocity is a result of the broad shock transition region over which cooling processes are effective (see Figure 1a). We have found that under quiescent cloud conditions:

1. Planar C-shocks produce narrow H_2 lines (v < 30 km s^{-1} (FWHM), Smith and Brand, 1990b).
2. Bow shocks, conical and other curved shocks can produce somewhat wider lines < 50 km s^{-1}, Smith and Brand, 1990c). Much of the emission is from warm undeflected gas.
3. Observed Herbig-Haro profiles (Zinnecker et al., 1989) are consistent with the H_2 profile shapes and widths from bow shocks and planar shocks (Smith and Brand, 1990c). Further high quality data on all aspects, including velocity imaging and mapping is needed.
4. H_2 line ratios depend essentially only on the shock shape, H_2O abundance and the density. Thus they are useful as C-shock (and J-shock) diagnostics (Smith et al., 1990a).

A new type of flow pattern emerges under "active" cloud conditions, i.e. where the magnetic field is amplified. The Alfvén speed may far exceed 10 km s^{-1} and the shock Alfvén number, M_A, is thus reduced (Figure 1b). The neutrals still

[1] DEPARTMENT OF PHYSICS, UNIVERSITY OF DURHAM, DURHAM DH1 3LE, U.K.
[2] NEW ADDRESS: INTERNATIONAL SCHOOL FOR ADVANCED STUDIES, STRADA COSTIERE 11, I-34014 TRIESTE-MIRAMARE, ITALY
[3] DEPARTMENT OF ASTRONOMY, UNIVERSITY OF EDINBURGH, ROYAL OBSERVATORY, EDINBURGH EH9 3HJ, U.K.
[4] DEPARTMENT OF MATHEMATICS, UMIST, MANCHESTER M60 1QD, U.K.

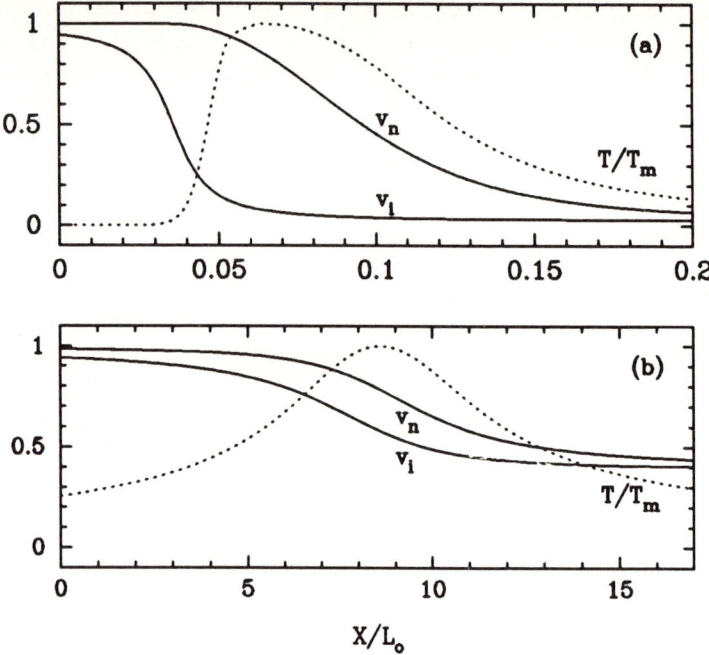

Figure 1: Sections through shocks with M_A = (a) 24.4 and (b) 2.1. Note the crucial difference in streaming speed (v_n-v_i). For details see Smith *et al.* (1990b)

lag behind the ions but only by a fraction of the shock velocity, the acceleration length scales being related by $L_i = L_n/M_A$. Therefore, in active regions, the ion-neutral streaming speed is lower for a given shock speed. High breakdown speeds are now found since it is the streaming speed which determines the dissociation rate (through both heating and impact-ionisation).

Molecular hydrogen can be accelerated to high speeds. Smith *et al.* (1990b) have shown that H_2 profiles from Peak 1 of OMC-1 can be interpreted as from a paraboloidal bow shock of velocity 250 km s^{-1} moving through dense gas in which $v_A = 46$ km s^{-1} (i.e. a field of 30 mG for an H_2 density of 10^6 cm^{-3}). Line ratios are quite well modelled for an H_2O abundance of $< 10^{-4}$ per hydrogen atom.

Are such high Alfvén speeds plausible? A number of arguments in favour are presented first.

1. Active star-forming regions do not proceed through a series of magnetic-gravitational equilibria. Rather, they can be shock-dominated. The shocks enhance the Alfvén speed through compression. Strong fields are predicted for H_2O masering clumps associated with dense post-shock layers (Elitzur *et al.*, 1989).

2. Ambipolar diffusion separates high and low v_A regions. MHD centrifugally-driven winds must sustain very high Alfvén speeds in order to drive the CO outflows.

3. High fields may be necessary to provide support for molecular clouds near

the Galactic centre (Güsten, 1989). High fields or high electron densities are required to explain the Faraday rotation measures of background sources in L1551 (Simonetti and Cordes, 1986).

Problem areas are as follows.

1. UV radiation from the J-type shock at the bow's leading edge pre-ionizes the gas approaching the oblique C-type bow tail section. Metal depletion onto grains and/or dust absorption are then required to maintain a low ionization level (Smith and Brand, 1990c).
2. If the ionization level is low far upstream, a C-type shock results unless (local) UV radiation produces a sufficiently ionized precursor. The precursor then forces the shock to be J-type, which, self-consistently, produces the high UV flux. Hence J- or C- type is sensitive to the shock history.
3. The high-v_A pre-shock medium proposed for OMC-1 cannot be contained by gravity. Instead, we relate it to the CO outflow or the cloud compression caused by the outflow. A second, higher-velocity, outflow drives the fast shocks. Two such distinct velocity components are quite common to bipolar sources, with the faster component often well-collimated.

References

Brand, P., Toner, M.P., Geballe, T.R. and Webster, A.S., 1989, Mon. Not. Roy. Astron. Soc., **237**, 1009.
Draine, B.T., Roberge, W.G. and Dalgarno, A., 1983, Astrophys. J., **264**, 485.
Elitzur, M., Hollenbach, D.J. and McKee, C.F., 1989, Astrophys. J., **346**, 983.
Güsten, R., 1989. In IAU Symp. 136 *The Galactic Centre*, ed. M. Morris, (Kluwer Academic, Dordrecht), p. 89.
Kwan, J., 1978, Astrophys. J., **216**, 713.
Simonetti, J. and Cordes, J., 1986, Astrophys. J., **303**, 659.
Smith, M.D. and Brand, P.W.J.L., 1990a, Mon. Not. Roy. Astron. Soc., **242**, 495.
Smith, M.D. and Brand, P.W.J.L., 1990b, Mon. Not. Roy. Astron. Soc., **243**, 498.
Smith, M.D. and Brand, P.W.J.L., 1990c, Mon. Not. Roy. Astron. Soc., **245**, 108.
Smith, M.D., Brand, P.W.J.L. and Moorhouse, A., 1990a and 1990b, Mon. Not. Roy. Astron. Soc., (in press).
Zinnecker, H., Mundt, R., Geballe, T.R. and Zealey, W.J., 1989, Astrophys. J., **342**, 337.

Hydromagnetic wave propagation, ambipolar diffusion and chemical fractionation in cold dark clouds

S.B. CHARNLEY and W.G. ROBERGE

PHYSICS DEPARTMENT, RENSSELAER POLYTECHNIC INSTITUTE, TROY, NEW YORK 12180, U.S.A.

Hydromagnetic waves are possibly a major source of line broadening in molecular clouds, and are also important in transporting angular momentum during gravitational collapse. Ambipolar diffusion between ions and neutrals determines the loss of magnetic flux from magnetically supported clumps of gas and thus plays a critical role in the initial stages of low-mass star formation. However, no *direct* evidence for the existence of hydromagnetic waves or the presence of ambipolar diffusion has been obtained to date.

As a representative model of MHD disturbances in dense clouds, we consider a Gaussian packet of planar shear Alfvén waves propagating parallel to a uniform magnetic field. The hydrodynamical structure and evolution of the wave packet are governed by Maxwell's equations, plus dynamical equations for the coupled fluids of neutral particles, ions, electrons, and dust grains (*e.g.* Draine, 1986, Mouschovias, 1987). The *linearized* forms of these equations give us hydrodynamical solutions for small amplitudes A (Roberge and Hanany, 1990).

Appreciable ion-neutral drift speeds can be attained in a propagating isothermal wave and these motions may drive some endothermic chemical reactions at enhanced rates, in a manner analogous to that which occurs in the magnetic presursors of C-type MHD shock waves (*e.g.* Shull and Draine, 1987). However, as the typical drift speeds of a few times 0.1 km s^{-1} are lower than those occurring in C-shocks (~ 5–10 km s^{-1}), only the rate coefficients of reactions which have small temperature barriers will be significantly affected. Four potentially important systems involve the reactions of (1) H_2D^+, (2) CH_2D^+, (3) C_2HD^+ and (4)N^+ with H_2, where reactions (1)–(3) fractionate deuterium and (4) leads to the formation of NH_3. Millar, Bennett and Herbst (1989) have shown that the fractionation chemistry is, as expected, sensitive to temperature and exhibits a strong time-dependence in the abundance ratios, $R(XH) = n(XD)/n(XH)$, of several species. We have estimated how the rate coefficient of each of the reactions (1)–(4) may vary within a wave packet. We used the rate coefficients listed in the compilation of Millar *et al.* for the rate coefficients of reactions (2), (3) and (4) respectively, and have followed Herbst (1982) to calculate the temperature dependence of (1). Figure 1 shows the rate coefficients of reactions (1)–(4) as a function of time; these may vary substantially in a hydromagnetic wave. The timescale over which a parcel of gas is affected by such a wave is greater than the chemical timescales of these reactions, and consequently the molecular chemistry may be altered by its presence.

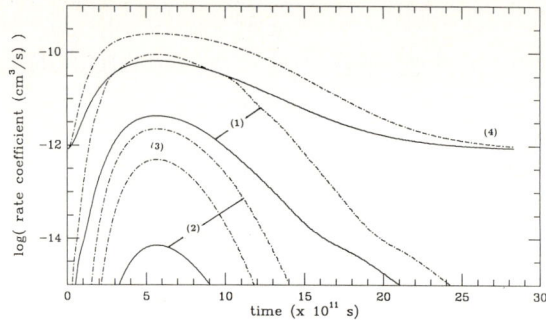

Figure 1. Variation of the rate coefficients of reactions (1)–(4) as a function of time at the centre of a Gaussian Alfvén wave packet, in which the field initially experiences a disturbance $\delta B = A B_0 \exp(-z^2/2 L_{ni}^2)$, where L_{ni} is the slowing-down length for a neutral particle streaming through the ions at the Alfvén speed (Spitzer, 1978). The density, temperature and fractional ionization of the gas were taken to be $10^4\,\mathrm{cm}^{-3}$, $12\,\mathrm{K}$ and 10^{-7} respectively. Solid curves are for the case where the maximum drift velocity in the wave is $0.52\,\mathrm{km\,s}^{-1}$; broken curves are for the $0.84\,\mathrm{km\,s}^{-1}$ case.

It is possible to speculate as to some possible effects. For example, destruction of H_2D^+ may be efficient in such a wave, leading to reduced fractionation in N_2H^+, HCO^+ and ammonia, whilst retaining high fractionation in H_2CO and HCN. The enhanced formation rate of NH^+ may also affect $R(NH_3)$. The results of Millar et al. show that $R(H_2CO)$ only varies slightly in time and is also little affected in warm gas below 70K. Thus, one possible chemical diagnostic of the presence of hydromagnetic waves in dark clouds could be low values of $R(NH_3)$ in gas for which $R(H_2CO)$ is high, and which is ostensibly cold (~ 10 K).

Acknowledgements

SBC was partially supported by the US Air Force under Grant AFOSR-89-0104. WGR is grateful to the Dudley Observatory for a Career Development Award.

References

Draine, B.T., 1986, Mon. Not. Roy. Astron. Soc., **220**, 133.
Herbst, E., 1982, Astron . Astrophys., **111**, 76.
Millar, T.J., Bennett, A. and Herbst, E., 1989, Astrophys. J. **340**, 906.
Mouschovias, T.Ch., 1987, in *Physical Processes in Interstellar Clouds*, ed. G.E. Morfill and
   M. Scholer, (Reidel), p. 491.
Roberge, W.G. and Hanany, S., 1990, (in preparation).
Shull, J.M. and Draine, B.T., 1987, in *Interstellar Processes*, ed. D.J. Hollenbach and
   H.A. Thronson Jr., (Reidel), p.283.
Spitzer, L. Jr., 1978, *Physical Processes in the Interstellar Medium*, (J. Wiley, New York).

Fluorescent HD emission from photon-dominated regions

AMIEL STERNBERG

MAX-PLANCK INSTITUT FÜR EXTRATERRESTRISCHE PHYSIK,

D-8046 GARCHING, FRG

Abstract Infrared fluorescent emission lines of the molecular isotope HD are produced in photon-dominated regions at the surfaces of molecular clouds exposed to far-ultraviolet radiation. The strongest emission lines may be detectable from dense clumps containing hydrogen densities equal to $\sim 5 \times 10^6 \, \text{cm}^{-3}$ which are exposed to intense radiation fields.

Introduction. Photon-dominated regions (PDR's) at the surfaces of molecular clouds exposed to far-ultraviolet (FUV) radiation are sources of a wide range of atomic and molecular line emission. In particular, PDRs emit near-infrared (1–5 µm) fluorescent lines of molecular hydrogen (H_2) which are excited by the direct absorption of FUV photons (Black and Dalgarno, 1976, Black and van Dishoeck, 1987, Dinerstein et al., 1988, Hasegawa et al., 1987, Hayashi et al., 1985, Israel et al., 1989, Sternberg, 1988, Sellgren, 1986). Dense PDRs which are heated by the FUV radiation may also be the source of thermal H_2 emission, particularly in the vicinity of intense UV fields (Sternberg and Dalgarno, 1989, Sternberg, 1989).

Many trace elements, including deuterium nuclei, are mixed with the interstellar hydrogen gas. The deuterium abundance is an important cosmological parameter. The detection (Wright and Morton, 1979) of ultraviolet HD absorption lines in diffuse clouds with small visual extinctions suggests that HD molecules are excited by FUV photons in PDRs with large visual extinctions, including those which radiate detectable H_2 emission lines. In this paper I describe computations (Sternberg, 1990) of the infrared fluorescent HD emission spectra that are produced in cold ($T \lesssim 500 \, \text{K}$) PDRs with hydrogen gas density n less than $\sim 10^7 \, \text{cm}^{-3}$.

Calculations. The ultraviolet radiation longward of 91.2 nm that penetrates into photon dominated regions of molecular clouds produces vibrationally excited HD and H_2 molecules by identical two-step pumping mechanisms. UV photon absorption is followed by transitions to excited vibrational levels of the ground electronic state. The H_2 and HD ultraviolet absorption lines are typically separated by ~ 0.1 nm. The intensities of the (optically thin) fluorescent HD emission lines are proportional to the total column densities of UV pumped molecules produced in the PDR. The local volume density of HD molecules varies with cloud depth due to changes in the rates of molecular formation and destruction. HD is formed (Black and Dalgarno, 1973, Watson, 1973, see Figure 4) by association on grain surfaces as well as by sequences of ion-molecule reactions initiated by cosmic-ray ionization of atomic and molecular hydrogen. HD is destroyed by UV photodissociation, gas phase reactions, and collisions with cosmic rays. Steady state plane parallel models may be employed to compute the local chemical abundances and populations of vibrationally excited HD (and H_2) molecules. Integration of the vibrationally excited level populations through the cloud yields excited state column densities and the fluorescent emission line intensities.

The intensities of the fluorescent HD emission lines depend on n, the intensity of the incident UV radiation, the fractional abundance of deuterium nuclei, the effective dust UV continuum cross section, the grain surface molecular formation rate coefficient, and the cosmic ray ionization rate. n ranges from $\sim 10^2\,\mathrm{cm}^{-3}$ in diffuse clouds to $\sim 10^8\,\mathrm{cm}^{-3}$ in star-forming regions. The deuterium abundance, x_D, is of order 5×10^{-5} in the solar neighbourhood and in many diffuse and dense molecular clouds (Dalgarno and Lepp, 1984). In some environments, such as in the vicinity of galactic nuclei, the deuterium abundance may be substantially enhanced by local production processes (Boyd et al., 1989). The UV intensity scaling factor χ ranges from ~ 1 in the average interstellar field to $> 10^5$ near hot stars or intense sources of nonthermal UV radiation. The effective grain UV continuum absorption cross section $\sigma = 1.0 \times 10^{-21}(\delta/\delta_0)\,\mathrm{cm}^2$ depends on the gas-to-dust mass ratio δ (with typical value $\delta_0 = 100$) and a factor k (of order unity) which depends on uncertain grain scattering properties. The grain surface molecular formation rate coefficient is $R = 3.0 \times 10^{-18}(\delta/\delta_0)\,T^{0.5}y_F\,\mathrm{cm}^3\,\mathrm{s}^{-1}$ where y_F is an uncertain efficiency factor with which HD molecules are formed on and ejected from the grain surfaces. The cosmic ray ionization rate is usually parameterized by ζ, the cosmic ray destruction rate of H_2, which is $\sim 5 \times 10^{-17}\,\mathrm{s}^{-1}$ in diffuse clouds and quiescent molecular clouds. It may be orders of magnitude larger near X-ray sources, supernova remnants and galactic nuclei.

In cold ($T < 500\,\mathrm{K}$) PDRs with $n \lesssim 10^7\,\mathrm{cm}^{-3}$ the relative intensities of the HD fluorescent emission lines are determined primarily by the branching ratios of the radiative rovibrational transitions which are molecular constants. The *relative* HD line intensities are therefore insensitive to the cloud parameters n, x_D, χ, R and ζ. However, the efficiency with which incident UV radiation is absorbed by HD molecules (rather than by dust grains) and converted into fluorescent emission lines depends critically on the cloud parameters and therefore so do the *absolute* intensities of the HD emission lines.

In PDRs the deuterium is primarily atomic or bound in HD molecules. For fixed values of σ and x_D it may be shown (Sternberg, 1990) that to a very good approximation the fraction of deuterium that is bound in HD molecules at any cloud depth is a function of two dimensionless ratios, $\alpha \equiv \chi D/Rn$ (where $D \sim 5 \times 10^{-11}\,\mathrm{s}^{-1}$ is the unattenuated molecular photodissociation rate in the average interstellar field) and $\beta \equiv \zeta/Rn$. It follows (Sternberg, 1990) that the efficiency, Y_{HD}, with which the incident UV radiation is absorbed by HD molecules (rather than by dust) and converted in fluorescent line emission also depends on these two parameters and that the intensity of a fluorescent emission line scales as

$$\frac{I(\chi, n, R, \sigma)}{I(\chi_0, n_0, R_0, \sigma_0)} = \frac{\chi}{\chi_0}\frac{Y(\alpha, \beta)}{Y(\alpha_0, \beta_0)}. \tag{1}$$

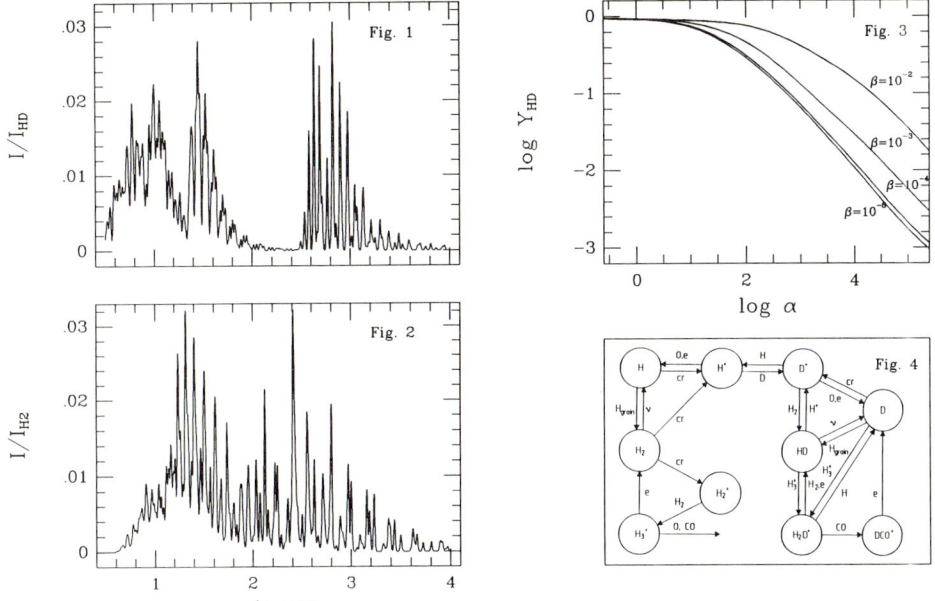

The efficiency $Y_{\rm HD}$ is displayed in Figure 3 as a function of α for various values of β. The curves are normalized to unity at the maximum of $Y_{\rm HD}$. When α is small the deterium is molecular throughout most of the PDR and the intensity of fluorescent emission is proportional to χ and independent of n, R and ζ. When α is sufficiently large the deuterium is primarily atomic throughout most of the PDR. $Y_{\rm HD}$ is then approximately proportional to $1/\alpha$ and the intensity of fluorescent emission becomes insensitive to χ. When β is small the formation of HD is dominated by grain surface reactions. $Y_{\rm HD}$ is then independent of β and when α is large the intensity of fluorescent emission is proportional to Rn and independent of ζ. When β is large the formation of HD is dominated by cosmic-ray driven gas phase reactions. $Y_{\rm HD}$ is then proportional to β and when α is large the intensity of fluorescent emission is proportional to ζ and independent of Rn. All of these regimes exist in interstellar environments.

The fluorescent HD emission line spectrum (figure 1) was computed for a PDR with $n = 10^5\,{\rm cm}^{-3}$, $X = 10^4$, $R = 3 \times 10^{-17}\,{\rm cm}^3\,{\rm s}^{-1}$, $\zeta = 5 \times 10^{-17}\,{\rm s}^{-1}$, $\sigma = 1.9 \times 10^{-21}\,{\rm cm}^2$ and $x_D = 5 \times 10^{-5}$. The line intensities are displayed as fractions of the total HD fluorescent intensity of $I_{\rm HD} = 1.9 \times 10^{-6}\,{\rm ergs\,s}^{-1}\,{\rm cm}^{-2}\,{\rm ster}^{-1}$. The value of $I_{\rm HD}$ for other values of the parameters χ, n, R and ζ may be estimated using equation (1). For comparison the H$_2$ fluorescent spectrum (figure 2) was computed for a cloud with $n = 10^3\,{\rm cm}^{-3}$, $\chi = 10^2$, $R = 3 \times 10^{-17}\,{\rm cm}^3\,{\rm s}^{-1}$ and $\sigma = 1.9 \times 10^{-21}\,{\rm cm}^2$. The line intensities are displayed as fractions of the total fluorescent H$_2$ intensity $I_{\rm H_2} = 7.4 \times 10^{-5}\,{\rm erg\,s}^{-1}\,{\rm cm}^{-2}\,{\rm ster}^{-1}$ produced in such a cloud. The differences between the *shapes* of the HD and H$_2$ spectra are due to the isotopic shifts of the HD rovibrational energy levels and the dipole (rather than quadrupole) selection rules which govern the HD radiative vibrational transitions.

It should be noted that the intensity ratio of HD and H_2 fluorescent emission from a PDR may, for various reasons, often greatly exceed the deuterium abundance x_D in them. When the gas is molecular the H_2 absorption lines become much more optically thick than those of the less abundant HD molecules so that the number of UV photons absorbed per HD molecule is much larger than the number of photons absorbed per H_2 molecule in the PDR. When the gas is primarily atomic the HD/H_2 density and fluorescent intensity ratios can become larger than x_D if, due to gas phase HD production processes, the HD formation rate exceeds the H_2 formation rate. Finally, because the HD radiative vibrational transition rates are about 100 times larger than those of H_2 a range of gas densities exists ($10^5 \, \text{cm}^{-3} < n < 10^7 \, \text{cm}^{-3}$) for which the fluorescent H_2 emission is collisionally quenched whereas the fluorescent HD emission is not. A combination of this latter effect with one of the first two can result in an HD/H_2 fluorescent emission intensity ratio that exceeds x_D by many orders of magnitude.

It follows from equation (1) and Figure 1 that the brightest HD lines in the H-band ($1.45-1.8 \, \mu m$) should be detectable from dense PDRs $n \sim 5 \times 10^6 \, \text{cm}^{-3}$ which are exposed to UV fields with $\chi \gtrsim 10^3$. Such regions may exist in the Orion bright bar, the Galactic centre and planetary nebulae and may be the source of the thermal H_2 emission which has been observed in these objects. The intensities of the fluorescent component of the H_2 and HD emission lines from these objects may be of comparable intensity.

References

Black, J.H. and Dalgarno, A., 1973, Astrophys. J., **184**, L101.
Black, J.H. and Dalgarno, A., 1976, Astrophys. J., **203**, 132.
Black, J.H. and van Dishoeck, E.F., 1987, Astrophys. J., **322**, 412.
Boyd, R.N., Ferland, G.J. and Schramm, D.N., 1989, Astrophys. J., **366**, L1.
Dalgarno, A. and Lepp, S., 1984, Astrophys. J., **287**, L47.
Dinerstein, H.L., Lester, D.F., Carr, J.S. and Harvey, P.M., 1988, Astrophys. J., **327**, L27.
Hasegawa, T., Gatley, I., Garden, R.P., Brand, P.W.J.L., Ohishi, M., Lightfoot, J.M., Hayashi, M. and Kaifu, N., 1987, Astrophys. J., **318**, L77.
Hayashi, M., Hasegawa, T., Gatley, I., Garden, R.P. and Kaifu, N., 1985, Mon. Not. Roy. Astron. Soc., **215**, 31P.
Israel, F.P., Hawarden, T.G., Wade, R., Geballe, T.R. and van Dishoeck, E.F., 1989, Mon. Not. Roy. Astron. Soc., **236**, 89.
Sellgren, K., 1986, Astrophys. J., **305**, 399.
Sternberg, A., 1988, Astrophys. J., **322**, 400.
Sternberg, A., 1989, in *Proc. 22nd Eslab Symposium on Infrared Spectroscopy in Astronomy* ed. B.H. Kaldeich, (ESA, Paris), p. 269.
Sternberg, A., 1990, Astrophys. J., **361**, 121.
Sternberg, A. and Dalgarno, A., 1989, Astrophys. J., **338**, 197.
Watson, D.W., 1973, Astrophys. J., **182**, L73.
Wright, E.L. and Morton, D.C., 1979, Astrophys. J., **227**, 483.

Models of the gas-grain interaction in dark clouds

P. D. BROWN[1] and S. B. CHARNLEY[2]

In the absence of an efficient desorption process, the heavy gas phase component of molecular clouds is expected to condense on to the grains on a timescale of a few million years. We have studied the chemistry in an accreting, reacting gas in which some desorption mechanisms operate, to seek possible signatures of grain chemistry. The gas-grain chemistry model has been described by Brown and Charnley (1990a). The temperature (10 K) and density ($3 \times 10^4 \, \text{cm}^{-3}$) are held constant throughout the calculation, and accretion occurs with unit sticking efficiency. Only activationless surface reactions involving H atoms are considered at present. We have modelled the chemistry of the dark cloud TMC-1, in particular that of the complex hydrocarbon molecules (Brown and Charnley, 1990b) and find good agreement for most species, including the cyanopolyynes and other complex hydrocarbons, at a model time which is consistent with estimates for the dynamical age of the cloud.

In gas phase models of cold cloud chemistry, ion-molecule reactions have difficulty in reproducing the observed abundances of several sulphur-bearing molecules (e.g. H_2S, C_2S and C_3S), and formation of some of these species by grain-surface reactions has often been suggested (e.g. Minh, Irvine and Ziurys, 1989). To study the possible role of grain chemistry we have incorporated the gas phase sulphur chemistry of Millar and Herbst (1990) into the basic gas-grain model.

Figures 1 and 2 show some results from two exemplary calculations. These are for one model which ignores accretion, and for one in which accretion occurs with volatile methane molecules being desorbed from the grains upon formation. We have used the model parameters of Brown and Charnley (1990b) with sulphur depleted from its cosmic abundance by a factor of 0.1. One notable point is that the evolution of the sulphur species when accretion operates is quite similar to that which occurs in the non-accretion model, until accretion dominates and the abundances fall almost monotonically. At times $\sim$ million years, accretion acts to keep the abundances of CS, OCS, SO and SO_2 lower than in the non-accretion model. The observed SO/SO_2 abundance ratio of greater than 5 is only reproduced in the accretion model. The peak H_2S abundance is relatively unaffected and lies just below its observed value in both models. The effects of methane desorption does lead to a slight enhancements in the abundances of C_2S, C_3S and H_2CS, compared to accretion models which have no CH_4 desorption (not shown). On the basis of these results, it is difficult to unambiguously ascribe the high organo-sulphur (and hydrocarbon) abundances in this region solely to a local methane enhancement due to grain chemistry. We shall present a more

---

[1] CANADIAN INSTITUTE FOR THEORETICAL ASTROPHYSICS, UNIVERSITY OF TORONTO, 60 ST. GEORGE STREET, TORONTO, ONTARIO M5S 1A1, CANADA.
[2] PHYSICS DEPARTMENT, RENSSELAER POLYTECHNIC INSTITUTE, TROY, NEW YORK 12180, U.S.A.

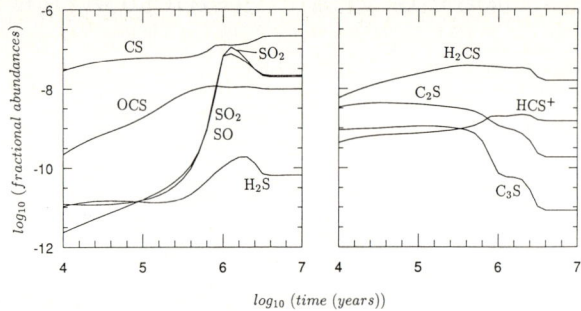

Figure 1. Chemical evolution for the case in which accretion is neglected.

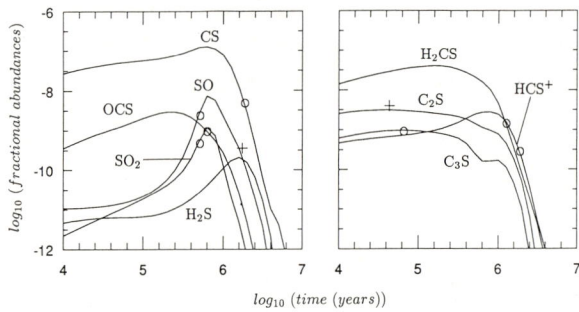

Figure 2. Chemical evolution for the case which includes accretion; methane molecules are desorbed after surface formation. For details see text.

detailed analysis of sulphur gas-grain chemistry in a forthcoming paper (Brown and Charnley 1990c).

Acknowledgments

PDB is grateful to SERC and NATO for a postdoctoral fellowship. SBC was partially supported by the US Air Force under Grant AFOSR-89-0104.

References

Brown, P.D. and Charnley, S.B., 1990a, Mon. Not. Roy. Astron. Soc., **244**, 432.
Brown, P.D. and Charnley, S.B., 1990b, Mon. Not. Roy. Astron. Soc., (in press).
Brown, P.D. and Charnley, S.B., 1990c, Mon. Not. Roy. Astron. Soc., (in preparation).
Irvine, W.M., Friberg, P., Kaifu, N., Kawaguchi, K., Kitamura, Y., Matthews, H.E., Minh, Y., Saito, S., Ukita, N. and Yamamoto, S., 1989, Astrophys. J., **342**, 871.
Millar, T.J. and Herbst, E., 1990, Astron. Astrophys. **231**, 466.
Minh, Y.C., Irvine, W.M. and Ziurys, L.M., 1989, Astrophys. J., **345**, L63.

The chemistry of the early universe at the epoch of recombination

WILLIAM B. LATTER[1] and JOHN H. BLACK[2]

We have re-examined in detail the atomic and molecular processes which may have taken place in the early Universe around the *epoch of recombination*. This calculation includes processes not previously considered, such as

$$H_{n=2} + H_{n=1} \to H_2 + h\nu \tag{1}$$

(n is the principal quantum number of the separated atoms; Latter and Black, 1991). A study of molecular abundances and their effects on cooling at early epochs has already been carried out by Lepp and Shull (1984), and the subject reviewed by Dalgarno and Lepp (1987).

Around the epoch of recombination, H_2 would have formed principally by the gas phase processes

$$H + H^+ \to H_2^+ + h\nu; \quad H_2^+ + H \to H_2 + H^+, \tag{2}$$

and

$$H + e^- \to H^- + h\nu; \quad H^- + H \to H_2 + e^-, \tag{3}$$

with a minor contribution early on from reaction (1). Destruction processes for H_2^+, which can slow the second of reactions (2), are photodissociation and dissociative recombination. A minor sink of $H_2^+(v > 3)$ is the reaction with He to form HeH^+, which is also a major source of HeH^+. Reactions (3) can be interrupted by photodetachment of H^-, or by the mutual neutralization of H^- and H^+.

The equations of ionization balance and molecular formation and destruction have been integrated through the epoch of recombination. We find excellent agreement with the earlier results of Lepp and Shull (1984). In the present study and that of Lepp and Shull, it was found that molecule formation must have been efficient, with H_2 reaching an asymptotic relative abundance $> 10^{-6}$ in all models. As discussed by Lepp and Shull, it is important to note that this is the minimum abundance of H_2 required to provide sufficient radiative cooling of gravitationally unstable regions for the initial collapse to continue. Collisional excitation of vibrational and rotational transitions in H_2, HD, HeH^+, and possibly even LiH^+ would have contributed significantly to the cooling of

---

[1] CITA, UNIVERSITY OF TORONTO, MCLENNEN LABORATORIES, TORONTO, ONTARIO M5S 1A1, CANADA

[2] STEWARD OBSERVATORY, UNIVERSITY OF ARIZONA, TUCSON, ARIZONA 85721, U.S.A.

the gas in the range $T_m \approx 10^2 - 10^4$ K, where the lowest excited states of atomic H and He are energetically inaccessible (T_m denotes the matter temperature). Without an initial abundance of molecular coolants, a collapsing region would be halted before renewed molecule formation could begin.

It is interesting to note that the helium bearing species He_2^+ and HeH^+ were apparently the first molecules to have formed out of the primordial plasma. The charge exchange reaction of H^+ with D determined the ionisation fraction of deuterium. Formation of the isotopic molecule HD took place by

$$D^+ + H_2 \to HD + H^+, \tag{4}$$

which led to a fractionation of the HD/H_2 ratio by a factor of ~ 13.6 over the primordial D/H ratio. Molecular cooling processes were considered in detail and found, not suprisingly, to have never competed with the adiabatic cooling of expansion. However, it was discovered that the persistence of a large ionized fraction after decoupling meant that Thomson heating continued to be significant. The matter temperature remains higher by a small factor than if only adiabatic cooling is considered.

References

Dalgarno A. and Lepp, S., 1987, in IAU Symposium **120**, *Astrochemistry*, ed. M.S. Vardya and S.P. Tarafdar (Reidel, Dordrecht), p. 109.

Latter, W.B. and Black, J.H., 1991, Astrophys. J., (in press).

Lepp, S. and Shull, M., 1984, Astrophys. J., **280**, 465.

Molecular processes in SN 1987A

J M C RAWLINGS and D A WILLIAMS

DEPARTMENT OF MATHEMATICS, UMIST, PO BOX 88, MANCHESTER
M60 1QD, U.K.

CO was detected in the ejecta of SN1987a as early as 112 days after the outburst. This is interesting for a number of reasons. Firstly it represents the first detection of a molecular species in a supernova. Secondly the region of emission is strongly hydrogen deficient so that we appear to be observing a non-hydrogen based chemistry and thirdly, the formation of CO has been shown to be a necessary pre-requisite for dust formation in novae (Rawlings and Williams, 1989). It may have a similar significance in supernovae. The CO was detected in the $\Delta v = 1,2$ lines (4.6 μm, 2.3 μm) and showed a steadily rising mass (Spyromilio et al., 1988). It has also been suggested by these authors that CO$^+$ was present. We find it very hard to maintain a high CO$^+$/CO ratio and consider the identification as weak. The physical processes in the ejecta are highly uncertain. We have adopted the mixed core models of Woosley (1988) as the physical basis of our work. In these models the source of the CO emission appears to be the cooler "core" region of the ejecta (where at t = 1 year; T = 2–3000 K, $n = 10^9$ cm^{-3} and the fractional abundances of hydrogen and oxygen are 10^{-4} and 0.33 respectively). The distinguishing features of our model are the nature of the radiation field (briefly discussed below), the energetic particle flux (estimated to be some 8 to 12 orders of magnitude greater than that in the ISM at t = 1 year) and a highly unorthodox chemistry appropriate to these physical and chemical conditions.

There are (at least) three possible contributing sources to the UV radiation field.

1. From the decay of the He(2^1S) excited by the degradation of gamma rays produced by ^{56}Co decay ($\lambda^{-1} \sim 114$ days) (c.f. Petuchowski et al., 1989).
2. A photospheric blackbody term (T $\sim$ 5500 K const.).
3. A compact X-ray source (?) causing the flux to level off after about 3 years.

Of these, (1) is intrinsic and highly dominant. The spectrum is also highly density dependent as a result of collisional coupling between He(2^1S) and He(2^1P). The absolute flux has been estimated (following Petuchowski et al.) using the FeIII/FeII ratio at t = 260 days as observed by Moseley et al. (1989). We note however that there are major uncertainties in this approach. The significance of the photospheric radiation field, (2) becomes apparent when considering the photodetachment of negative ions for which processes cross-sections often extend to wavelengths larger than 1 μm.

Results from the time-dependent model are briefly discussed below:
- CO forms via a variety of routes and thus is not sensitive to the physical conditions.
- In a chemistry limited to the elements H,He,C,N and O, no molecular species other than CO, CO^+, C_2 and O_2 achieves a fractional abundance of greater than 10^{-10}.
- The CO abundance rises sharply from 10^{-11} to 10^{-6} over the period of 1–3 years after the outburst as the UV field both becomes weaker and shifts to longer wavelengths.
- The observations indicate that the CO abundance rose substantially prior to one year. A detailed study has shown that this can only be modelled if the UV field is some 10^{-3} times smaller than in the "standard" case. The UV has probably been overestimated by a factor of 10 because of inconsistency with the oxygen ionization level, but there is still a large discrepancy. This could imply that either some sort of shielding is occurring (e.g. in clumps at the interface of the ejecta core with the envelope), or the CO emission cannot be identified with the bulk of the metal-rich core, or the FeIII 22.92 μm line identification by Woosley is incorrect.
- In the "standard" case the CO destruction and hence abundance is largely determined by the He^+ abundance for times greater than a few years. In the "reduced" UV run (UV$\times 10^{-3}$) this is true throughout the chemical evolution (as in the model of Lepp, Dalgarno and McCray, 1990). The CO formation is, however, partly dependent on the abundances of C^+ and C^- and thus the abundance is related to the UV field strength.
- In the "standard" model $CO^+/CO \ll 1$ for $t > 100$ days. In the "reduced" UV model $CO^+/CO \ll 1$ at all times.
- If the physical conditions do not change we may expect the CO mass to rise to about $10^{-3} M_\odot$ (dependent on the behaviour of the high energy particle flux at late times). Increasing ejecta transparency to the UV and the high energy particle flux will further enhance the CO abundance. If there is no compact source then a further enhancement factor of between 6 and 10 may be expected. The lack of detection of CO at late times is probably due to effective cooling of the core.
- In a simple model of the mantle-envelope region, CO formation also appears to be efficient.

References

Lepp, S., Dalgarno, A. and McCray, R., 1990, Astrophys. J., **358**, 262.
Moseley, S.H., Dwek, E., Silverberg, R.F., Glaccum, W.J., Graham J.R. and Loewenstein, R.F., 1989, Astrophys. J., **329**, 1119.
Petuchowski, S.J., Dwek, E., Allen, J.E. Jnr. and Nuth, III, J.A., 1989, Astrophys. J., **342**, 406.
Rawlings, J.M.C. and Williams, D.A., 1989, Mon. Not. Roy. Astron. Soc., **240**, 729.
Spyromilio, J., Meikle, W.P.S., Learner, R.C.M. and Allen, D.A., 1988, Nature, **334**, 327.
Woosley, S.E., 1988, Astrophys. J., **330**, 218.

Monte Carlo simulation of molecular cloud fragmentation

MAREK WOLF

DEPARTMENT OF ASTRONOMY AND ASTROPHYSICS, CHARLES UNIVERSITY PRAGUE, 150 00 PRAHA, ŠVÉDSKÁ 8, CZECHOSLOVAKIA.

Abstract The hierarchical fragmentation of a molecular cloud is modelled as a random process by the Monte Carlo method. It is assumed that the probability of fragmentation in each step is a simple function of fragment mass. The critical fragment mass is introduced and the role of the initial temperature and surface density of the molecular cloud on the resulting mass function is discussed. The computed mass function is compared with the mean mass function in open clusters and associations. The present fragmentation model is also applied to possible bimodal star formation.

1 Introduction The fragmentation of molecular clouds is one of the most complicated physical processes involved in galaxy evolution. The exact solution of this problem leads to the three-dimensional magneto-hydrodynamic equations, where all existing phenomena, like rotation, turbulence, fragment interaction, etc. have to be accounted for. Only particular problems have been solved and a few perturbation theories derived. Therefore the stochastic approach to the fragmentation process is natural.

In the literature we can find approximately three main approaches to the description of the fragmentation process: (a) particle simulations of rotating and/or self-gravitating clouds, (b) hierarchical and multiplicative fragmentation models and (c) simulations by the Monte Carlo method.

2 Fragmentation Model The present fragmentation model is based on the Monte Carlo simulation of hierarchical fragmentation proposed by Elmegreen and Mathieu (1983). As in our previous papers (Wolf and Vanysek, 1986, Wolf, 1988) the three following probability distributions are assumed:

1) The integer number of fragments N_f which is formed in every fragmentation step is a random variable with normal probability distribution

$$P_f(N_f) = \frac{1}{\sigma_f \sqrt{2\pi}} \int_{N_{f-1}}^{N_f} \exp\left[\frac{-(N - N_{fo})^2}{2\sigma_f^2}\right] dN,$$

where N_{fo} is the mean number of fragments per event and σ_f is the standard deviation.

2) The relative masses m_i/m_f of all N_f fragments were allowed to be random

variables with Gaussian distribution

$$P_m\left(\frac{m_i}{m_f}\right) = P_0 \exp\left[-\frac{(m_i/m_f - 1)^2}{2\sigma_m{}^2}\right], \quad m_i > 0,$$

where σ_m is the mass dispersion.

3) The probability of fragmentation of each fragment is a simple function of its mass. This probability for a fragment of mass m_f is given by the following distribution

$$P_f(m_f) = \frac{1}{(m_f/m_c)^x + 1}, \quad x < 0,$$

where x is a free parameter and m_c is the so-called "critical" fragment mass discussed in the next section.

3 The Critical Mass For every molecular cloud, depending on its properties, it is possible to derive several critical masses. The role of this parameter is like the limiting fragmentation factor: a fragment with mass greater than the value of the critical mass will fragment into subfragments with higher probability than a fragment with mass less then the critical mass.

In the first place as the critical mass for gravitational instability of isothermal molecular clouds we can consider the well-known Jeans mass $m_c = 10\,T^{1.5}\,n^{-0.5}$, where n is the density (in atoms cm^{-3}), T is the temperature in Kelvins and m_c is expressed in solar masses. Homogeneous and isotropic turbulence in a molecular cloud characterized by a velocity dispersion v also changes the critical mass so that (Chandrasekhar, 1958) $m_c = 10\,(T + 100\,v^2)^{1.5}\,n^{-0.5}$. Larson (1985) derived the critical masses of flattened or filamentary star-forming regions in the following simple form: $m_c = 2.4\,T^2/\mu$, where μ is the cloud surface density (expressed in M$_\odot$ pc^{-2}). The numerical coefficient is a funtion of the polytropic index and has approximately the same value for sheets and filaments. For magnetically supported clouds Blitz (this volume, p. 49) defines the critical mass in the form $m_c = 70(B/10\,\mu\mathrm{G})$ where B is the magnetic field intensity.

4 Results Each simulation was initiated by choosing the mean number of fragments N_{fo}, the standard deviation σ_f, the slope x and the temperature and surface density of the primary molecular cloud.

The Monte Carlo iteration procedure was as follows. During each event a value of the fragmentation probability was chosen for each existing fragment. The fragmentation process was halted when the probability of next fragmentation was less than a random number from the interval (0,1). The number of subfragments and their relative masses were computed according to the distributions given above. The final number of fragments created by this procedure was always less than 10^4. As the last step the mass function for the whole set of fragments was computed. In this study we are concerned about the influence of temperature and density on the resulting mass function. These changes for different parameters are illustrated in the figures.

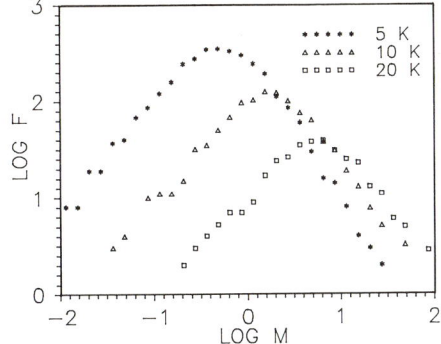
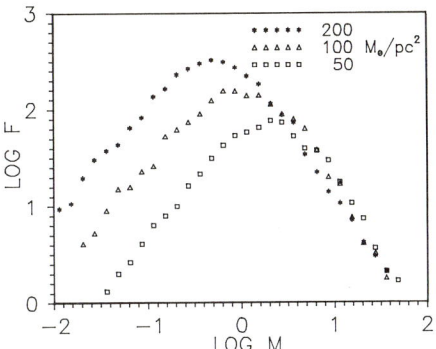

Figure 1: The comparison of the resulting mass function for three different values of the initial cloud temperature (5, 10 and 20 K). In these simulations the surface density of 100 $M_\odot \, pc^{-2}$ was assumed.

Figure 2: The comparison of the resulting mass function for the three different values of the initial cloud surface density (50, 100 and 200 $M_\odot \, pc^{-2}$). In this case the temperature was 10 K.

Figure 1 shows the influence of the initial cloud temperature on the resulting mass function. In Figure 2 we compare the mass functions for three different surface densities. From these pictures one can see that the maximum of the mass function is strongly dependent on these parameters. On the other hand, the slope of this function in the well-studied mass interval 1–10 $M_\odot$ has approximately the same value and varies between -1.3 and -1.6. Figure 3 shows the best fit of our model with the composite mass function in open clusters and associations according to Taff (1974). The critical mass for this simulation was 0.6 $M_\odot$, the initial cloud mass was 3000 $M_\odot$.

5 Bimodal Star Formation

An idea which has become quite popular in recent years is that there are two distinct mechanisms for star formation. This suggestion, which has become known as "bimodal star formation" would have important implications for star formation and galaxy evolution. From the observations of the nearest molecular clouds and star formation regions it follows that low-mass stars are dominant in small cold dark clouds (as in the Taurus-Auriga region), while massive stars appear to form in massive and relatively hot molecular clouds (such as Orion).

The fragmentation model presented here was also applied to the idea of bimodal star formation. Our Monte Carlo procedure was iterated as in our previous cases, except that each mode of star formation had its own initial parameters. For the two sets of created fragments we computed the mass function. As an example, Figure 4 shows the bimodal mass function resulting from such a simulation. The low-mass mode is modelled for initial cloud mass 2000 $M_\odot$ and critical masses 2 $M_\odot$ and 3 $M_\odot$ respectively. In this picture is also plotted the Initial Mass Function (IMF) according to Larson (1986). The division of this curve into two parts for masses greater than 1 $M_\odot$ corresponds to two probably extreme assumptions about the time-dependence of the star formation rate (SFR). For

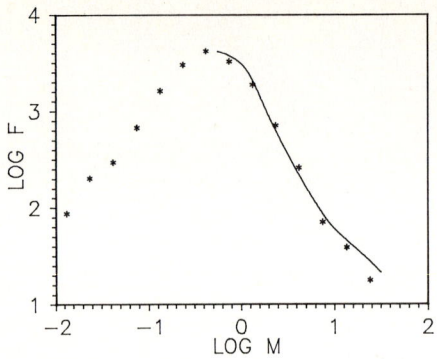 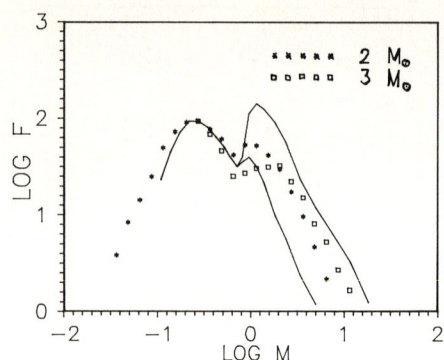

Figure 3: The best fit of our model with the composite mass function in open clusters and associations according to Taff (1974). The model parameters were: critical mass $0.6\,M_\odot$ and initial cloud mass $3000\,M_\odot$.

Figure 4: Simulation of the bimodal IMF (see text for details).

the upper part the SFR is assumed to decay with time as $\exp(-t/t_0)$, where $t_0 = 4.15\,\text{Gyr}$; for the lower part the SFR is constant with time ($t_0 \to \infty$).

6 Conclusions Although this model is a simple numerical study of the fragmentation process, we can make the two following conclusions:

1. The role of the critical mass as a limiting fragmentation parameter is real condition of every fragmentation process.
2. The behaviour of the mass function in clusters and associations depends particularly on the physical properties of initial molecular cloud.

References

Chandrasekhar S., 1958, Proc. Roy. Soc. A., **246**, 301.
Elmegreen, B.G. and Mathieu, R.D., 1983, Mon. Not. Roy. Astron. Soc., **203**, 305.
Larson, R.B., 1985, Mon. Not. Roy. Astron. Soc., **214**, 379.
Larson, R.B., 1986, Mon. Not. Roy. Astron. Soc., **218**, 409.
Taff, G.L., 1974, Astron. J., **79**, 1280.
Wolf, M., 1988, Proc. Xth IAU European Regional Astronomical Meeting, Prague, ed. J. Palous, p. 97.
Wolf, M. and Vanysek, V., 1986, Astrophys. and Space Sci., **128**, 229.

Clump formation in a non-uniform ultraviolet radiation field

JEAN-PIERRE CHIÈZE and CONSTANCE DE BOISANGER

COMMISSARIAT À L'ENERGIE ATOMIQUE, CENTRE D'ETUDES DE

BRUYÈRES-LE-CHÂTEL, SERVICE PTN,

F-91680 BRUYÈRES-LE-CHÂTEL BP 12, FRANCE

Abstract The non-uniform exposure of molecular cloud envelopes to ultraviolet radiation can trigger the formation and destruction of dense regions. The influence of the ultraviolet radiation field on the gas pressure is examined in order to study the gas response to local shielding. We show that gas condensation or expansion by more than one order of magnitude can be induced by extinction variations of about 1 magnitude in the visible. Along with density variations, mildly supersonic collective motions are induced which can feed turbulence. The short time scales involved in density variations, governed by cooling or heating processes, favour non equilibrium molecular chemistry.

1. Introduction. Many clumps, especially in cloud envelopes, have masses far below the local Jeans mass and exhibit features, such as clear cut sharp edges, that are irreconciliable with a strict gravitational origin. We suggest here that an irregular exposure of cloud material to the UV radiation field is likely to play an important role in gas dynamics, resulting in clump formation or destruction.

2. Time dependent gas condensation. The local intensity of the UV interstellar radiation field dominates, at low extinctions, the thermal state of the gas through the dust grain photoelectric heating rate, the abundance of the major coolant C^+ and the dissociation of H_2 molecules.

2.1. Chemical network and thermal balance. The time-dependent chemical evolution is followed based on the dark cloud chemistry described by Pineau des Forêts *et al.* (1988). It is modified to be representative of both envelopes and cores of molecular clouds.

Molecular gas cooling is mainly due to the excitation of C^+, CI, OI, CO and H_2. The dominant heating processes included in this work are photoelectric effect on dust grains, ionization by cosmic rays, photoionization of CI and photodissociation of H_2. The heating and cooling rates are listed in Chièze and Pineau des Forêts (1987) except for H_2.

We have selected three cases corresponding to an energy deposition $\epsilon_1 = 2.24\,\text{eV}$ (equipartition of the released H_2 binding energy between molecules and grains), $\epsilon_2 = 0.2\,\text{eV}$ (Williams, 1987) and finally $\epsilon_3 = 0$. The heating rate is then $\Gamma_F(H_2) = 1.6 \times 10^{-12} \epsilon(H_2) n(HI) R_F(H_2)\,\text{erg cm}^{-3}\,\text{s}^{-1}$. The equilibrium gas pressure governs the time scale of late time variations of the thermodynamic quantities (Boisanger and Chièze, 1990).

2.2. Application to gas condensation. Since the equilibrium molecular gas pressure is a sensitive decreasing function of the extinction, we analyse the gas response to a variation ΔA_v of the visual extinction, in a region of scale l in which the sound speed is c_s. Define the cooling time as $t_{cool} = \rho\varepsilon/|L|$ where ε is the internal energy of the gas and L the net cooling rate. For small scale perturbations, *ie:* $l \ll c_s t_{cool}$, local cooling will only slightly alter the large scale hydrostatic pressure field (if any), resulting in *isobaric* cooling conditions. To follow the time dependent condensation, we calculate the condensation factor $\eta(t) = n_H(t)/n_{H_0}$. Assuming a constant pressure P_0, the gas density evolution following the extinction perturbation ΔA_v, is described by the equation:

$$\frac{1}{n}\frac{dn}{dt} = \frac{2}{5}\frac{L}{P_0} + \frac{1}{n}\sum_i \left(\frac{dn_i}{dt}\right)_{ch}$$

where n is the total number density of particles and the index "ch" denotes purely chemical transformations. An example of the evolution of the condensation $\eta(t)$ for initial extinction $A_v = 0.25$ mag is presented in Figure 1 in the two cases $\phi_{UV} = \phi_0$ and $\phi_{UV} = 10\phi_0$.

The initial condensation rate is determined by the gas pressure and the dominant cooling term (C^+). As the gas begins to condense, C^+ recombination increases the fraction of neutral carbon in low density regions and allows for CO formation in the denser ones: these two species then dominate cooling, so that, after 0.4 to 1 Myr the cooling rate is approximately constant. Meanwhile, different heating processes in shielded regions lead to a decrease in the condensation rate, until thermal equilibrium is restored. This requires less than 0.6 Myr.

3. Discussion and conclusion. Non-uniform exposure of the molecular gas to UV field is efficient in producing non gravitationally driven gas condensation by more than one order of magnitude. The amplitude of the condensation exceeds two orders of magnitude in low density gas and density contrasts of about 10 are achieved for $n_H \sim 100\,\text{cm}^{-3}$. An increase of the UV intensity strengthens condensation.

The condensation growth time is always quite short. The cooling time, t_{cool}, defines a length scale $l = c_s t_{cool}$, where c_s is the sound speed, on which the condensation is nearly isobaric. It is a good approximation to the condensation time scale if the shielded region has extension less than l. Thus on scales $l \approx 0.03-0.2$ pc, condensations by a factor of 10 may be achieved by $t \approx 0.1-0.2$ Myr $\approx 1\%$ of the free fall time.

Two dimensional hydrodynamical simulations including the relevant heating and cooling terms, show that condensations actually grow with such short time scales. Figure 2 represents the density contours achieved after 0.5 Myr, in a cloud envelope perturbed by a UV gaussian screening of $\Delta A_v = 1$ mag, moving parallel to the initial surface of the hydrostatic cloud with velocity $v = 0.2\,\text{km s}^{-1}$. Density enhancements by more than a factor of 20 are achieved (Boisanger, Chièze and Gambart 1990). The nature of this mechanism favors the spontaneous growth of globule or "elephant-trunk" shaped condensations.

3.1 Induced collective motions and gravitational collapse. In response to the pressure variation, a mildly supersonic velocity field is soon established, with ve-

Clump formation in a non-uniform ultraviolet radiation field

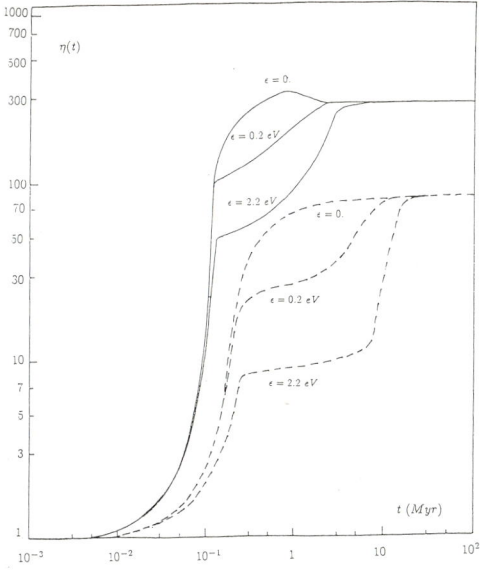

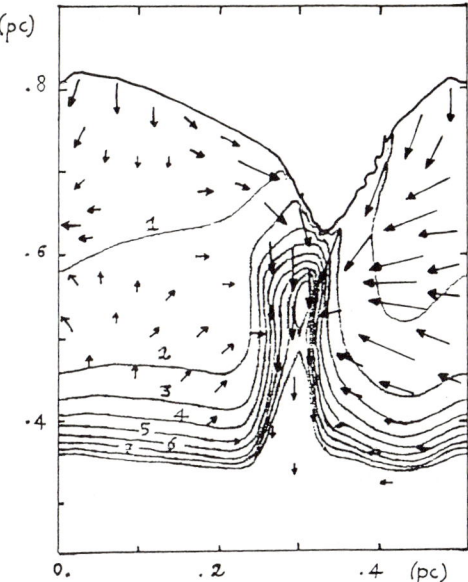

Figure 1: Time dependent evolution of the isobaric condensation factor $\eta(t)$, induced by an increase $\Delta A_v = 2\,\mathrm{mag}$ of the visual extinction in a cloud envelope of density $n_{H_0} = 20\,\mathrm{cm}^{-3}$ and mean, unperturbed, visual extinction $A_{v0} = 0.25\,\mathrm{mag}$. The dashed lines are for an incident radiation field $\phi_{UV} = \phi_0$, and the solid lines for $\phi_{UV} = 10\phi_0$. In each set, the three curves correspond to different assumed values of the energy (in eV) released by H_2 formation on dust grains.

Figure 2: Two dimensional hydrodynamical simulation showing the density contours and the velocity field which develop in a cloud envelope subject to variations of the incident UV field (see text). The contours 1 – 9 are equally spaced in density from $n_H = 50\,\mathrm{cm}^{-3}$ to $n_H = 850\,\mathrm{cm}^{-3}$. Arrows are scaled to the gas velocity, between 0.1 and 0.7 km s^{-1}.

locities in the range $u = 0.3 - 0.7\,\mathrm{km\,s^{-1}}$ (Figure 2). Supersonic motions can be induced in this way only over scales $l \gg c_s t_{cool}$. Furthermore, local condensations by a factor $\eta \gg 1$ induce buoyant forces which trigger collective motions. During their motions, condensed clumps eventually escape from the shielded regions. Recovering a normal UV heating, they get warmer and expand. These UV driven turnover motions are likely to feed turbulence in clouds envelopes.

The time scale for this condensation-expansion cycle of molecular cloud material is of the order of 0.3 Myr. A small fraction of the mass flux may leak out of the cycle due to gravitational collapse.

During an isobaric condensation, the Jeans mass varies as η^{-2} and may decrease locally by two or three orders of magnitude. This can induce gravitational collapse if the length scale is shorter than the turbulence correlation length.

3.2 Role of the H_2 formation heating rate. The late condensation time scale depends strongly on the energy $\epsilon(H_2)$ released by H_2 formation, though in the initial stages of condensation the dependence is weak (Figure 1). After about $t \sim 0.1\,\mathrm{Myr}$, the H_2 heating rate slows down the gas condensation. At this time, the condensation factors range between 3 (high initial density, low UV field, $\epsilon(H_2) = 2.24\,\text{eV}$) and 100 (low initial density, high UV field, $\epsilon(H_2) = 0.$). Further low rate condensation brings the gas to its equilibrium density, which depends only on the initial conditions, and not on $\epsilon(H_2)$.

3.3 Non equilibrium chemistry. Non equilibrium chemical effects are associated with this cycle. Chièze and des Forêts (1989) have found that significant departures from equilibrium chemistry are obtained for mixing rates $t_{mix} < 0.3\,\text{Myr}$. The condensation-expansion process examined in the present study may offer the required time scales. As an example, we followed the C/CO ratio in the condensation and expansion regimes (but not allowing for mixing) – such a cycle leads to C/CO ratios of the order of 0.1.

The present study suggests that both local clump condensation or destruction and collective motions may be the molecular gas response to fluctuations in the interstellar UV radiation field intensity. For isobaric transformations, extinction variations of about two magnitudes in molecular cloud envelopes result in density contrasts of one or two orders of magnitude in less than 0.5 Myr. These unusually short time scales reflect the strong dependence of the gas pressure on the UV field intensity. In more realistic conditions, large pressure gradients lead to supersonic motions through radiative shocks. The isobaric condensation of shielded regions decreases the Jeans mass and can trigger gravitational collapse. Mutual shielding of moving clumps may result in time dependent variations of the UV intensity; previously shielded regions may then recover normal UV heating and expand, giving rise to convection-like motions. These UV-driven motions favor non equilibrium chemistry and high C/CO ratios.

Acknowledgements

We would like to acknowledge G. Pineau des Forêts and D.R. Flower for providing us the chemical tools and network used in this work.

References

de Boisanger, C. and Chièze, J.P., 1990, Astron. Astrophys., (in press).
de Boisanger, C., Chièze, J.P. and Gambart, J., 1990, (in preparation).
Chièze, J.P. and Pineau des Forêts, G., 1987, Astron. Astrophys., **183**, 98.
Chièze, J.P. and Pineau des Forêts, G., 1989, Astron. Astrophys., **221**, 89.
Heck, L., Flower, D.R. and Pineau des Forêts, G., 1990, Comp. Phys. Comm., **58**, 169.
Mathis, J.S., Mezger, P.G. and Panagia, N., 1983, Astron. Astrophys., **128**, 212.
Pineau des Forêts, G., Flower, D.R. and Dalgarno, A., 1988, Mon. Not. Roy. Astron. Soc., **235**, 621.
Williams, D.A., 1987, in *Physical Processes in Interstellar Clouds*, ed. G. Morfill and M. Scholer, (Reidel, Dordrecht), p. 377.

Interstellar gas cycling powered by star formation

A. LIOURE and J-P. CHIÈZE

COMMISSARIAT À L'ENERGIE ATOMIQUE,
CENTRE D'ETUDES DE BRUYÈRES-LE-CHÂTEL, SERVICE PTN,
F-91680 BRUYÈRES-LE-CHÂTEL BP 12, FRANCE.

Introduction. Although the estimated amount of mass involved in star formation is small ($3 - 10\,M_\odot\,yr^{-1}$), young stars are very efficient in destroying molecular clouds, reducing the efficiency of star formation to $\sim 5\%$. This induces a mass flow which can be calculated from the observed star formation rate (R_{SF}) and a typical value of the efficiency (ϵ): $\phi_0 \sim R_{SF}/\epsilon \sim 60 - 200\,M_\odot\,yr^{-1}$. This considerable amount of low density gas is injected in the diffuse phase of the interstellar medium, and should condense back, if one assumes a stationary state, and a cycling process is initiated. The aim of the model is to account quantitatively for the mass distribution of interstellar gas throughout the whole density spectrum. We argue that this is tractable if one supposes that for each value of the density, one can identify a dominant physical process associated with a well defined condensation rate $\omega(n_H) = n_H^{-1}\,dn_H/dt$. In a stationary state the mass distribution function $g(n_H, t) = \partial M/\partial Log(n_H)$ is equal to $\phi_0/\omega(n_H)$ where ϕ_0 is the constant mass flux throughout the density spectrum.

Thermal cascade. We assume: isobaric evolution; the heated gas enters the cycle at a temperature of about $8000\,K$ with a density convenient for global pressure balance; a mean interstellar pressure of $3800\,K\,cm^{-3}$; a cooling function, $B(n_H, T)$. The gas evolution is completely described by the condensation rate $\propto$ the heat exchange rate $\omega_{th} \equiv d(\log n_H)/dt = -0.4(B/P_0)$. For a given interstellar pressure the condensation rate depends only upon the density n_H. We assume that the gas always lies in the cooling region so most of the the gas heated by young stars cools to its HI equilibrium temperature. The isobaric evolutionary path is partly unstable in the sense that small density perturbations grow exponentially at a rate $\omega_F = (2/5P_0)(B_T T - B_{n_H} n_H)$, when positive (this is the same as the Field criterion). Due to this cooling instability, we suggest that the gas is already fragmented in small cloudlets when it enters the thermally stable HI phase, where the continuity of the condensation flow is carried out by the collisional processes.

The major contribution to the total mass arises naturally from the regions where the condensation rate is low: $M_\omega = 8 \times 10^8\,M_\odot$. Accordingly, most of the warm gas lies in the vicinity of the low density bump of the thermal equilibrium curve, in the so-called warm phase.

Collisional cascade. Since the HI gas in clouds is in thermal equilibrium, it can only evolve through mechanical interactions dominated by collisions. Chièze and Lazareff (1980) self-consistently derive a stationary cloud mass spectrum,

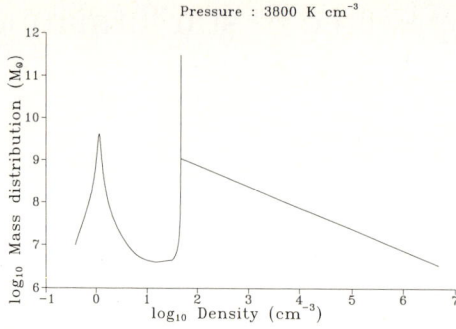
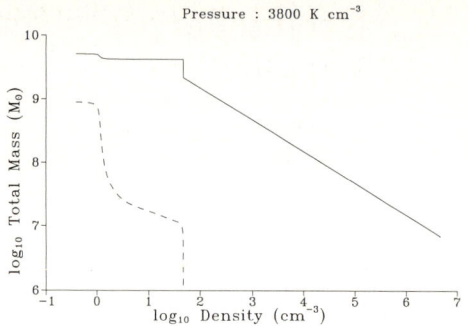

Figure 1. Mass distribution of gas in the Galaxy obtained by the cycle model. Most of the mass is contained in warm and HI phases, as well as in the low-density molecular gas.

Figure 2. Total mass of gas in the Galaxy. This curve is obtained by integrating the one in Figure 1 over density, from high ones to low ones. The discontinuity represents the mass locked in the HI phase. The dashed line shows the separate integration for warm gas, indicating the predicted amount of lukewarm gas.

driving a net mass flux carrying the mass distributed among low mass clouds into more massive clouds, compatible with the one calculated above: $\phi_{coll} = 1.3 \times 10^7 \, M_\odot \, \text{Myr}^{-1}$. We suggest that the high-mass-end clouds make the transition from HI clouds to individual molecular clouds ($3 \times 10^2 \, M_\odot$). Moreover, the total HI mass of these colliding clouds roughly fits the observation, most of it being locked in the massive ones: $M_{HI} \sim 2 \times 10^9 \, M_\odot$.

Gravitational cascade. We suppose here that gravitation is the main physical process at work inside molecular clouds. Some features are in favour of this hypothesis, even if it is well known that the naive picture of free-falling molecular material grossly overestimates the observed star formation rate: fragmentation and high density contrasts inside individual clumps. Hence, we choose to scale the condensation rate with the usual local Jeans rate and use an effective rate $\omega = z\omega_J = z(4\pi G \rho)^{0.5}$, where $0 < z \leq 1$ is a measure of the efficiency of the gravitational cascade. We find: $M(H_2) = 1.5 \times 10^9 \, (z/0.1)^{-1} \, (n_{HI}/100\,\text{cm}^{-3})^{-0.5} \, M_\odot$.

Conclusion. This model has been built making as few assumptions as possible (stationary state, pressure equilibrium) and emphasizing some conspicuous observational facts (the prominent role of molecular clouds of 1 – 3 pc and 100 – 300 $M_\odot$, the importance of gravity inside these clouds, the presence of gas at various temperatures, the power of star formation in ionizing molecular gas). The results are displayed in Figures 1 and 2 - see also Lioure and Chièze (1990).

References

Chièze, J.P. and Lazareff, B., 1980, Astron. Astrophys., **91**, 290.
Lioure, A. and Chièze, J.P., 1990, Astron. Astrophys., **235**, 379.

A self-regulated state for the warm, cool neutral and molecular Galactic gas

A. PARRAVANO[1,2] and J. MANTILLA CH.[2]

We have calculated the self-regulated star formation rate and the value of various variables related to the state of the interstellar medium assuming that two regulating mechanisms act simultaneously. One is the "Heating-Regulated Warm Gas Condensation" (HRWC, Parravano, 1989), that acts as a large scale self-regulating mechanism of star formation. This mechanism tends to maintain the warm gas in a critical state over which a phase transition resulting in the formation of small neutral clouds will occur. The state of the warm gas is very sensitive to the UV energy density, and so this mechanism determines the "equilibrium" formation rate of massive stars.

The other regulating mechanism is "Photoionisation-Regulated Star Formation" (PIRSF, McKee, 1989) that acts at the scale of single molecular clouds but depends on the ambient conditions. This mechanism tends to maintain the cloud mass fraction where cosmic rays dominate ionisation and low mass star formation takes place, near the value for energy balance. PIRSF specifies (under some assumptions) the efficiency of the stellar formation process and thereby the amount of molecular gas necessary to maintain the "equilibrium star formation rate".

The two mechanisms used here tend to maintain the star formation rate (SFR) near its equilibrium value because when the SFR is higher (lower) than a critical value then the rate of condensation of warm gas and the star formation efficiency in molecular clouds are both reduced (increased). The HRWC mechanism controls the SFR at large scale; the PIRSF mechanism assures the rapid response of the system to departures from equilibrium.

Given the radial dependence of the gas metallicity and warm gas density, the HRWC assumption permits us to calculate the radial dependence of the formation rate of massive stars, the thermal pressure and the temperature and ionisation degree of the warm and cool neutral gases.

Given the radial dependence of the primary cosmic ray ionisation rate, the PIRSF assumption permits us to calculate the radial dependence of the minimum visual extinction of star forming molecular clouds and the mass in star forming regions.

There are large quantitative differences in the literature for the radial dependence of the gas metallicity and warm gas density, but qualitatively they have similar radial behaviour. Then, assuming that the two regulating mechanisms control the SFR, the results might be seen only as indicators of the radial dependence of the SFR and of the ISM state. The resulting equilibrium state

[1] DEPARTMENT OF ASTRONOMY, UNIVERSITY OF EDINBURGH, ROYAL OBSERVATORY EDINBURGH, BLACKFORD HILL, EDINBURGH EH9 3HJ, SCOTLAND
[2] UNIVERSIDAD DE LOS ANDES, DEPARTAMENTO DE FISICA, MERIDA, VENEZUELA

compares reasonably well with observations for $R > 6\,\text{kpc}$, but for lower galactocentric distances there is an increasing discrepancy. This discrepancy might be explained in terms of a saturation of the HRWC mechanism. That is: at the equilibrium SFR the mass in the warm gas phase is insufficient to supply the net consumption of the star formation process and consequently the SFR will be lower than the equilibrium value. In these saturated regions the highest possible fraction of molecular gas is expected because the warm gas does not reach thermal equilibrium.

Acknowledgments

We are very grateful to Dr. Peter Brand who suggested that we couple the two regulating mechanisms. We would like to thank the Commission of the European Communities and the CDCHT of the Universidad de Los Andes for their support.

References

McKee, C.F., 1989, Astrophys. J., **345**, 782.
Parravano, A., 1989, Astrophys. J., **347**, 812.

Stabilization of the ISM by fluctuations

A. JUST, B. M. DEISS AND W. H. KEGEL

INSTITUT FÜR THEORETISCHE PHYSIK, ROBERT MAYER-STRASSE 10,

POSTFACH 11 19 32, D-6000 FRANKFURT AM MAIN 11, F.R.G

Usually the effect of small scale fluctuations, especially of turbulent gas motions, on the large scale dynamics is taken into acount by introducing an additional force term $u^2 \nabla \rho$, where u is the velocity dispersion of the gas. The term $u^2 \rho$ may be a good guess for the turbulent pressure, but the pressure gradient $\nabla(u^2 \rho) = u^2 \nabla \rho + \rho \nabla u^2$ is very sensitive to the parameter dependence of u^2, which is simply neglected in the expression given above. Taking e.g. the scale dependence $u^2 \propto L$, $\rho \propto L^{-1}$ from Larson's relations for the molecular cloud hierarchy (Larson, 1981), one finds that turbulent pressure is independent of scale. This result gives no direct information for the pressure gradient on a fixed scale but it suggests that the parameter dependence may be quite different.

Here we are especially investigating the influence of fluctuations on the large scale stability of the two component system ISM – stars (Just and Deiss, 1990). For this question all fluctuation terms, e.g. turbulent heating, thermal pressure enhancement, dynamical friction, and mass flow in the fluctuations, occuring in the averaged equations (Just et al., 1986, Deiss et al., 1990), must be taken into account.

Two main aspects of stability can be seen in the limiting cases of large and small values of the wavenumber k. Firstly, the conditions for thermal stability (large values of k) are influenced in a more involved way. One condition, however, namely that the denominator in the generalized Jeans criterion (equation 2 below) has to be positive, can be seen directly.

Secondly, the variation of the Jeans criterion, i.e. stability on large scales, can be investigated. In this context one finds from the general structure that dynamical friction and the mass flow in the fluctuations lead to a nonvanishing critical frequency $\omega_c \neq 0$, so the critical Jeans wavenumber k_c corresponds to an oscillating mode.

In previous papers we investigated small scale fluctuations induced via gravitation by randomly moving stars (Kegel and Völk, 1983, Deiss and Kegel, 1986, Jacobi et al., 1990) and calculated large scale effects in quasilinear approximation (Just et al., 1986, Deiss et al., 1990). With respect to the large scale stability we find that the main contribution comes from the term

$$A = \frac{1}{3}\rho_{g0}\langle V_1^2\rangle + \kappa c_V \langle \rho_{g1}T_1\rangle \tag{1}$$

which is the total pressure enhancement by fluctuations. Thus the Jeans mode

k_c is a collapse mode $\omega_c \approx 0$ and the modified Jeans criterion is

$$k_c^2 = k_s^2 + k_g^2 \frac{1 - A_s}{1 + A_g - \delta(1 + A_T)} \tag{2}$$

where k_s and k_g are the critical wavenumbers of the stellar system and an isothermal gas respectively (Deiss, 1989). A correction term of $(1 - \delta)$ in the denominator is the result of heating and cooling processes (Kegel and Traving, 1976). The terms A_s, A_g, and A_T are the normalized partial derivatives of A with respect to stellar density, gas density, and gas temperature, respectively.

In thermally stable systems usually the quantities A_s and A_g are positive and A_T is negative, so that all terms lead to a stabilization of the system. Computing the correction terms induced by the grainy structure of the stellar system (Kegel and Völk, 1983) in quasilinear approximation leads to stabilization factors up to about seven for the Jeans length. As an example we find $k_c^2 = k_s^2 + k_g^2/12$ for the parameters $\rho_{s0} = 2.5 \times 10^{-21}\,\mathrm{g\,cm^{-3}}$, $\sigma = 1\,\mathrm{km\,s^{-1}}$, $\rho_{g0} = 4 \times 10^{-21}\,\mathrm{g\,cm^{-3}}$, $T = 10\,\mathrm{K}$, $\tau_{cool} = 100\,\mathrm{yr}$ typical for molecular clouds in the disk. The Jeans length is increased by a factor of 2.7 to 1.1 pc due to fluctuations.

References

Deiss, B.M., 1989, Ph.D. Thesis, (Frankfurt).
Deiss, B.M., Just, A. and Kegel, W.H., 1990, Astron. Astrophys. (in press).
Deiss, B.M. and Kegel, W.H., 1986, Astron. Astrophys., **161**, 23.
Jacobi, S., Just, A., Deiss, B.M. and Kegel, W.H., 1990, Astron. Astrophys., **237**, 461.
Just, A. and Deiss, B.M., 1990, (in preparation).
Just, A., Kegel, W.H. and Deiss, B.M., 1986, Astron. Astrophys., **164**, 337.
Kegel, W.H. and Traving, G., 1976, Astron. Astrophys., **50**, 137.
Kegel, W.H. and Völk, H.J., 1983, Astron. Astrophys., **119**, 101.
Larson, R.B., 1981, Mon. Not. Roy. Astron. Soc., **194**, 809.

Comets and molecular clouds: the sink and the source

M. E. BAILEY

DEPARTMENT OF ASTRONOMY, UNIVERSITY OF MANCHESTER,

MANCHESTER M13 9PL, U.K.

Abstract This paper reviews the rôle of molecular clouds as both a sink and source of comets. On timescales greater than a few hundred Myr, close encounters with molecular clouds cause the largest single changes of cometary orbits in the outer Oort cloud ($a \gtrsim 2 \times 10^4\,\mathrm{AU}$), whilst on longer timescales ($t \gtrsim 2\,\mathrm{Gyr}$) molecular cloud encounters also dominate the mean energy transfer rate. The additional possibility of comet formation within molecular clouds (whether or not associated with star and planet formation) highlights further links between comets and molecular clouds. Consideration of hierarchical grain aggregation processes within clouds and protostellar systems may provide clues both to the nature of cometary dust and the structure of the icy nucleus.

1 Introduction

The discovery during the early nineteen-seventies that diffuse gas in the Galaxy exists mainly in the form of massive cool molecular clouds, and that the clouds themselves clump into dense complexes of cold gas known as giant molecular clouds (with masses in excess of $10^5\,\mathrm{M}_\odot$) has had a profound influence on cometary theory. Biermann (1978) concluded that virtually all comets initially with aphelia beyond about $5 \times 10^4\,\mathrm{AU}$ would be removed by molecular cloud perturbations during the age of the solar system, while Napier and Clube (1979), reporting the results of calculations by themselves and Staniucha (Napier and Staniucha, 1982), suggested that cloud perturbations would be even more damaging. These results caused the survival of a primordial comet cloud to be called into question. Since this time, the dynamical effects of molecular clouds have been discussed by many authors, though there is no consensus concerning the overall degree of disruption.

In addition to these dynamical effects, the study of molecular clouds has also impinged on theories of comet formation. Most stars and planets are presumably formed within molecular clouds, and hence (on almost any theory) molecular clouds are probably where most comets originate (Clube and Napier, 1982, 1984, 1985, Napier 1982, Clube 1985). However, whether comets are born in protostellar or protoplanetary regions, or simply in ordinary (but still dense) "interstellar" regions within clouds is not known, and the many proposed theories together encompass virtually all possibilities. This paper reviews the status of molecular clouds, first as perturbers of the Oort cloud, and secondly as potentially important sources of cometary material, both for stellar systems and the wider interstellar medium.

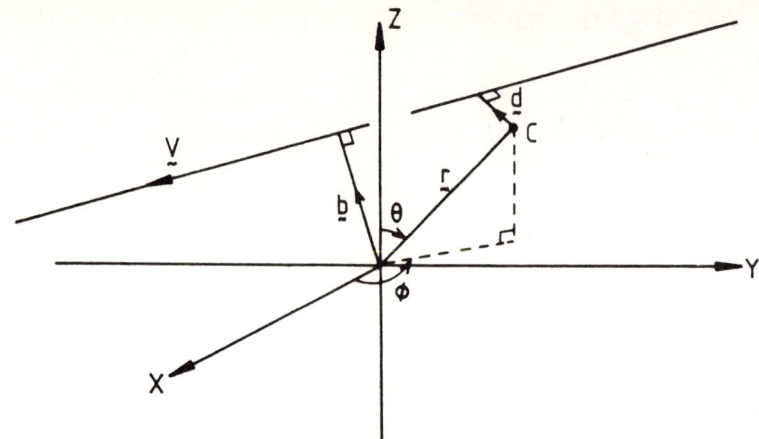

Figure 1: The encounter geometry for the calculation of $\Delta\mathbf{v}$ is shown. The sun lies at the origin, the comet has coordinates $\mathbf{r} = (r, \theta, \phi)$, and the perturber of mass M has relative velocity $\mathbf{V}$. The points of closest approach of the perturber to the sun and comet are $\mathbf{b}$ and $\mathbf{d}$ respectively.

2   Evolution of the Oort cloud

Figure 1, taken from Bailey (1986a), defines the geometry of an encounter between the solar system and a perturber of mass M moving with relative velocity $\mathbf{V}$ and impact parameter $\mathbf{b}$ with respect to the sun. The hyperbolic orbit of the perturber has been approximated by a straight line, while neglecting the orbital motion of the comet about the sun a comet with heliocentric position $\mathbf{r}$ follows essentially the same path but with impact parameter $\mathbf{d} = \mathbf{b} - \mathbf{r} + r(\hat{\mathbf{r}}.\hat{\mathbf{V}})\hat{\mathbf{V}}$. Using the impulse approximation (Sekanina, 1968, Rickman, 1976, Bailey, 1983, 1986a), the comet's velocity change $\Delta\mathbf{v}$ with respect to the sun may be expressed in the form

$$\Delta\mathbf{v} = \frac{2GM}{dV}\hat{\mathbf{d}} - \frac{2GM}{bV}\hat{\mathbf{b}} = \frac{2GM}{bV}\left\{\left(\frac{b^2}{d^2}-1\right)\hat{\mathbf{b}} - \frac{rb}{d^2}\left[\hat{\mathbf{r}} - (\hat{\mathbf{r}}.\hat{\mathbf{V}})\hat{\mathbf{V}}\right]\right\} \quad (1)$$

This implies

$$(\Delta\mathbf{v})^2 = \left(\frac{2GM}{V}\right)^2 \frac{r^2[1-(\hat{\mathbf{r}}.\hat{\mathbf{V}})^2]}{b^2 d^2} \quad (2)$$

where $\hat{\mathbf{b}} = \mathbf{b}/|\mathbf{b}|$, $\hat{\mathbf{r}} = \mathbf{r}/|\mathbf{r}|$, and $\hat{\mathbf{V}} = \mathbf{V}/|\mathbf{V}|$ are unit vectors in the directions of $\mathbf{b}$, $\mathbf{r}$, and $\mathbf{V}$, respectively. For completeness, we note that the radial and transverse components of $\Delta\mathbf{v}$, Δv_r and Δv_t respectively, are

$$\Delta v_r = \frac{2GM}{bV}\left\{\left(\frac{b^2}{d^2}-1\right)\hat{\mathbf{b}}.\hat{\mathbf{r}} - \frac{rb}{d^2}\left[1-\left(\hat{\mathbf{r}}.\hat{\mathbf{V}}\right)^2\right]\right\} \quad (3)$$

and

$$\Delta v_t = \frac{2GM}{bV} \left\{ \left(\frac{b^2}{d^2} - 1\right)^2 \left[1 - \left(\hat{\mathbf{r}}.\hat{\mathbf{b}}\right)^2 - \left(\hat{\mathbf{r}}.\hat{\mathbf{V}}\right)^2\right] \right.$$
$$\left. + \frac{r^2}{d^2} \left(\hat{\mathbf{r}}.\hat{\mathbf{V}}\right)^2 \left[1 - \left(\hat{\mathbf{r}}.\hat{\mathbf{V}}\right)^2\right] \right\}^{1/2} \quad (4)$$

In general, since the perturbation is assumed to be impulsive in nature, the mean change in specific energy of the orbit is

$$\Delta E = \mathbf{v_0}.\Delta \mathbf{v} + \frac{1}{2}(\Delta \mathbf{v})^2 \quad (5)$$

where $\mathbf{v_0}$ is the orbital velocity of the comet about the sun at the time of the encounter. Since $\mathbf{v_0}.\Delta \mathbf{v}$ averages to zero for a randomly distributed set of cometary orbits, the mean energy transferred during a single encounter is

$$\langle \Delta E \rangle = \left\langle \frac{1}{2}(\Delta \mathbf{v})^2 \right\rangle \quad (6)$$

Appropriate averages of these equations over an Oort cloud containing comets with a particular set of orbital parameters (*e.g.* for given values of the semi-major axis a and eccentricity e) then determine whether comets may be directly ejected during a single encounter, and how rapidly their orbital energies change towards higher and lower values. The results of such calculations are usually expressed in one of several forms depending on the context: for example, as a disruption half-life, a mean energy transfer rate, or a net Δv either per encounter or summed over the possibly many encounters which occur during a given time interval t. In the remainder of this section we first consider the stellar disruption half-life, and then give approximate expressions for the mean energy transfer rate and the r.m.s. velocity change and change in perihelion distance per encounter, treating the perturbers (stars or clouds) as point masses. Combining these results allows the overall effect of molecular clouds to be compared with that of stars and the disruption half-life due to both kinds of perturber to be determined.

2.1   Stellar disruption half-life

It is readily verified that single stellar perturbations cause a negligible change in orbital energy unless the impact parameter with respect to either the sun or the comet is much less than the semi-major axis. For such encounters, equation (1) implies $\Delta v \simeq 2GM/bV$ or $\Delta v \simeq 2GM/dV$, depending on whether the star passes closest to the sun or the comet. The former possibility may be ignored, since the survival of the Oort cloud (assumed primordial) and the regularity of the outer planetary system both show that exceptionally close star-sun encounters have not yet occurred (Morris and O'Neill, 1988). In order for a single stellar encounter to eject a comet of semi-major axis a, the impact parameter with respect to the comet must be less than some minimum value $d_{min,*}$. This is obtained by setting $\Delta E > GM_\odot/2a$ in equation (5), and to a first approximation may be estimated by ignoring the comet's orbital motion. This leads to

$$d_{min,*} = 2\left(\frac{GM_\odot a}{V^2}\right)^{1/2}\left(\frac{M}{M_\odot}\right) \tag{7}$$

A more detailed calculation including the effect of the $\mathbf{v_0}.\mathbf{\Delta v}$ terms (Heggie, 1975, Hut, 1983) yields

$$\pi d^2_{min,*} = \frac{20\pi}{3}\frac{GM^2 a}{M_\odot V^2} \tag{8}$$

This is larger than the equivalent result derived from equation (7) by a factor 5/3, though we note that equation (8) is smaller by a factor of 2 than the cross-section given by Hut (1983, his equation 5.1') because we have here considered close encounters only to one member of the sun-comet binary, namely the comet.

Considering the frequency of encounters with stars having number density n, relative velocity V and impact parameters smaller than $d_{min,*}$, the number of comets with semi-major axes a may be shown to decrease exponentially with a half-life $t_{d,1/2} = \ln 2/\langle n\pi d^2_{min,*} V\rangle$, i.e.

$$t_{d,1/2} = \frac{3}{5}\ln 2 \frac{GM_\odot}{a}\bigg/4\pi G^2 \langle nM^2/V\rangle \tag{9}$$

Substituting $\langle V^{-1}\rangle = (2/\pi)^{1/2}\sigma^{-1}$ for an isotropic Maxwellian distribution of one-dimensional velocity dispersion σ together with $\langle nM^2/\sigma\rangle \simeq 1.6 \times 10^{-3} M_\odot^2$ pc^{-3}/km s^{-1} (appropriate to the present solar neighbourhood), this expression reduces approximately to $t_{d,1/2} = 1.2 \times 10^{11} a_4^{-1}$ yr, where $a_4 = a/10^4$ AU. For comparison, the numerical value for $t_{d,1/2}$ obtained from the Monte Carlo simulation described by Heisler et al. (1987, see their Table III) is about 35% smaller.

The half-life for direct ejection may also be compared with the corresponding timescale for removal of comets by weaker, more distant stellar perturbations, though still mostly involving stars passing *through* the Oort cloud. Following Bailey (1986a) this half-life may be written in the form (*cf.* equation (22))

$$t_{1/2,*} = \frac{1}{4.732\, I_*}\frac{GM_\odot}{a}\bigg/4\pi G^2 \langle nM^2/V\rangle \tag{10}$$

where I_* is a logarithmic factor on the order of 10. Evaluating this, as above, for an isotropic Maxwellian distribution with parameters appropriate to the solar neighbourhood yields $t_{1/2,*} = 6 \times 10^9 a_4^{-1}$ yr, in good agreement with the results presented by Bahcall et al. (1985). In this way we find $t_{d,1/2} \simeq 20\, t_{1/2,*}$, showing that evolution of the Oort cloud through direct ejection of comets by stars is generally negligible compared to the more gradual effect of weak, more distant encounters. Despite this, it is worth noting that about 2.5% of comets initially with semi-major axes $a_0 = 10^4$ AU will be directly ejected by stars over the age of the solar system, the fraction increasing to 5%, 7% and 10% for $a_0 = 2 \times 10^4$, 3×10^4 and 4×10^4 AU respectively (*cf.* Nezhinski, 1972, Weissman, 1980). Since power-law models of the Oort cloud usually have many more comets with semi-major axes less than 10^4 AU than with $a \gtrsim 3 \times 10^4$ AU, the majority of comets directly eliminated by stellar perturbations will probably come from the denser, inner regions of the Oort cloud.

2.2 Energy transfer rate and mean velocity change

The determination of the half-life for removal of comets initially with semi-major axes a due to a sequence of stellar or molecular cloud perturbations may also be approached using the mean energy transfer rate $\dot{\varepsilon}(t)$. This is the average rate of increase of cometary energies due to external perturbations over a time interval t. Approximate expressions for $\dot{\varepsilon}$ when the perturbers are treated as point masses have been given by Bailey (1986b, c), namely

$$\dot{\varepsilon}(t) = \frac{4\pi G^2 M^2 n}{V} \begin{cases} (a/a_c)^2 & a < a_c \\ 2\ln(a/a_c) + 1 & a > a_c \end{cases} \tag{11}$$

where $a_c = \sqrt{\frac{12}{7}}\, b_{\min}$ and $b_{\min} = (2\pi nVt)^{-1/2}$ is the most probable minimum impact parameter during a time interval t. In the notation of equation (10) the factor $2\ln(a/a_c) + 1$ corresponds to the logarithmic factor I_*. The total energy change expected during time t is then $\sum \Delta E = \dot{\varepsilon} t = \frac{1}{2}(\Delta v_{tot})^2$ where Δv_{tot} is the equivalent total change in velocity. In a similar way, the mean relative velocity change experienced in a single encounter may be expressed in the approximate form

$$\Delta v = \frac{2GM}{bV} \begin{cases} \sqrt{2} & b < \sqrt{7/12}\, a \\ \sqrt{7/6}\, a/b & b > \sqrt{7/12}\, a \end{cases} \tag{12}$$

The extra factor of $2^{1/2}$ in the first part of this expression occurs because perturbations of both the sun and the comet contribute to the net energy exchange per encounter (Weinberg *et al.*, 1986, Weinberg, 1990), while the numerical factor in the second part of the expression arises from averaging equation (2) in the limit $b \simeq d \gg r$ over all orientations (giving $[1 - (\hat{\mathbf{r}}.\hat{\mathbf{V}})^2] = 2/3$), then around an elliptical orbit of eccentricity e (giving $\langle r^2 \rangle = a^2[1 + 3e^2/2]$), and finally over all eccentricities for an isotropic velocity distribution (giving $\langle e^2 \rangle = 1/2$).

2.3 Comparison between stars and clouds

2.3.1 *Maximum velocity change*

Over relatively short timescales ($t \lesssim 5$–$30\,\mathrm{Myr}$) the closest stellar encounter expected during a given time interval has a minimum impact parameter greater than the semi-major axes $a \simeq 1$–$3 \times 10^4\,\mathrm{AU}$ typical for comets in the outer Oort cloud. Over these short timescales, the maximum velocity change due to stars or clouds is obtained from equation (12) by substituting $b_{\min} = (2\pi n V t)^{-1/2}$ in the limit $b \gtrsim a$; this gives

$$\Delta v_{\max} \simeq 4\pi (7/6)^{1/2}\, G\rho a t \tag{13}$$

where $\rho = nM$ is the mass density of perturbers. Since the mean stellar mass density in the solar neighbourhood ($\rho_* \simeq 0.05\,M_\odot\,\mathrm{pc}^{-3}$) exceeds that of molecular clouds ($\rho_c \simeq 0.015\,M_\odot\,\mathrm{pc}^{-3}$), the strongest individual perturbations experienced by comets over short timescales are those due to stars. Evaluating equation (13) gives the approximate result $\Delta v_{\max} \simeq 4.3\,(a/3 \times 10^4\,\mathrm{AU})(t/10\,\mathrm{Myr})\,\mathrm{m\,s^{-1}}$.

Assuming that such a velocity change leads to a corresponding change in the perihelion distance of a nearly parabolic orbit on the order of $\Delta q \simeq 2a^2 (\Delta v)^2 / GM_\odot$ we find $\Delta q_{\max}$ due to the single strongest stellar perturbation during time t is on the order of $38\,(a/3 \times 10^4\,\mathrm{AU})^4 (t/10\,\mathrm{Myr})^2\,\mathrm{AU}$, comparable with the change in perihelion distance introduced by the galactic tide *per revolution* (see Bailey et al., 1990 for a recent review). In general, significant fluctuations in the arrival rate of new comets may occur when comets with $a \lesssim 2 \times 10^4\,\mathrm{AU}$ receive changes in the perihelion distance Δq comparable with the radius, $q_{\mathrm{lc}} \simeq 10$–$15\,\mathrm{AU}$, of the loss cylinder imposed on the velocity distribution by planetary perturbations (Bailey et al., 1987, Fernández and Ip, 1987, Heisler et al., 1987, Heisler, 1990).

Over longer timescales the minimum impact parameter due to clouds remains in the range $b_{\min} \gg a$, but that due to stellar perturbations gradually decreases towards the limit $b_{\min} \ll a$. The ratio of the maximum change in velocity due to stars and clouds is then

$$\frac{\Delta v_*}{\Delta v_c} = \sqrt{\frac{12}{7}}\,\frac{\rho_*}{\rho_c}\,\frac{b_{\min,*}}{a}$$

$$\simeq 0.9 \left(\frac{0.01\,M_\odot\,\mathrm{pc}^{-3}}{\rho_c}\right)\left(\frac{3 \times 10^4\,\mathrm{AU}}{a}\right)\left(\frac{200\,\mathrm{Myr}}{t}\right)^{1/2} \tag{14}$$

showing that the single strongest molecular cloud perturbation is comparable with the strongest stellar perturbation at $a = 3 \times 10^4\,\mathrm{AU}$ at a time on the order of $200\,\mathrm{Myr}$. For times longer than this (depending on a), occasional strong molecular cloud perturbations become dominant.

For example, if we consider a particular encounter geometry (*cf.* Figure 1) in which the molecular cloud has velocity **V** directed parallel to the negative z axis of a rectangular heliocentric coordinate system, so that the point of closest approach cuts the (x, y) plane at an azimuthal angle ϕ_c, then $\hat{\mathbf{r}}.\hat{\mathbf{V}} = -\cos\theta$ and $\hat{\mathbf{r}}.\hat{\mathbf{b}} = \sin\theta\cos(\phi_c - \phi)$, where the comet's position vector is $\mathbf{r} = r(\sin\theta\cos\phi, \sin\theta\sin\phi, \cos\theta)$. Equation (4) then implies

$$(\Delta v_t)^2 = \left(\frac{2GM}{bV}\right)^2 \sin^2\theta \left\{\left(\frac{b^2}{d^2} - 1\right)^2 \sin^2(\phi_c - \phi) + \frac{r^2}{d^2}\cos^2\theta\right\} \tag{15}$$

Expanding this in the limit $b \simeq d \gg r$, using $d^2 = b^2 + r^2[1 - (\hat{\mathbf{r}}.\hat{\mathbf{V}})^2] - 2\mathbf{r}.\mathbf{b}$, and assuming $\Delta q = (r\,\Delta v_t)^2/2GM_\odot$ yields

$$\Delta q = \frac{2GM_\odot}{V^2}\left(\frac{M}{M_\odot}\right)^2 \frac{r^4 \sin^2\theta}{b^4}\left[1 - \sin^2\theta \cos^2\psi\right] \tag{16}$$

where $\psi = 2(\phi_c - \phi)$. Lastly, averaging this expression around an elliptical orbit of eccentricity e, giving $\langle r^4 \rangle = a^4(1 + 5e^2 + 15e^4/8)$, and setting $e \simeq 1$ for a nearly parabolic orbit, we obtain

$$\Delta q = \frac{2GM_\odot}{V^2}\left(\frac{M}{M_\odot}\right)^2 \frac{63}{8}\frac{a^4}{b^4}\sin^2\theta \left[1 - \sin^2\theta \cos^2\psi\right] \tag{17}$$

which may be compared with the corresponding expression derived by Torbett (1986, equation 9) on the assumption that the comet remains at aphelion throughout the perturbation and $\phi_c = \phi$. We note that inclusion of the ϕ dependence shows that the signature of a recent molecular cloud encounter – a non-uniform distribution of cometary aphelia over the sky – is more complex than a simple $\sin\theta\cos\theta$ factor (cf. Torbett, 1986, Fernández and Ip, 1990).

Finally, substituting $b = b_{min}$ in equation (17), the largest single change in perihelion distance due to molecular cloud encounters expected over a timescale $t \gtrsim 200\,\mathrm{Myr}$ is $\Delta q_{max} \simeq 1000\,(a/3 \times 10^4\,\mathrm{AU})^4\,(t/200\,\mathrm{Myr})^2$ AU. Molecular cloud encounters therefore produce comet showers with an intensity corresponding to filling the loss cylinder to $a \simeq 10^4$ AU about once every 200 Myr or so. Over these timescales and longer, occasional molecular cloud perturbations represent the dominant process which randomizes cometary orbits.

2.3.2 Total energy change

In a similar way we may use equation (11) to estimate the total energy transferred to the comet cloud by stellar and molecular cloud perturbations over time t. On short timescales the mean energy transfer is negligible, the perturbations being dominated by the term $\mathbf{v_0}.\Delta\mathbf{V}$ in equation (5). Over longer times equation (11) reveals that stellar perturbations initially dominate the energy transfer rate, but for $t \gtrsim 10^9\,\mathrm{yr}$ the cumulative effect of molecular clouds becomes more important. The ratio is approximately

$$\frac{\sum \Delta E_*}{\sum \Delta E_c} \simeq 1.3 \left(\frac{0.01 M_\odot \, \mathrm{pc}^{-3}}{\rho_c}\right)^2 \left(\frac{3 \times 10^4 \, \mathrm{AU}}{a}\right)^2 \left(\frac{10^9 \, \mathrm{yr}}{t}\right) \qquad (18)$$

again demonstrating that over the longest timescales the evolution of the outer Oort cloud is dominated by encounters with massive molecular clouds.

2.4 The survival problem

Whereas the original Oort cloud theory assumed that stellar perturbations were dominant on all timescales with Δq and $\sum \Delta E$ both governed by close stellar encounters, the understanding achieved during the course of the past ten years leads to a more complicated picture. On short timescales ($t \lesssim 10\,\mathrm{Myr}$) changes in perihelion distance are dominated by the galactic tide and occasional stellar perturbations, while $\sum \Delta E$ is dominated by stars. On longer timescales ($10 \lesssim t \lesssim 200\,\mathrm{Myr}$) occasional stellar perturbations become increasingly important in randomizing the cometary orbits, while for $t \gtrsim 200\,\mathrm{Myr}$ the strongest perturbations at $a \approx 3 \times 10^4\,\mathrm{AU}$ are those due to molecular clouds. Despite this, changes in $\sum \Delta E$ continue to be dominated by stars up to about $t \approx 10^9\,\mathrm{yr}$ (depending on a), until on the longest timescales occasional molecular cloud encounters dominate both $\Delta v_{\max}$ and $\sum \Delta E$. These qualitative statements and transition timescales depend on a (equation 18), but the general conclusion is clear: on timescales $t \gtrsim 200\text{--}1000\,\mathrm{Myr}$ the outer Oort cloud is severely disturbed by close encounters with nebulae. As suggested by Napier and Clube (1979) this leads to a severe survival problem for the outer Oort cloud.

Following Bailey (1986a) and the form of equation (11), the mean energy transfer rate due to both stars and clouds may be written in the form

$$\dot{\varepsilon} = \dot{\varepsilon}_* + \dot{\varepsilon}_c \simeq A_* + A_c a^2 \qquad (19)$$

where

$$A_* \simeq 4\pi G^2 \langle n_* M_*^2 / V_* \rangle I_* \simeq 10^{-13} \, \mathrm{m}^2 \, \mathrm{s}^{-3} \qquad (20)$$

and (allowing for finite radii R_c of the clouds and the likelihood of compact internal substructure; Bailey, 1986a, Hut and Tremaine, 1985)

$$A_c \simeq 4\pi G^2 \langle n_c M_c^2 / V_c \rangle \frac{7}{12 R_c^2} g_{\mathrm{tot}} \approx 10^{-44} \, \mathrm{s}^{-3} \qquad (21)$$

Here, the numerical factor $g_{\mathrm{tot}} \simeq 5$ depends on the internal density distribution and degree of clumping of the molecular clouds; its value, and those of the other parameters in equations (20) and (21) evaluated for average conditions in the solar neighbourhood, are discussed in Bailey (1986a). For completeness, we note that g_{tot} is related to the clumping factor C introduced by Hut and Tremaine (1985) by $g_{tot} = 2C$ (see their equations A11–A14). For a homogeneous cloud C is on the order of unity, but Hut and Tremaine (1985) argue that clumpiness probably increases the disruptive effect of clouds by a factor of about 2–3, consistent with the above value of g_{tot}.

Given these expressions for the mean energy transfer rate, it is straightforward to calculate the half-life due to stellar perturbations or molecular clouds (Bailey, 1986a, Hut and Tremaine, 1985). Following Bailey (*loc. cit.*, equations 11 and 21), the respective results are

$$t_{1/2,*} = \frac{1}{4.732} \frac{GM_\odot}{A_* a} \simeq 6 \times 10^9 \left(\frac{10^4 \, \text{AU}}{a}\right) \, \text{yr} \tag{22}$$

(*cf.* equation 10) and

$$t_{1/2,c} = \frac{1}{8.190} \frac{GM_\odot}{A_c a^3} \simeq 1.5 \times 10^{10} \left(\frac{10^4 \, \text{AU}}{a}\right)^3 \, \text{yr} \tag{23}$$

Thus, over the age of the solar system, the majority of comets originally formed with semi-major axes $a \gtrsim 2 \times 10^4$ AU will by now have been lost from the Oort cloud. Combining the half-lives indicates that stellar and molecular cloud perturbations together remove roughly half the comets originally formed with semi-major axes $a = 10^4$ AU (Weinberg *et al.*, 1986, Shoemaker and Wolfe, 1986, Staniucha and Banaszkiewicz, 1988, Staniucha, 1990).

Parenthetically, we note that occasional molecular cloud perturbations are equally important in discussions of the evolution of open star clusters (Terlevich, 1987) and wide binaries. In the latter case, Abt (1986, 1988) has argued that binaries of solar age have maximum separations on the order of 5×10^3 AU, although the exact figure may be strongly influenced by observational selection effects (*e.g.* Giannuzzi, 1988, Weinberg, 1990). Nevertheless, the available studies agree that there are relatively few wide binaries in the solar neighbourhood with semi-major axes $a \gtrsim 2 \times 10^4$ AU (*e.g.* Retterer and King, 1982, Latham *et al.*, 1984, Mazeh and Latham, 1988). This confirms the importance of external perturbations in determining the fate of such weakly bound systems (including the Oort cloud), although the surprisingly large variation in the number of wide binaries in different parts of the sky (Saarinen and Gilmore, 1989) and the complexity of realistic theoretical modelling of the quantities involved mean that one cannot yet convincingly argue that one or another class of perturber (*e.g.* molecular clouds or hypothetical dark matter) must be negligible (Bahcall *et al.*, 1985, Weinberg *et al.*, 1987, Wasserman and Weinberg, 1987). At the very least, these studies confirm the necessity in studies of Oort cloud evolution to consider both stars and molecular clouds (Weinberg *et al.*, 1987, Wasserman, 1988, Weinberg, 1990).

These results indicate the severity of the survival problem for the Oort cloud. Unless the parameter $A_c \simeq 10^{-44} \, \text{s}^{-3}$ in equation (21) has been seriously overestimated (which, it should be emphasized, may be the case in view of observational uncertainties in the average masses, radii and degree of clumping of molecular clouds, and the uncertainties involved in averaging these quantities over the encounter history of the solar system), those comets now entering the inner planetary region for the first time with $a \gtrsim 2 \times 10^4$ AU cannot have resided in such weakly bound orbits for the age of the solar system. Either they were formed in a hypothetical dense inner core of the Oort cloud, or they originated further out (in interstellar space) and have subsequently been captured, possibly during a past encounter of the solar system with a massive molecular cloud. These alternatives have been reviewed in detail by Bailey *et al.* (1990). In either case, observed comets may be regarded as the tip of an iceberg: in order for capture to be efficient there must be huge numbers of interstellar comets (presumably located close to their sites of formation in dense clouds), while on the more conventional solar system hypothesis the inner core must be of sufficient mass to replenish the dynamically unstable outer cloud for the entire age of the solar system.

In this way, the discovery of molecular clouds has had a considerable influence on cometary theories. In the future, there is a need to improve the accuracy of the molecular cloud half-life $t_{1/2,c}$ (for example, by reducing the uncertainty in molecular cloud masses through factors such as the CO/H_2 ratio, the internal structure and possible magnetic support of clouds, and the assumed distance to the centre of the Galaxy). On the dynamical side it is important to determine the effects of stellar and molecular cloud encounters on self-consistent evolutionary models of the Oort cloud (for example, to look for observational signatures of the process or to assess the viability of different solar system hypotheses). Finally, if our solar system is indeed ordinary in having lost a substantial mass of cometary material, the total number of unbound comets produced by stellar systems over the age of the Galaxy (particularly in the central regions, Fogg, 1988, 1990) could be sufficient to have important implications for a wide range of studies (*e.g.* Bailey, 1988, 1990, McGlynn and Chapman, 1989, Stern and Shull, 1988, 1990).

3   Molecular cloud comets

Two important lines of argument can be traced in answer to the question whether comets originate within molecular clouds. On the one hand, the primordial comet cloud originally envisaged by Oort (1950) would by now have been almost completely dispersed by the combination of stellar and molecular cloud perturbations over the age of the solar system. This result, together with other arguments, has led a number of authors to suggest that comets may not be primordial solar system bodies but instead are interstellar objects recently captured. Possible interstellar formation processes have been discussed, for example, by McCrea (1975), Yabushita (1983), Clube (1985), Napier and Humphries (1986), Bailey (1987a) and Napier (1990), these all implying that comets are a potentially important component of molecular clouds.

On the other hand, the dynamically unstable outer Oort cloud might be replenished from within, in which case a primordial solar system hypothesis for comets could be retained provided the Oort cloud contains a massive inner core. This implies a much larger original cometary mass than formerly envisaged and an enhanced rôle for comets in the early solar system. In particular, since the origin of comets on these theories is related to star or planet formation, the fact that most stars seem to originate within molecular clouds suggests that comets too are probably formed in such systems. Moreover, if the formation of the Oort cloud is inefficient (as, for example, on the planetesimal hypothesis, in which $\approx 90\%$ of the original mass of planetesimals is thought to be hyperbolically ejected), this view of comet formation requires the contemporaneous production of many interstellar comets.

These arguments, including observational evidence for volatile species in comet nuclei (indicating formation of ices in a very cold "molecular cloud" environment; *e.g.* Yamamoto *et al.*, 1983, A'Hearn and Feldman 1985), suggest that molecular clouds should contain many comets. In addition to the observed gas and dust, the latter mostly of submicron size with volatile icy mantles, molecular clouds are thus expected to contain solid bodies with dimensions in the range 1–100 km, not to mention a predicted population of "giant" grains as well (*e.g.* Bailey, 1987b, and references therein).

A third line of argument (*e.g.* Greenberg, 1982, 1988) approaches the question of comet formation from the perspective provided by interstellar dust. Here, the main result is the realization that a typical interstellar grain passes through many different phases of the interstellar medium during its life, from a hot intercloud phase to cold atomic and molecular clouds, and back again. The final grain eventually incorporated into a new star, meteorite parent body or comet should show isotopic and chemical signatures of this complex history (*e.g.* Clayton, 1982, 1988, Clayton *et al.*, 1989), providing potentially important constraints on theories of galactic chemical evolution and the interstellar medium (*e.g.* Ming *et al.*, 1989).

This picture of the evolution of interstellar dust (which has additional consequences for the chemical evolution of dense clouds; *e.g.* Williams, 1987, 1988), implies that cometary material is best viewed as an aggregate of ice-covered interstellar particles, and that the *pre-cometary* dust grain probably has a structure characterized by one or more submicron-sized silicate grains glued together with a mantle of volatile icy material or less volatile organic residue modified by ultraviolet irradiation, sputtering and grain-grain collisions (Greenberg, 1982, 1988, Greenberg and Hage, 1990).

Interstellar grains are thus expected to undergo a variety of condensation, coagulation and accumulation processes until they eventually grow to sizes comparable to observed comets. The precise details of this growth naturally depend on the underlying assumptions about the final site of comet formation, but virtually all proposed theories point to early grain aggregation occurring within a dense molecular cloud. Subsequently, depending on whether comets are presumed to form entirely in interstellar space (Napier, 1990, and references therein), in the outer parts of a collapsing protostellar cloud (Hills, 1981, 1982), in an expanding wind-driven shell around a young star (Bailey, 1987a), in an extended protoplanetary nebula (Biermann and Michel, 1978, Cameron, 1978) or deep within the protoplanetary zone itself (Safronov, 1969), different aspects of the grain accumulation hierarchy may be realized.

3.1 The hierarchical grain aggregate model

This section briefly illustrates how this picture of interstellar grain growth may be applied to the case of comet formation in a protoplanetary disc. The idea is an extension of work by previous authors (Daniels and Hughes, 1981, Greenberg, 1982, 1986, Yamamoto, 1985, Fechtig and Mukai, 1985, Meakin and Donn, 1988, Donn, 1990), each of whom have separately emphasized both the porous "fluffy" nature of interstellar or interplanetary dust particles and the importance of considering the interstellar/protostellar link and the possible condensation of molecules from the surrounding gaseous medium, whether a molecular cloud or protoplanetary disc.

First, refractory interstellar grain cores and volatile icy mantles experience what may be described as a prolonged phase of "cooking" in the interstellar medium (Greenberg, 1978, 1982, Williams, 1988). During this stage of evolution they frequently pass between different parts of the interstellar medium, suffering a variety of physical and chemical changes due to collisions, sputtering, condensation, and ultraviolet or cosmic-ray irradiation. This phase of evolution lasts on the order of 10^8–10^9 yr.

Eventually, the grains in a particular location enter a giant molecular cloud for the last time, and the refractory grain cores and any surviving organic residue become coated with a volatile icy layer of condensed gases, principally molecules such as H_2O, CO, CH_4, and CO_2 (Greenberg and d'Hendecourt, 1985, Knacke and Larson, 1990). These mantled grains may persist in the molecular cloud for times up to about 10^7 yr before finally becoming part of a cool, dense clump eventually destined to form a new star. In such a region the collision timescale t_c for grains of radius a_g material density ρ_g and mean mass density $\rho_d = n_d m_d$, where $m_d = \frac{4}{3}\pi \rho_g a_g^3$, is on the order of $a_g \rho_g / \rho_d v_{\rm rel}$, i.e.

$$t_c \approx 5 \times 10^5 \left(\frac{a_g}{0.1\,\mu{\rm m}}\right) \left(\frac{\rho_g}{1\,{\rm g\,cm^{-3}}}\right) \left(\frac{10^4\,{\rm cm^{-3}}}{n_{H_2}}\right) \left(\frac{10\,{\rm m\,s^{-1}}}{v_{\rm rel}}\right) \,\,{\rm yr} \qquad (24)$$

where we have assumed a dust-to-gas ratio by mass on the order of 1%. Here n_{H_2} is the molecular hydrogen number density and $v_{\rm rel}$ is the relative velocity of different grains. Following Scalo (1977), Völk et al. (1980) and Hills (1982), we assume that turbulence in the gas, possibly aided by effects due to differential radiation pressure, leads to relative grain-grain velocities on the order of 1–10 m s^{-1}. In this way we expect a modest degree of coagulation to occur during the final molecular cloud phase of precometary grains, leading to the formation of particles with sizes on the order of 0.5–1 μm by the time they enter a collapsing fragment. The grains produced in such a process will tend to have a porous open structure with mean densities rather less than 1 g cm^{-3} (Daniels and Hughes, 1981, Donn and Hughes, 1986, Meakin and Donn, 1988, Brooks, 1990), though they are *not* generally expected to have a simple fractal structure (Brooks, 1990, *cf.* Donn, 1990). If the grain temperatures are sufficiently low, the spaces within them may also be partially filled by condensation of molecules from the gas phase. This leads to a heterogeneous "interstellar grain aggregate", in which volatile and non-volatile compounds are intimately mixed down to scales comparable with those of the original interstellar particles.

Once these aggregates have become part of a protostellar cloud, they are either accreted on to a growing protostar or become incorporated into a protoplanetary disc, part of which eventually forms planets. The high dust density within the disc allows grain growth to proceed much more rapidly than before, and it is plausible to assume that both the porosity and the types of icy compound that condense within the aggregate are now different than before: not only are the fundamental building blocks different, but the temperature and chemistry of the surrounding gas is almost certain to have changed. Moreover, the processes governing grain-grain collisions in the disc are bound to differ from the formerly considered interstellar or protostellar environments (*e.g.* Mizuno, 1989). We thus envisage that aggregates formed by accretion in the protoplanetary disc should be qualitatively different from those formed during the earlier interstellar phase of evolution, though they should still have a weakly bound, low-density, fluffy structure.

Following Greenberg *et al.* (1984) and Bailey (1990), for example, the rate of grain growth during this protoplanetary phase of evolution may be estimated from the expression

$$a_b \simeq \frac{\sqrt{3}\xi_c}{8} \frac{\Sigma_d \Omega}{\rho_b} t \approx 10\, \xi_c \left(\frac{0.5 \text{ g cm}^{-3}}{\rho_b}\right) \left(\frac{t}{10^5 \text{ yr}}\right) \quad \text{m} \qquad (25)$$

where ξ_c is the mean sticking probability per collision, Σ_d is the surface density of dust in the disc, Ω is the rotational angular velocity of the disc, a_b is the radius of the ensuing "boulders" at time t, and ρ_b is their mean material density. The interstellar grain aggregates thus accumulate into boulder-sized bodies with diameters ranging up to a few tens of metres on a timescale on the order of 10^5–10^6 yr, assuming parameters typical of those in the Uranus-Neptune zone of the sun's protoplanetary disc (i.e. $\Sigma_d \simeq 5 \text{ kg m}^{-2}$ and a heliocentric distance $r \simeq 25 \text{ AU}$).

The later stages of accumulation, on timescales ranging up to 10^8 yr or more and leading to planet formation in the case of the few largest bodies, are again expected to differ from the previous phases, perhaps because the remaining cool gas has been lost from the system (Yamamoto and Kozasa, 1988) and certainly because the velocity dispersion induced by gravitational effects will lead to significant local heating and compression of particles at points of impact. Local particle adhesion may also occur as a result of processing by energetic ions (Johnson, 1985), possibly associated with an early protostellar wind.

In this way, a hierarchy of accretion processes leads finally to bodies, presumably comet nuclei, with diameters ranging up to some tens or hundreds of kilometres. The process has not been described in detail, but it is important to emphasize that the model predicts that cometary dust and comet nuclei should show evidence for several distinct phases of accumulation: the "interstellar grain aggregate" phase (associated with grain growth in the parent molecular or protostellar cloud), the protoplanetary disc phase, and finally the comet nucleus phase. Each of these episodes, if they occur, should leave a mark in the form of detailed physical and chemical substructure within the aggregated nucleus (Weissman, 1986, 1988), and their presence or otherwise in particular cases may be used to test different theories of comet formation. (The particular scenario described above applies, of course, only to the planetesimal hypothesis.) The general picture implies that comet nuclei are a heterogeneous collection of grains of widely varying sizes, comprising volatile and non-volatile components intimately mixed down to the smallest scales. The degree to which cometary grains and grain aggregates with signatures characteristic of a particular mode of accretion are dominant should provide useful constraints on the ultimate site of comet formation.

4 Conclusions

This paper has briefly considered the impact of molecular clouds on the development of cometary theory during the course of the past ten years or so. These massive clouds have probably exerted a significantly disruptive effect on the Oort cloud over the age of the solar system, to the extent that comets now observed entering the inner planetary region with semi-major axes $a \gtrsim 2 \times 10^4$ AU are unlikely to have originated in such weakly bound orbits.

This result has led to a growing awareness that a primordial solar system hypothesis for cometary origin probably requires the existence of a massive inner core to replenish the dynamically unstable outer region, and hence a greater mass of comets. Alternatively, comets might be very numerous within molecular clouds allowing one or another proposed capture mechanism to provide enough comets to explain the observations. Nevertheless, even a solar system hypothesis for comet formation requires that molecular clouds should contain many comets. Moreover, the initial accumulation of grains within clouds is almost certainly the first stage in a process that ultimately ends by forming a comet nucleus.

A greater understanding of molecular clouds is thus fundamental to theories of cometary origin, whilst consideration of grain accumulation processes in different environments leads to a picture in which comet nuclei are essentially *hierarchical* aggregates. The proposed sequence leads finally to kilometre-sized bodies, incorporating particles which on the smallest scales are of interstellar dust dimensions and contain clues to the earliest interstellar and protostellar phases of accretion. These tiny grains (which may be observed as the ultimate disintegration product of cometary dust) are incorporated into larger bodies which should also contain signatures of their accretion history, for example of a possibly warmer protoplanetary disc phase. These particles are finally combined into more massive aggregates (boulder-sized objects and larger) which eventually make up the observed comet nucleus. An understanding of the physics and chemistry of each phase of accretion should provide a useful diagnostic for the structure of the comet nucleus and the final site of comet formation.

Acknowledgments

The results on cometary grain growth are based on a continuing investigation in collaboration with A. Brooks and F.D. Kahn. This work was supported by the SERC.

References

Abt, H.A., 1986, Astrophys. J., **304**, 688.
Abt, H.A., 1988, Astrophys. J., **331**, 922.
A'Hearn, M.F. and Feldman, P.D., 1985, in *Ices in the Solar System*, ed. J. Klinger, D. Benest, A. Dollfus and R. Smoluchowski, (Reidel, Dordrecht) p. 463.
Bahcall, J.N., Hut, P. and Tremaine, S., 1985, Astrophys. J., **290**, 15.
Bailey, M.E., 1983, Mon. Not. Roy. Astron. Soc., **204**, 603.
Bailey, M.E., 1986a, Mon. Not. Roy. Astron. Soc., **218**, 1.
Bailey, M.E., 1986b, in *Asteroids, Comets, Meteors II*, ed. C.-I. Lagerkvist, B.A. Lindblad, H. Lundstedt and H. Rickman (Reprocentralen HSC, Uppsala), p. 207.
Bailey, M.E., 1986c, Nature, **324**, 350.
Bailey, M.E., 1987a, Icarus, **69**, 70.
Bailey, M.E., 1987b, Q. J. Roy. Astron. Soc., **28**, 242.
Bailey, M.E., 1988, *Dust in the Universe*, ed. M.E. Bailey and D.A. Williams, (Cambridge University Press.) p. 113.
Bailey, M.E., 1990 in *Baryonic Dark Matter*, ed. D. Lynden-Bell and G. Gilmore, (Kluwer, Dordrecht) p. 7.
Bailey, M.E., Clube, S.V.M. and Napier, W.M., 1990, *The Origin of Comets*. (Pergamon, Oxford).
Bailey, M.E., Wilkinson, D.A. and Wolfendale, A.W., 1987, Mon. Not. Roy. Astron. Soc., **227**, 863.
Biermann, L., 1978, in *Astronomical Papers Dedicated to Bengt Strömgren*, ed. A. Reiz, and T. Anderson, (Copenhagen University Observatory) p. 327.
Biermann, L. and Michel, K.W., 1978, Moon and Planets, **18**, 447.
Brooks, A., 1990, in *Asteroids, Comets, Meteors III*, ed. C.-I. Lagerkvist, H. Rickman, B.A. Lindblad and M. Lindgren, (Reprocentralen HSC, Uppsala) p. 25.
Cameron, A.G.W., 1978, Moon and Planets, **18**, 5.
Clayton, D.D., 1982, Q. J. Roy. Astron. Soc., **23**, 174.
Clayton, D.D., 1988, in *Dust in the Universe*, ed. M.E Bailey and D.A. Williams, (Cambridge University Press) p. 145.
Clayton, D.D., Scowen, P. and Liffman, K., 1989, Astrophys. J., **346**, 531.
Clube, S.V.M., 1985, in *Dynamics of Comets: Their Origin and Evolution*, ed. A. Carusi, and G.B. Valsecchi, IAU Coll. No. 83, (Reidel, Dordrecht) p. 19.
Clube, S.V.M. and Napier, W.M., 1982, Q. J. R. Astron. Soc., **23**, 45.
Clube, S.V.M. and Napier, W.M., 1984, Mon. Not. Roy. Astron. Soc., **208**, 575.
Clube, S.V.M. and Napier, W.M., 1985, Icarus, **62**, 384.
Daniels, P.A. and Hughes, D.W., 1981, Mon. Not. Roy. Astron. Soc., **195**, 1001.
Donn, B., 1990, Astron. Astrophys., **235**, 441.

Donn, B. and Hughes, D.W., 1986, 20th ESLAB Symposium on *The Exploration of Halley's Comet*, ed. B. Battrick, E.J. Rolfe and R. Reinhard, ESA SP-250, Vol. III, (ESA Publications, ESTEC, Noordwijk) p. 523.
Fechtig, H. and Mukai, T., 1985, in *Ices in the Solar System*, ed. J. Klinger, D. Benest, A. Dollfus and R. Smoluchowski, (Reidel, Dordrecht) p. 251.
Fernández, J.A. and Ip, W.-H., 1987, Icarus, **71**, 46.
Fernández, J.A. and Ip, W.-H., 1990, in *Comets in the Post-Halley Era*, IAU Coll. No. 116, ed. R.L. Newburn, J. Rahe and M.M. Neugebauer, (in press, Kluwer, Dordrecht).
Fogg, M.J., 1988, Earth, Moon, Planets, **43**, 123.
Fogg, M.J., 1990, Comments Astrophys., **14**, 357.
Giannuzzi, M.A., 1988, Astrophys. Space Sci., **142**, 241.
Greenberg, J.M., 1978, in *Cosmic Dust*, ed. J.A.M. McDonnell, (Wiley-Interscience, Chichester), p. 187.
Greenberg, J.M., 1982, in *Comets*, ed. L. Wilkening, IAU Coll. No. 61 (University of Arizona Press, Tucson) p. 131.
Greenberg, J.M., 1986, in *Asteroids, Comets, Meteors II*, ed. C.-I. Lagerkvist, B.A. Lindblad, H. Lundstedt and H. Rickman, (Reprocentralen HSC, Uppsala), p. 221.
Greenberg, J.M., 1988, in *Dust in the Universe*, ed. M.E. Bailey and D.A. Williams, (Cambridge University Press), p. 121.
Greenberg, J.M. and d'Hendecourt, L.B., 1985, in *Ices in the Solar System*, ed. J. Klinger, D. Benest, A. Dollfus and R. Smoluchowski, (Reidel, Dordrecht), p. 185.
Greenberg, J.M. and Hage, J.I., 1990, Astrophys. J., **361**, 260.
Greenberg, R., Weidenschilling, S.J., Chapman, C.R. and Davis, D.R., 1984, Icarus, **59**, 87.
Heggie, D.C., 1975, Mon. Not. Roy. Astron. Soc., **173**, 729.
Heisler, J., 1990, Icarus, (in press).
Heisler, J., Tremaine, S. and Alcock, C., 1987, Icarus, **70**, 269.
Hills, J.G., 1981, Astron. J., **86**, 1730.
Hills, J.G., 1982, Astron. J., **87**, 906.
Hut, P., 1983, Astrophys. J., **268**, 342.
Hut, P. and Tremaine, S., 1985, Astron. J., **90**, 1548.
Johnson, R.E., 1985, in *Ices in the Solar System*, ed. J. Klinger, D. Benest, A. Dollfus and R. Smoluchowski, (Reidel, Dordrecht), p. 337.
Knacke, R.F. and Larson, H.P., 1990, Astrophys. J., (in press).
Latham, D.W., Tonry, J., Bahcall, J.N., Soneira, R.M. and Schechter, P., 1984, Astrophys. J., **281**, L41.
Mazeh, T. and Latham, D.W., 1988, Astrophys. Space Sci., **142**, 131.
McCrea, W.H., 1975, Observatory, **95**, 239.
McGlynn, T.A. and Chapman, R.D., 1989, Astrophys. J., **346**, L105.
Meakin, P. and Donn, B., 1988, Astrophys. J., **329**, L39.
Ming, T., Anders, E., Hoppe, P. and Zinner, E., 1989, Nature, **339**, 351.
Mizuno, H., 1989, Icarus, **80**, 189.
Morris, D.E. and O'Neill, T.G., 1988, Astron. J., **96**, 1127.
Napier, W.M., 1982, in *Sun and Planetary System*, ed. W. Fricke and G. Teleki, (Proc. Sixth European Regional Astronomy Meeting; Astrophys. Space Sci. Lib, **96**) (Reidel, Dordrecht), p. 375.
Napier, W.M., 1990, in *Dusty Objects in the Universe*, ed. E. Bussoletti and A.A. Vittore, (Kluwer, Dordrecht), p. 103.
Napier, W.M. and Clube, S.V.M., 1979, Nature, **282**, 455.
Napier, W.M. and Humphries, C.M., 1986, Mon. Not. Roy. Astron. Soc., **221**, 105.
Napier, W.M. and Staniucha, M., 1982, Mon. Not. Roy. Astron. Soc., **198**, 723.
Nezhinski, E.M., 1972, in *The Motion, Evolution of Orbits, and Origin of Comets*, ed. G.A. Chebotarev, E.I. Kazimirchak-Polonskaya and B.G. Marsden, IAU Symp. No. 45, (Reidel, Dordrecht), p. 335.
Oort, J.H., 1950, Bull. Astron. Inst. Neth., **11**, 91.

Retterer, J.M. and King, I.R., 1982, Astrophys. J., **254**, 214.
Rickman, H., 1976, Bull. Astron. Inst. Czechosl., **27**, 92.
Saarinen, S. and Gilmore, G., 1989, Mon. Not. Roy. Astron. Soc., **237**, 311 and Microfiche 237/1.
Safronov, V.S., 1969, Translated by Israel program for scientific translations, (Jerusalem, 1972).
Scalo, J.M., 1977, Astron. Astrophys., **55**, 253.
Sekanina, Z., 1968, Bull. Astron. Inst. Czechosl. **19**, 291.
Shoemaker, E.M. and Wolfe, R.F., 1986, in *The Galaxy and the Solar System*, ed. R. Smoluchowski, J.N. Bahcall and M.S. Matthews, (University of Arizona Press, Tucson), p. 338.
Staniucha, M.S., 1990, in *Asteroids Comets Meteors III*, ed. C.-I. Lagerkvist, H. Rickman, B.A. Lindblad and M. Lindgren, (Uppsala University). p. 439.
Staniucha, M.S. and Banaszkiewicz, M., 1988, in *The Few Body Problem*, ed. M.J. Valtonen, IAU Coll. No. 96, (Kluwer, Dordrecht), p. 201.
Stern, S.A. and Shull, J.M., 1988, Nature, **332**, 407.
Stern, S.A. and Shull, J.M., 1990, Astrophys. J., **359**, 506.
Terlevich, E., 1987, Mon. Not. Roy. Astron. Soc., **224**, 193.
Torbett, M.V., 1986, Mon. Not. Roy. Astron. Soc., **223**, 885.
Völk, H.J., Jones, F.C., Morfill, G.E. and Röser, S., 1980, Astron. Astrophys., **85**, 316.
Wasserman, I., 1988, Astrophys. Space Sci., **142**, 267.
Wasserman, I. and Weinberg, M.D., 1987, Astrophys. J., **312**, 390.
Weinberg, M.D., 1990, in *Baryonic Dark Matter*, ed. D. Lynden-Bell and G. Gilmore, (Kluwer, Dordrecht), p. 117.
Weinberg, M.D., Shapiro, S.L. and Wasserman, I., 1986, Icarus, **65**, 27.
Weinberg, M.D., Shapiro, S.L. and Wasserman, I., 1987, Astrophys. J., **312**, 367.
Weissman, P.R., 1980, Nature, **288**, 242.
Weissman, P.R., 1986, Nature, **320**, 242.
Weissman, P.R., 1988, in *Comet Halley 1986: Worldwide Investigations, Results and Interpretations*, ed. J. Mason and P. Moore, (in press, Ellis Horwood Publ. Co.)
Williams, D.A., 1987, in *Physical Processes in Interstellar Clouds*, ed. G.E. Morfill and M. Scholer, (Reidel, Dordrecht), p. 377.
Williams, D.A., 1988, in *Dust in the Universe*, ed. M.E. Bailey, and D.A. Williams, (Cambridge University Press), p. 391.
Yabushita, S., 1983, Astrophys. Space Sci., **89**, 159.
Yamamoto, T., 1985, in *Ices in the Solar System*, ed. J. Klinger, D. Benest, A. Dollfus and R. Smoluchowski, (Reidel, Dordrecht), p. 205.
Yamamoto, T., Nakagawa, N. and Fukui, Y., 1983, Astron. Astrophys., **122**, 171.
Yamamoto, T. and Kozasa, T., 1988, Icarus, **75**, 540.

Molecular cloud isotopic chemistry and comets

V. VANYSEK

DEPARTMENT OF ASTRONOMY AND ASTROPHYSICS, CHARLES UNIVERSITY, PRAGUE, SVÉDSKÁ 8. 150 00 PRAHA 5, CZECHOSLOVAKIA.

Abstract The relation between chemical composition of comets to that of dense cool molecular interstellar clouds could be interpreted as an indirect evidence that the comets contain pristine material of the protosolar nebula with different history of evolution and formation. In this material may be inherited signatures of isotopic ratios related to the different environments in the interstellar matter and cool clouds – precursors of the solar nebula. The results concerning the isotopic abundances in comets indicate that the isotopic ratio in cometary material is essentially terrestrial. However, there is an exception. The deuterium-to-hydrogen ratio in cometary water is enhanced by a factor about 100 relative to the D/H ratio in diffuse ISM and much larger enhancement could be expected in other volatile species. Also the ratio of carbon stable isotopes 12/13 of the volatile phase seems to be significantly lower in contrast to the ratio derived from mass spectroscopy of the refractory material, which tends to be terrestrial. This effect could be explained by the gas-phase or by gas-grain chemistry in dense ISMC.

Isotopic ratios in comets. It is almost generally accepted that comets are relics of the protosolar nebula, composed from material partly formed in the interstellar environment and possibly containing information about processes in the dense molecular cloud (or ISMC) precursor(s) of the solar nebula. However, the abundances of molecular species in comets and in the ISMC are only partly known and the number of possible channels involved in the interstellar chemistry is very large. Straightforward comparison of molecular abundances in comets and the ISMC could be misleading. On this problem we may shed the light of studies concerning the abundances of some stable isotopes. Although the available data about isotopic ratios in comets are still very scarce and limited only to a few stable isotopes, they may provide hints about the evolution of the ISMC and presolar nebula.

Cometary nuclei contain refractory and volatile material of different history which may exhibit compositon signatures of different environments in ISMC, whereas dust grains keep their original composition virtually unchanged. The abundance of stable isotopes in dust particles is therefore determined by the chemical composition of interstellar diffuse matter or by the abundances of heavier elements in circumstellar environments. In molecular compounds of the cometary ice the isotopic ratios may be influenced by fractionation processes. Important channels for the isotopic exchange between interstellar molecules are the ion-neutral reactions at low temperature.

Isotopic ratios are known for a few comets for the stable isotopes of ^{12}C and ^{13}C, derived from ground based observations of isotopically shifted features in molecular bands. For Comet Halley we have data for more isotopes obtained by mass spectroscopy *in situ*. However all these results are to some extent hampered by observational or instrumental artifacts. A review about isotopes in comets with a long list of references was very recently published by Vanysek (1990).

Available data are insufficient for quantitative analysis of the isotopic abundance in comets. The abundances of stable isotopes heavier than carbon, *i.e.* as oxygen or sulphur in the gas phase as well as in cometary dust, appear to be almost identical with the solar system data (here denoted as terrestrial). The same holds to some extent for Si, Ca and perhaps also for Mg. No conclusion can be made for ^{14}N, ^{15}N. Further unbiased but indirect information about the isotopic ratio in comets can be inferred from the analysis of dust particles collected in the stratosphere, some of which are likely to be of cometary origin. The data (Geiss, 1988) show that for heavier elements the isotopic ratio is essentially solar. However, this is not true for the D/H ratio and the ^{12}C/^{13}C ratio also deviates from the solar value. The D/H ratio inferred from the mass-spectroscopic data obtained *in situ* for a ratio HDO/H$_2$O ranges from 5×10^{-5} to 5×10^{-4} (Eberhardt et al., 1987) and the average is almost the same as in the solar system, but deuterium enhancement in distant objects seems to be a real effect. If we accept the lowest value as representative, the deuterium abundance in cometary water is enhanced by a factor of 500 relative to the ISM or 100 relative to the presolar nebula (Kunde et al., 1982, Geiss and Reeves, 1981). The average D/H ratio is enhanced by a factor 100 relative to that assumed for the diffuse ISM. Larger enhancements may be expected for other volatile species in cometary nuclei. The D/H ratio in comets covers the range of D/H in outer planets and Titan and tends to be larger than in the SMOW (standard mean ocean water).

The ^{12}C/^{13}C ratio is the only isotopic ratio obtained for more than one comet. The range of the ^{12}C/^{13}C ratio obtained by various methods varies between ~ 20 and several hundred. The low ratios however are due partly to mass ambiguities caused by ^{12}CH$^+$. Results outlined here indicate that the ratio of ^{12}C/^{13}C as derived from dust and from diatomic carbon molecules bands tends to be solar or even higher, but in CN molecules ^{13}C seems to be enhanced or ^{12}C depleted.

Isotopic Fractionation Effect. Chemical processes in interstellar clouds and the protosolar nebula may lead to isotopic fractionation in molecular species at low temperatures. The deuterium enrichment in outer planets has been discussed in terms of the equilibrium fractionation (Geiss and Reeves, 1981). These enrichments may be a result of equilibrium partitioning reactions *i.e.* deuterium will be partitioned into heavier molecules as the temperature of the solar nebula drops (Reeves and Bottinga, 1972). However the efficiency of such a process depends on the ratio of time-scales of the partition process t* and the characteristic "free-fall" time of the solar nebula is $t \sim 10^6$ years. For reactions such as

$$CH_4 + HD \Leftrightarrow CH_3D + H_2 ; \quad \text{or} \quad H_2O + HD \Leftrightarrow HDO + H_2$$

(Beer and Taylor, 1974) the time-scale for $T < 500\,\text{K}$ is $t^*/t = \tau, \tau > 10^3$ so partitioning reactions are ineffective. Even the surface chemistry (Grinspoon and Lewis, 1987) requires extremely long equilibration times. Vanysek and Vanysek (1985) and Ip (1985) have proposed another process for deuterium enhancement, based on the assumption that relatively fast ion-neutral molecule reactions are important in redistributing deuterium among the various species. One example is

$$D^+ + H_2 \Leftrightarrow H^+ + HD$$

which is exothermic to the right and is very efficient as a deuteration process at low temperatures. Such reactions have been extensively studied as main channels for deuteration of interstellar molecules (*e.g.* Watson, 1976, Smith, 1987, Herbst, 1988, Millar *et al.*, 1989). In dense and cool ISMC the deuterated species with higher D/H ratios might condense on the dust particles before the cometesimals formation. Deuteration is virtually terminated by the burial of the molecules in grains (Geiss, 1988), so the enrichment of condensates or organic refractory by the deuterium depends on time scales of the saturation as well as the condensation process. Quasi-time dependent deuteration models for

$$H_2 \Rightarrow HD, \quad H_2O \Rightarrow HDO, \quad HCN \Rightarrow DCN, \quad NH_3 \Rightarrow NH_2D$$

and other species were computed by Brown and Rice (1986) assuming $n_H \sim 2 \times 10^4$ cm and ionization coefficient (by cosmic rays) $10^{-17}\,s^{-1}$ for a dark cloud with T = 10 K. Such a model suits well the properties of the dark clouds as TMC-1. The results may be representative also for the solar nebula if most of heavier elements (Mg, Si, Fe) were contained in the dust grains. The calculated evolution of deuterium enrichment in this model shows that the time required for the saturation in all considered species is about 3×10^5 years, or $\tau = 0.3$. The enrichment of a compound XH by D, without stochiometric reduction, can be defined as

$$R(X) = \frac{n(XD)}{n(XH)} \left(\frac{H}{D}\right)$$

where $(H/D) = 6.7 \times 10^4$ is the average conventional value for the ISM. For almost all cases, with the exception of HD/H_2, $R(X) > 100$ is reached at $\tau = 0.1$. The condensable species such as H_2O, HCN, H_2CO and CH are highly deuterated so the abundance of D could be significantly enhanced in cometary ices even if the formation of cometary nuclei is relatively fast *cf.* the free-fall time scale of the presolar nebula. Owen *et al.* (1986) pointed out that the observed enhancement of deuterium in methane in the atmospheres of giant planets is evidence that the gases from ices which were trapped in planets and have mixed with the gaseous envelope were enriched in deuterium prior to planetary formation. Recently Lutz *et al.* (1990) estimated the H/D ratio that may be present in these primitive ices in the protosolar nebula. They developed simple but limiting models which constrain the amount of deuterated volatile species diluted in the original hydrogen envelopes of giant planets. These suggest that a D/H ratio in primitive ices between 10^{-4} and 10^{-3}, corresponding to $R(X) \sim 10 - 100$. The enhancement may be significantly larger if the condensed volatiles form stable mantles on the dust particles in the dark clouds before a very early stage of the solar nebula. The gas-dust grain interaction concerning the deuterium interstellar chemistry has been discussed recently by Brown and Millar (1989). In their time-dependent models the deuterated species are together with the corresponding hydrogen compounds almost all accreted on the dust particles within a time-interval $\tau \simeq 1$.

A similar effect can be expected in the distribution of ^{13}C in carbon bearing species. Fractionation of ^{13}C occurs via the ion-molecule reaction which may lead to enrichment in CO, but will be efficient only at temperatures of 10 to 30 K in the periphery of a dense cloud. If the above mentioned differences between the isotopic ratios of the molecular carbon and CN in comets are real (Wyckoff *et al.*, 1989), they indicate chemical fractionation processes in the environment where the cometary volatile material was formed.

References

Beer, R. and Taylor, F.W., 1974, Astrophys. J. **178**, 309.
Brown, P.D. and Millar, T.J., 1989, Mon. Not. Roy. Astron. Soc., **237**, 661.
Brown, R.D. and Rice, E.H.N., 1986, Mon. Not. Roy. Astron. Soc., **223**, 429.
Eberhardt, P., Krankowsky, D., Schulte, W., Dolder, U., Laemerzahl, P., Berthelier, J,J., Woweries, J., Stubbemann, U., Hodges, R.R., Hoffmann, J.H. and Illiano, J. M., 1987, Astron. Astrophys., **187**, 435.
Geiss, J. and Reeves, H., 1981, Astron. Astrophys., **93**, 189.
Geiss, J., 1988, in *Review of Modern Astronomy 1 (Cosmic Chemistry)*, ed. G. Klare, (Springer Verlag, Heidelberg) p. 1.
Grinspoon, D.H. and Lewis, J.S., 1987, Icarus, **61**, 430.
Herbst, E., 1988, in *Review of Modern Astronomy 1 (Cosmic Chemistry)*, ed. G. Klare, (Springer Verlag, Heidelberg) p. 114.
Ip, W.-H., 1985, in *Ices in the Solar System*, ed. J. Klinger, D. Benest, A. Dollfus and R. Smoluchowski, (Reidel, Dordrecht) p. 389.
Kunde, V., Honel, R., Maguire, W., Gautier, D., Baliteau, J.P., Marten, A., Chedin, A., Husson, N. and Scott, N., 1982, Astrophys. J., **263**, 443.
Lutz, B.L., Owen, T. and de Bergh, C., 1990, Icarus, **86**, 329.
Millar, T.J., Bennet, A. and Herbst, E., 1989, Astrophys. J., **340**, 906.
Owen, T., Lutz, B.L. and de Bergh, C., 1986, Nature, **320**, 244.
Reeves, H. and Bottinga, V., 1972, Nature, **232**, 1.
Smith, D., 1987, Phil. Trans. Roy. Soc. London A **323**, 269.
Vanysek, V. and Vanysek, P., 1985, Icarus, **61**, 57.
Vanysek, V., 1990, in *Comets in Post-Halley Era*, ed. R. Newburn, J. Rahe and M.M. Neugebauer, (Kluwer, Dordrecht, in press).
Watson, W.D., 1976, Rev. Mod. Phys., **48**, 513.
Wyckoff, S., Lindholm, E., Wehinger, P.A., Peterson, B.A., Zucconi, J-M. and Festou, M.C., 1989, Astrophys. J., **339**, 488.

Dust in molecular clouds

D. A. WILLIAMS

DEPARTMENT OF MATHEMATICS, UMIST, MANCHESTER M60 1QD, U.K.

Abstract The nature of molecular mantles on dust in molecular clouds is briefly described. It is argued that the efficiency of mechanisms continuously limiting mantle growth on dust is relatively low, and that mantle growth is limited by intermittent events such as star formation. The consequences of high depletions of heavy molecules on to dust are explored. A cyclic model of molecular clouds, indicated by observations and appropriate to some objects, is described. It is shown that winds of low mass stars driving a circulation of gas can affect both the chemistry and the growth and composition of mantles.

1 Introduction and Survey

For the purposes of this discussion, I shall regard interstellar clouds with extinction $A_v \lesssim 1$ as diffuse, and those with $A_v \gtrsim 1$ as molecular. This division is arbitrary; molecules are clearly present in many diffuse clouds. However, the chemistry in interstellar clouds becomes much more well-developed and molecules contain a significant fraction of elemental abundances when the effective extinction is more than a couple of magnitudes, and such clouds deserve the epithet "molecular". Such clouds incorporate both quiescent regions and more active localities where, for example, star formation is occurring, or a strong interaction between cloud and star is present. Dust plays important parts in both diffuse and molecular clouds. Since the nature of molecular clouds is different from that of diffuse clouds, so too the roles of dust are different.

The presence of dust in molecular clouds is directly apparent from infrared extinction studies which – on the assumption of a "standard" interstellar extinction curve – often imply very large visual extinctions. Infrared continuum emission detected by IRAS from warm dust in localized regions and from cooler dust in the general interstellar medium reveals the widespread nature of interstellar dust. The silicate features in emission and absorption, near 10 and 20 μm, are also detected in a wide variety of sources and indicate that silicates are a significant component of dust in molecular clouds (as is the case in diffuse clouds; see Whittet, 1988). Dust in molecular clouds has a range of sizes which may be systematically larger than in diffuse clouds because of the accretion of mantles of molecular ices. The size distribution certainly extends to small sizes (radius $\lesssim 100\text{Å}$) but the lower cut-off (if any) is unknown. The *widespread* existence of polycyclic aromatic hydrocarbons (PAHs) in molecular clouds has been postulated (see, e.g., reviews by Allamandola, Tielens and Barker, 1989, Puget and Léger, 1989), but this suggestion remains controversial (cf. Jones, Duley and Williams, 1990, Cossart-Magos and Leach, this volume, p. 319).

The major roles of dust in both diffuse and molecular clouds are summarized in Table 1. The optically related effects (extinction, polarization, scattering, luminescence, and photoelectric effect) may be significant in molecular clouds especially in regions of high radiation field (such as star forming regions). The depletion of refractory elements such as C, Mg, Si and other metals observed in diffuse clouds also applies to molecular clouds. In addition, the molecular ices detected in molecular clouds indicate that there is also significant further depletion of C and O from the gas phase.

Molecular hydrogen is formed efficiently at the surfaces of dust grains in diffuse clouds. It is not known whether this process is equally efficient in molecular clouds. Surface reactions on dust may contribute to diffuse cloud chemistry, although the significance of the contribution cannot yet be determined from observations. In molecular clouds, both direct and indirect contributions to the chemistry are likely to occur. Direct routes imply immediate ejection of the products of surface reactions to the gas, whereas indirect routes imply retention of the products as part of a molecular mantle, to be released at some later time.

Mantles of molecular ice are widespread in molecular clouds. They may contain significant fractions of the interstellar elemental abundances. Mantles may be destroyed continuously, or abruptly in a single event. In either case, the injection of the molecules may significantly modify the gas phase chemistry. Unrestricted growth of molecular mantles would ultimately lead to the depletion of all heavy species from the gas phase, with consequences for the gas temperature and ionization. Various scenarios have been proposed to address this question.

The presence of dust grains in interstellar shocks may have important consequences. Sputtering and grain-grain collisions may erode or destroy both the mantles and – possibly – the refractory cores, thus modifying the abundance and size distribution of the dust, and contributing to gas phase processes in the shock. In MHD C-type shocks charged grains can form a significant additional fluid which contributes to the ion-neutral drag and may modify the shock structure. In dense clouds where the ionization is low and dust grains are the dominant charge carriers these effects may become pronounced (Pilipp, Hartquist and Havnes, 1990).

In this paper I shall describe in detail several of these aspects of dust behaviour in molecular clouds. In Section 2 I discuss some recent work on molecular mantles. Section 3 deals with mechanisms which may limit mantle growth. Section 4 describes the chemistry and physics in collapsing cores in which accretion of molecules on to dust is occurring. This suggests the need for cyclic models of molecular clouds, and these are briefly described in Section 5. Section 6 makes some conclusions.

2 Dust and mantles in molecular clouds

Detailed reviews of these subjects have been published recently by Whittet (1988) and by Tielens (1989), and the reader is referred to these papers and to many others appearing in the two books "Dust in the Universe" (1988) (ed. Bailey and Williams) and "Interstellar Dust" (1989) (ed. Allamandola and Tielens) for further information. I draw particular attention to the paper by Allamandola and Sandford (1988). In brief, near infrared observations of molecular clouds

Table 1. *Dust grains in diffuse and molecular clouds*

Dust grain property	Diffuse	Molecular
Depletions of refractory elements	✓	✓
Absorption, scattering, polarization of starlight	✓	✓
Luminescence	✓	✓ (cloud edges)
IR - bands	✓ weak absorption	✓ strong emission and absorption
- continuum	✓ (inc. cirrus)	✓
Heat gas by photoelectric effect	✓	✓ (cloud edges)
Carriers of electric charge	✓ (+ and −)	✓ (weak −)
Effects in shocks		
- sputtering, grain-grain collisions	✓	✓ (inc. loss of mantles)
- modify shock structure	−	✓ (very dense gas)
Molecular Mantles		
- detection	−	✓ ($H_2O: A_v \gtrsim 3$; $CO: A_v \gtrsim 5$)
- solid state chemistry	−	✓ (e.g. $N \rightarrow NH_3$)
- photochemistry	−	✓ (e.g. CH_3OH formation)
Limitation or inhibition of mantle growth		
- continuous	✓ (photo-desorption?)	✓ (photodesorption, cosmic rays)
- intermittent	−	✓ (associated with star formation)
Effects of mantle growth	−	✓ (ionization, ambipolar diffusion time, cooling, star formation)

show a number of absorption features attributed to dust and molecular mantles on dust. See Table 2 for a list of observed absorption features and the proposed assignments.

Table 2. *Observed Interstellar Absorption Features attributed to dust and mantle material*

Wavelength (μm)	Proposed Identification
2.9	NH_3
3.0	H_2O
3.4	C–H stretch
3.5	CH_3OH
3.5	H_2CO
3.9	H_2S
4.6	OCN^-, SiH
4.7	CO
4.9	OCS
6.0	H_2O
6.8	NH_4^+, CH_3OH
9.7	Silicate
15	CO_2
19	Silicate
21	Iron oxide
42	H_2O

(after Whittet, 1988, Tielens, 1989)

The presence of the silicate features suggest that molecular cloud dust is, in fact, diffuse cloud dust acting as a low temperature substrate for the deposition of molecular ices from the gas phase. The water ice feature at 3.0 μm varies from source to source, apparently in response to local conditions, but it can be well fitted to laboratory data (cf. Tielens and Allamandola, 1987a, b). Water ice is the most abundant constituent of the ice mantle. Absorption by solid CO is also widely detected. Adamson *et al.*, (1988) show that the onsets of mantles of H_2O and CO in the Taurus dark clouds occur at extinctions of about 3 and 5 magnitudes, respectively. The onset of H_2O mantles in ρ Oph sources, however, is at extinction of 10–15 magnitudes (Tanaka *et al.*, 1990).

There is no doubt that an extensive solid-state photochemistry is occurring in the icy mantles on dust in molecular clouds (Grim *et al.*, 1989, Allamandola and Sandford, 1988, Tielens, 1989). This is illustrated, for example, in the work of Baas *et al.* (1989) in assigning a feature near 3.5 μm to CH_3OH. Further examples will be found in the work of Grim *et al.*, (1989), who demonstrate, *inter alia*, the photochemical production of trapped solid state ions OCN^- and NH_4^+ in the laboratory. These ions provide good fits to features observed in W33A near 4.62 and 6.86 μm.

Theoretical studies of the limited gas and solid chemistries have also been made. Tielens (1983) demonstrated that the two compositions may be quite different in character. Brown (1990) has investigated the formation of complex molecules in the solid state, and has suggested that the reactions of H-atoms with other species must be suppressed if the complex species are to form (see also Pickles and Williams, 1977). Brown and Millar (1989a, b) have described surface chemistries leading to high fractions of singly and multiply deuterated species.

Various models of individual molecular clouds, incorporating time dependent gas phase chemistry, and freeze-out without return of heavy species on the surfaces of dust grains, and some surface chemistry have been studied in the context of a static, uniform dense gas. The well observed quiescent object, TMC-1, has been modelled in this way by Brown and Charnley (1990) who find that it is possible to match the observed molecular abundances, to predict an "age" of the cloud, and to describe the present composition of the molecular mantles. In the absence of a suitable background or embedded source, the mantle composition cannot be observed. Such studies do not, however, deal with the physical development of such objects (see Section 4).

If molecular mantles can be released into the gas, then the rich solid-state chemistry can significantly affect the observed gas phase abundances. d'Hendecourt et al. (1985) investigated a model in which mantles were efficiently and continuously recycled into the gas. Mantle return may, however, occur abruptly, creating transient "anomalous" populations of molecules (cf. Wilson, 1985). Such a situation may be occurring in the Orion Hot Core and similar objects (see, e.g. Pauls et al., 1983). In the Orion Hot Core the molecules NH_3, HDO and CH_3OH are detected with abundances greatly enhanced with respect to values expected for cold clouds. Brown, Charnley and Millar (1988) have explored this situation in some detail. In their model, free-fall isothermal collapse of a cloud, accompanied by gas phase chemistry and accretion of heavy species on to dust grains with some surface chemistry, is assumed to be halted by the birth of a nearby bright star. The dust is heated, and the mantles are abruptly evaporated, returning both processed and unprocessed material to the gas phase. Brown et al. showed the Hot Core abundances do not reflect the local conditions; rather, the situation is transient. Some species, such as NH_3, could be formed in high abundance in the cold collapsing cloud via surface reactions, while other species – again, formed in low temperature chemistry – such as HCN, HC_3N, H_2CO are merely preserved in the mantle and released in the Hot Core. The "anomalous" abundances are maintained for about 10^4 years, until the unusually low ion abundance in the core increases in response to the changed conditions. Brown and Millar (1989a, b) predict high abundances for OD, can explain the observed enhanced abundances of NH_2D and HDO, and propose that multi-deuterated species such as NHD_2 should be detectable in hot cores.

I conclude that the chemical composition of molecular clouds is significantly affected by the presence of dust, through the processes of mantle growth, solid-state chemistry, and destruction. I shall illustrate below (Section 4) that the *physical* properties of the molecular cloud may also be significantly modified by these same processes.

3 The gas-grain interaction

The detection of molecular mantles in many molecular clouds indicates that the sticking of heavy atoms and molecules at the surfaces of cold dust grains is efficient. The conclusion is also expected from theoretical considerations (Leitch-Devlin and Williams, 1984). The significant thickness of observed mantles seems to rule out the possibility that the sticking efficiency declines as successive monolayers are accreted (cf. Stecher and Williams, 1967).

The time required for substantial mantles to form (assuming no return of material from solid to the gas) is t_m where

$$t_m(yr) \simeq 1.5 \times 10^5 \left(\frac{10^4 cm^{-3}}{n}\right)\left(\frac{1}{S}\right)\left(\frac{2.1 \times 10^{-21} cm^2}{\pi a_0^2 n_g/n}\right)\left(\frac{10}{T}\right)^{\frac{1}{2}}$$

where n is the total hydrogen number density, S is the effective sticking probability for the main component of the mantle (e.g. H_2O, CO, etc.) on grains which have effective radius a_0 and number density n_g. The gas temperature is T. The normalization is to the grain surface area per hydrogen atom for dust causing the visible and UV extinction (Duley and Williams, 1984). It may be that only the larger grains accrete mantles. If so, t_m will of course be larger. If small grains can accrete mantles, and if there is a large population of very small grains, then t_m would be smaller; but this seems unlikely.

For typical molecular clouds, t_m could be short compared to both the expected age of such clouds and to the time needed to achieve chemical steady-state, and could be comparable to the free-fall time. Thus, unless mechanisms exist to return mantle material efficiently to the gas, no chemical steady-state is possible, and both the chemistry and the physical development of the cloud will be affected by mantle growth. Because these implications are so significant, mechanisms limiting mantle growth have been proposed and developed in detail.

Léger et al. (1985) showed that the passage of the iron nuclei in cosmic rays through dust grains dominates the cosmic ray heating of grains. This heating causes the evaporation of CO from sufficiently small grains and partial evaporation from larger grains. However, H_2O, NH_3 and other polar molecules are not removed from the mantles. Léger et al. deduced that while H_2O should be almost totally frozen-out in dense, long-lived quiescent regions, an equilibrium abundance of gas phase CO could be maintained by this process.

However, equilibrium cannot be attained. The gas phase reactions

$$He^+ + CO \longrightarrow He + C^+ + O$$

and

$$C^+ + H_2 \longrightarrow CH_2^+ + h\nu$$

drive the formation of hydrocarbons which – if they strike a grain before being converted to CO will be permanently retained in the molecular mantle

Dust in molecular clouds

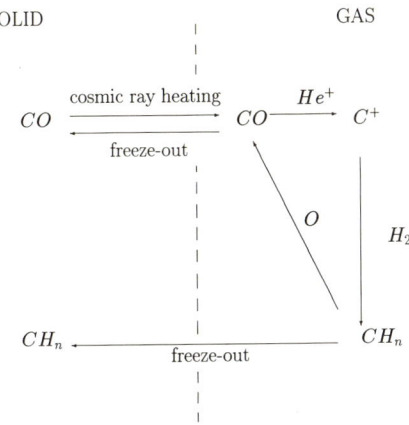

Figure 1: A simplified scheme illustrating the loss of gas phase CO.

(Hartquist and Williams, 1990), see Figure 1. For $n \gtrsim 10^6$ cm^{-3} the gas phase CO abundance decays with time approximately exponentially:

$$n_{gas}(\text{CO}) \propto exp\left[-t/\{2 \times 10^6 yr\, (n/10^6 cm^{-3})\}\right]$$

At lower densities, the situation is slightly more complicated in that the loss of CO is affected more severely by several factors including sticking to grains, the He$^+$ production, and the reaction with He$^+$. Thus, the CO abundances calculated by Léger et al. are appropriate at high densities only for relatively short times.

Photodesorption may provide an additional mechanism for limiting mantle growth. Duley et al. (1989) discuss the response of amorphous dust to the ambient radiation field near molecular cloud edges, and show that photodesorption is capable of accounting for the onsets of H$_2$O and CO mantles at distinct values of A_v (see Whittet et al., 1988, 1989). In addition, Duley et al. considered the possibility that photodesorption may occur deep in cloud interiors by the UV radiation field initiated in the interaction of cosmic rays with H$_2$ (Prasad and Tarafdar, 1983). The details of the photochemical processes involved have been further considered by Hartquist and Williams (1990), who relied on relevant experimental work such as that of Nishi, Shinohara and Okuyama (1984). Direct photodesorption and photodissociation leading to the ejection of products can both occur. Hartquist and Williams (1990) conclude that in the case of H$_2$O the latter process is the more important, and the consequence is that the photodissociation products (especially OH) will not be lost from the mantle unless the dissociation occurs in the outermost layers of the mantle. This leads to a yield per UV photon of about 0.1. Then, the ratio of the desorption rate to the rate at which heavy molecules stick to grains is about

$$\left(\frac{X}{10^{-7}}\right)^{-1} \left(\frac{n}{10^5 cm^{-3}}\right)^{-1} \left(\frac{\zeta}{10^{-17} s^{-1}}\right)$$

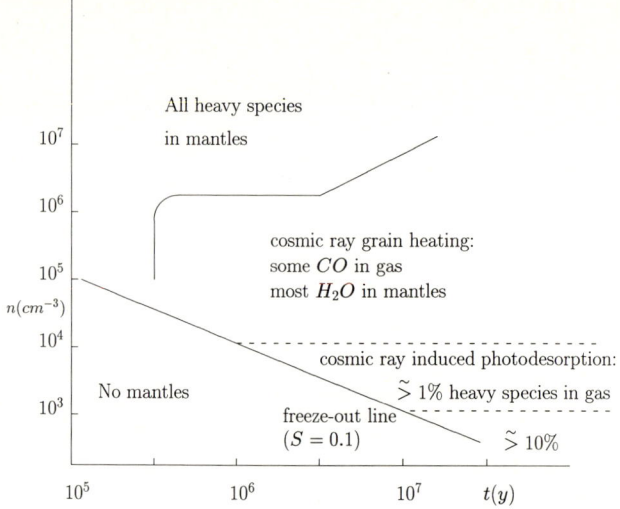

Figure 2: The figure illustrates semi-schematically the effects of cosmic ray grain heating and cosmic ray induced photodesorption in controlling mantle growth. Below the freeze-out line (illustrated for a sticking probability of 0.1) no mantles occur in this density (n)/time (t) plot. Above the freeze-out line, the growth of mantles is inhibited to some extent by the cosmic ray mechanisms described in the text.

where X is the fractional abundance of heavy species in the gas phase (Hartquist and Williams, 1990). Thus, for densities greater than about 10^3 cm^{-3}, the freeze-out of molecules may be expected to dominate photodesorption in cloud interiors.

Figure 2 indicates semi-schematically the effects of the two cosmic ray induced desorption mechanisms discussed above. Young, relatively low density clouds (lower left) initially without mantles may evolve variously. If clouds remain at low density, then ageing takes them ultimately into a region (lower right) where mantles occur. The régime described by Léger et al. occupies the central zone in this n,t diagram. Here, some gaseous CO is present while most H$_2$O is in ice. At higher densities, we expect almost all heavy species to be in mantles. We explore the consequences of such a situation in the next section.

The collapse of clumps in the process of star formation is obviously one important type of dynamical evolution that can occur. It is worth noting that the NH$_3$ line profiles for dense cores with embedded infrared sources (Benson and Myers, 1983, Menten et al., 1984, Menten and Walmsley, 1985) do not show the expected high velocity wings associated with the collapsing gas. Menten et al. (1984) suggested that the NH$_3$ may be depleted by a large factor in this densest and most rapidly infalling material. The NH$_3$ is clearly evident in low density material, and it has a high abundance. This situation supports the view that the densest material will – in the absence of mantle evaporation or removal – tend to be the most heavily depleted and that the timescales associated with such dense regions must normally be very short.

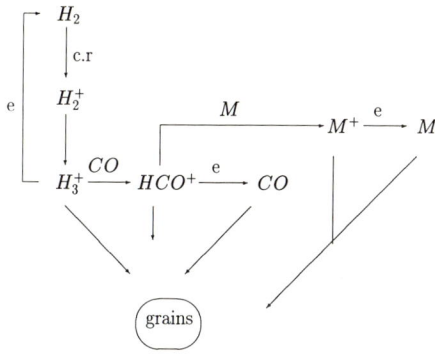

Figure 3: A simplified scheme illustrating the processes controlling the level of ionization in molecular clouds

4 Dust, chemistry, and the evolution of molecular clouds

In this section we consider the fate of clumps of gas in which it is assumed that freeze-out of heavy species is unrestricted.

Magnetic fields almost certainly play an important part in supporting molecular clouds and clumps against gravity. The fields are coupled to the gas through the ions and electrons which are constrained to move with the field. Collisions between ions and neutrals causes a friction which limits the speed of movement between the field and the neutral gas. Ambipolar diffusion – the ion-neutral relative motion – is controlled by this friction, and the ambipolar diffusion time, t_D, is a measure of the period for which magnetic support may be significant. The time t_D may be shown to be proportional to the level of ionization $x_i = n_i/n$ (where n_i is the number density of ions) and is, approximately

$$t_D(yr) \sim 10^6 \, (x_i/10^{-8})$$

(Mouschovias, 1979, Hartquist and Williams, 1989). The ionization is controlled by the chemistry. A simplified chemistry containing the essential features for dark clouds is illustrated in Figure 3, in which M represents heavy metallic elements (e.g. Mg, Fe). The recombinations of H_3^+ and M^+ with electrons are relatively slow compared to the dissociative recombination of HCO^+. Thus, in molecular clouds in which gas phase CO is abundant, fast recombinations of HCO^+ ions with electrons maintain the level of ionization at relatively low values, $x_i \sim 10^{-8} - 10^{-7}$.

If, however, CO is strongly depleted from the gas then recombination will be slower and the level of ionization will increase, with a consequent increase in t_D.

Hartquist and Williams (1989) compare the time required for freeze-out of CO to proceed far enough to effect a significant increase in fractional ionization with the time required for ambipolar diffusion to weaken the magnetic field in a (partially) magnetically supported clump. They show that events are determined according to density. If collapse to form a star is well established while $n < n_{crit}$, a critical density, then ambipolar diffusion occurs, magnetic support diminishes, and the collapse continues. On the other hand, if $n > n_{crit}$ then the loss of heavy molecules with consequent increase in x_i occurs before ambipolar diffusion can reduce the field strength, so that the field is frozen-in and magnetic support continues. Hartquist and Williams find that $n_{crit} \sim 10^7$ cm^{-3}.

Thus, the presence of dust and the phenomenon of freeze-out of heavy molecules may have a direct bearing on the process of star formation in molecular clouds. In addition, the discussion of Hartquist and Williams suggests the possibility that magnetically supported clumps within molecular clouds may exist as examples of arrested star formation, a metastable state (cf. Williams, 1985). These clumps would have densities $\gtrsim 10^7$ cm^{-3}, sizes $\sim 10^{-2}$ pc, with very high depletions of all heavy species leading to high fractional ionizations, long ambipolar diffusion times, and frozen-in magnetic fields. Obviously, the dust grains in such objects would have the maximum possible mass of molecular mantles. The abundances of H_3^+ and H_2D^+ should be high, but – unfortunately – neither molecule has yet been detected in interstellar clouds. The molecule CH should have an anomalously high abundance since the He^+, CO reaction ultimately feeds CH, whose loss with O atoms is now suppressed. These clumps will also be unusually warm, since most coolants have been removed. Temperatures will be controlled ultimately by H_2 and limited to 50–100 K.

5 Dynamical models of molecular clouds

Large scale turbulence-driven mass re-circulation of gas (Boland and de Jong, 1982) or grain-grain collisions (Greenberg, 1979) have been invoked as mantle limitation mechanisms. Both such processes require a substantial energy input, which is – presumably – derivable ultimately from stellar energies. Observations of a nearby molecular cloud, Barnard 5 (B5) by Goldsmith, Langer and Wilson (1986), suggest that winds of newly formed low mass stars (detected as IRAS sources) may drive a mass circulation of the gas. Goldsmith *et al.* propose that the winds of newly formed stars are powerful enough to disrupt and destroy the clumps from which the stars were formed. The clump gas is eroded by these winds, and the mass-loaded wind is eventually brought to rest by a reverse shock to form a new clump from which a new star will later be born. The cycle then repeats. Goldsmith *et al.* suggest that the cycle time is short, possibly $< 10^6$ years. If so, the chemistry is always out of steady-state, and there is a natural process limiting mantle growth. This kind of cyclic model has similarities with that proposed by Norman and Silk (1980). The behaviour of the chemistry and the dust in a cyclic type of model was discussed in outline by Williams and Hartquist (1984), and in a Norman and Silk picture by Charnley *et al.* (1988a). The chemical behaviour of B5 in the dynamical picture of Goldsmith *et al.* has been studied by Charnley *et al.* (1988b). The chemistry in this model has now been considerably extended by Nejad, Williams and Charnley (1990).

The general characteristics of these dynamical models are interesting, and may resolve difficulties associated with static equilibrium or pseudo-time-dependent models. These characteristics include the following features:

1. After a few cycles in which the final abundances from one cycle form the input for the next, the chemistry repeats precisely, and the molecular abundances are true tracers of the position in the cycle (the "limit cycle" effect). Knowledge of the initial values arbitrarily adopted for the start of the first cycle is lost after one cycle.
2. The chemistry never attains a steady state since the dynamical time is less than the chemical time. Calculated abundances always represent a "young" chemistry.
3. Growth of mantles on dust in molecular clouds subject to cyclic events is limited by the periodic ablation and shocks associated with star formation.
4. The models predict the chemistry in both gas and solid phases as a function of time and space. Hence, detailed point-by-point comparisons with observational maps are possible.
5. The cycling process requires molecular gas to pass repeatedly through "interface" zones, which may have characteristic observables (Charnley et al., 1990).

Nejad et al. (1990) studied in particular the chemistry of nitrogen-bearing molecules. Their model indicates, in particular, that substantial abundances of ammonia should be present in the dense clump phase. While a variety of gas phase chemical routes feed the NH_3 chemistry, in this cyclic model it appears that surface reactions followed by ejection of NH_3 etc. during the passage through a shock provide the major source. In addition to the gas phase abundances, these cyclic models also make predictions concerning the distribution of molecular mantles. The detection of ice mantles at IRS1 in B5 (Charnley, Whittet and Williams, 1990) is consistent with the cyclic model. The model predicts that the cycle time must be comparable to the freeze-out time if – as is observed – comparable amounts of molecular material are in gas and solid phases.

6 Conclusions

In addition to the conventional effects attributed to dust (see Table 1), it appears that dust in molecular clouds plays a significant, possibly dominant role in the evolution of those clouds. Although the precise nature of the underlying dust is relatively unimportant, the dust must be sufficiently refractory to survive repeated cycles involving shocks and heating. The dust must be able to provide a stable cold substrate, preferably with a broad phonon spectrum, on which the freeze-out of gas phase species can occur. The dust models proposed for the diffuse interstellar medium, involving silicates and carbon, provide suitable substrates, and such dust is the carrier of the near infrared emission bands seen in regions of high excitation. We infer that molecular cloud dust is diffuse cloud dust with accreted molecular mantles.

The mantles are widely detected in molecular clouds, principally through absorption features of H_2O and CO ices. These mantles contain a substantial fraction of the elements C, O, N available. Solid-state chemistry occurs in the mantles. At its simplest, this consists of H-addition reactions. Several species (CH_3OH, NH_4^+, OCN^-) provide good fits to observed infrared features and indicate the presence of a complex photochemistry.

Release of mantles (e.g. from heated grains) can provide local molecular overabundances; these effects are transient and persist only until the chemical network is able to readjust to the local conditions.

The processes which may continuously return mantle material to the gas have limited efficiency: some clouds may evolve into conditions in which freeze-out of heavy molecules is almost complete. A major consequence of significant freeze-out is the rise in ionization and increase in the ambipolar diffusion time. Thus, trajectories which lead to star formation may be impeded by these processes. The possibility arises that in some circumstances the star forming collapse may be arrested and that a magnetically supported clump is formed.

Observations indicate that in some molecular clouds a recycling of material into and out of dense regions is driven by winds of low mass stars. Such clouds can sustain their own physical existence for many cycles, develop a permanently "young" chemistry, and limit the mantle growth. Thus, in these clouds, the timescales of depletion and of clump collapse must be comparable, with comparable gas and solid fractions of heavy molecules.

Acknowledgement

This review includes a report of work done in collaboration with Dr T. W. Hartquist, to whom I am grateful for many stimulating discussions.

References

Adamson, A.J., Whittet, D.C.B., Bode, M.F., Longmore, A.J., Roche, P.F., McFadzean, A.D., Geballe, T.R. and Aitken, D.K., 1988, in *Dust in the Universe*, ed. M.E. Bailey and D.A. Williams, Cambridge University Press, p.61.
Allamandola, L.J. and Sandford, S.A., 1988, in *Dust in the Universe*, ed. M.E. Bailey and D.A. Williams, Cambridge University Press, p.229.
Allamandola, L.J. and Tielens, A.G.G.M., 1989, ed. *Interstellar Dust*, Kluwer Academic Publishers, Dordrecht.
Allamandola, L.J., Tielens, A.G.G.M. and Barker, J.R., 1989, Astrophys. J. Suppl. **71**, 733.
Baas, F., Grim, R.J.A., Geballe, T.R., Schutte, W. and Greenberg, J.M., 1988, in *Dust in the Universe*, ed. M.E. Bailey and D.A. Williams, Cambridge University Press, p.55.
Bailey, M.E. and Williams, D.A., 1988, eds. *Dust in the Universe*, Cambridge University Press.
Benson, P.J. and Myers, P.C., 1983, Astrophys. J., **270**, 589.
Boland, W. and de Jong, T., 1982, Astrophys. J., **261**, 110.
Brown, P.D., 1990, Mon. Not. Roy. Astron. Soc. **243**, 65.
Brown, P.D., Charnley, S.B. and Millar, T.J., 1988, Mon. Not. Roy. Astron. Soc., **231**, 409.
Brown, P.D. and Charnley, S.B., 1990. Mon. Not. Roy. Astron. Soc., **244**, 432.

Brown, P.D. and Millar, T.J., 1989a, Mon. Not. Roy. Astron. Soc., **237**, 661.
Brown, P.D. and Millar, T.J., 1989b, Mon. Not. Roy. Astron. Soc., **240**, 25P.
Charnley, S.B., Dyson, J.E., Hartquist, T.W. and Williams, D.A., 1988a, Mon. Not. Roy. Astron. Soc., **231**, 269.
Charnley, S.B., Dyson, J.E., Hartquist, T.W. and Williams, D.A., 1988b, Mon. Not. Roy. Astron. Soc., **235**, 1257.
Charnley, S.B., Dyson, J.E., Hartquist, T.W. and Williams, D.A., 1990, Mon. Not. Roy. Astron. Soc. **243**, 405.
Charnley, S.B., Whittet, D.C.B. and Williams, D.A., 1990, Mon. Not. Roy. Astron. Soc., **245**, 161.
d'Hendecourt, L.B., Allamandola, L.J. and Greenberg, J.M., 1985, Astron. Astrophys., **152**, 130.
Duley, W.W. and Williams, D.A., 1984, *Interstellar Chemistry*, Academic Press, London.
Duley, W.W., Jones, A.P., Whittet, D.C.B. and Williams, D.A., 1989, Mon. Not. Roy. Astron. Soc. **241**, 699.
Goldsmith, P.F., Langer, W.D. and Wilson, R.L., 1986, Astrophys. J. **303**, L11.
Greenberg, J.M., 1979, in *Stars and Star Systems*, ed. B.E. Westerland (D. Reidel, Dordrecht) p. 173.
Grim, R.J.A., Greenberg, J.M., de Groot, M.S., Baas, F., Schutte, W.A., and Schmitt, B., 1989, Astron. Astrophys. Suppl., **78**, 161.
Hartquist, T.W. and Williams, D.A., 1989, Mon. Not. Roy. Astron. Soc., **241**, 417.
Hartquist, T.W. and Williams, D.A., 1990, Mon. Not. Roy. Astron. Soc. (in press).
Jones, A.P., Duley, W.W. and Williams, D.A., 1990, Quart. J. Roy. Astron. Soc. (in press).
Léger, A., Jura, M. and Omont, A., 1985, Astron. Astrophys., **144**, 147.
Leitch-Devlin, M.A. and Williams, D.A., 1984, Mon. Not. Roy. Astron. Soc., **210**, 577.
Menten, K.M. and Walmsley, C.M., 1985, Astron. Astrophys., **146**, 369.
Menten, K.M., Walmsley, C.M., Krügel, E. and Ungerechts, H., 1984, Astron. Astrophys., **137**, 108.
Mouschovias, T. Ch., 1979, Astrophys. J., **228**, 475.
Nejad, L.A.M., Williams, D.A. and Charnley, S.B., 1990, Mon. Not. Roy. Astron. Soc., **246**, 183.
Nishi, N., Shinohara, H. and Okuyama, T., 1984, J. Chem. Phys., **80**, 3898.
Norman, C. and Silk, J., 1980, Astrophys. J., **238**, 158.
Pilipp, W., Hartquist, T.W. and Havnes, O., 1990, Mon. Not. Roy. Astron. Soc., **243**, 685.
Prasad, S.S. and Tarafdar, S.P., 1983, Astrophys. J., **267**, 603.
Puget, J.L. and Léger, A., 1989, Ann. Rev. Astron. Astrophys., **27**, 161.
Stecher, T.P. and Williams, D.A., 1968, Nature, **219**, 1349.
Tanaka, M., Sato, S., Nagata, T. and Yamamoto, T., 1990, Astrophys. J., **352**, 724.
Tielens, A.G.G.M., 1983, Astron. Astrophys., **119**, 177.
Tielens, A.G.G.M., 1989, in *Interstellar Dust*, ed. L.J. Allamandola and A.G.G.M. Tielens, Kluwer Academic Publishers, Dordrecht, p.239.
Tielens, A.G.G.M. and Allamandola, L.J., 1987a, in *Physical Processes in Dense Clouds*, ed. G.E. Morfill and M. Scholer, D. Reidel, Dordrecht, p.333.
Tielens, A.G.G.M. and Allamandola, L.J., 1987b, in *Interstellar Processes*, ed. D. Hollenbach and H. Thronson, D. Reidel, Dordrecht, p.397.
Whittet, D.C.B., 1988, in *Dust in the Universe*, ed. M.E. Bailey and D.A. Williams, Cambridge University Press, p.25.
Whittet, D.C.B., Bode, M.F., Longmore, A.J., Adamson, A.J., McFadzean, A.D., Aitken, D.K. and Roche, P.F., 1988, Mon. Not. Roy. Astron. Soc., **233**, 321.
Whittet, D.C.B., Adamson, A.J., Duley, W.W., Geballe, T.R. and McFadzean, A.D., 1989, Mon. Not. Roy. Astron. Soc., **241**, 707.
Williams, D.A. and Hartquist, T.W., 1984, Mon. Not. Roy. Astron. Soc., **210**, 141.
Wilson, T.L., 1985, Comments Astrophys., **11**, 83.

Solid CO and CO$_2$ in grain mantles

D.C.B. WHITTET[1,2] and H.J. WALKER[3]

Abstract The infrared spectral absorption feature of solid CO near 4.7 μm provides an important diagnostic of the chemical composition and thermal evolution of grain mantles, and leads to the prediction that CO$_2$ is also present in solid form. Recent attempts to confirm this by observations of the CO$_2$ feature at 15 μm in IRAS low-resolution spectra have produced somewhat conflicting results. Detection of the intrinsically-stronger CO$_2$ feature at 4.27 μm will be a major objective for the Infrared Space Observatory.

1 The detection of solid CO and its significance

The presence of solid carbon monoxide in interstellar grain mantles was anticipated by Duley (1974) and confirmed observationally by Lacy *et al.* (1984), who detected the vibrational feature of CO at 4.67 μm wavelength in several molecular clouds. Subsequent observations have shown that solid CO is ubiquitous in the cold, quiescent molecular cloud environment: in some lines of sight, the degree of CO depletion onto grains may be sufficient to reduce the gas phase CO abundance appreciably, a result of considerable astrophysical significance. In addition to its importance as a tracer of molecular material, gas phase CO is vital to the production of many polyatomic molecules by gas phase reaction schemes. The loss of CO onto grains could have a drastic effect on the carbon budget in dense clouds, altering or perhaps even controlling the abundances of larger carbon-bearing molecules (Mitchell *et al.*, 1988), but many models for interstellar chemistry discussed in recent years neglect the freeze-out process.

Observations of solid CO and their implications for gas phase abundances and the chemical and thermal evolution of grain mantles have been reviewed in depth by Whittet and Duley (1990), and the reader is referred to this paper for further discussion. We wish to draw attention here to the fact that details of the 4.67 μm CO feature profile (position, width and shape) provide important constraints on the chemical composition of the host grain mantle in which the CO is contained. Laboratory studies have shown, for example, that the width of the feature is a function of the dipole moment of the host (Sandford *et al.*, 1988): ices containing a substantial fraction of polar molecules such as H$_2$O *cannot* explain the sharpness of the feature, and this provides the first direct observational evidence for the presence of grain mantles in which H$_2$O is not the dominant constituent. One mixture studied in the laboratory which appears to give a more acceptable fit to the observations is CO:CO$_2$.

[1] SCHOOL OF PHYSICS AND ASTRONOMY, LANCASHIRE POLYTECHNIC, PRESTON PR1 2TQ, U.K.
[2] LABORATORY FOR SPACE RESEARCH, PO BOX 800, 9700 AV GRONINGEN, THE NETHERLANDS.
[3] RUTHERFORD APPLETON LABORATORY, CHILTON, DIDCOT, OXON OX11 0QX, U.K.

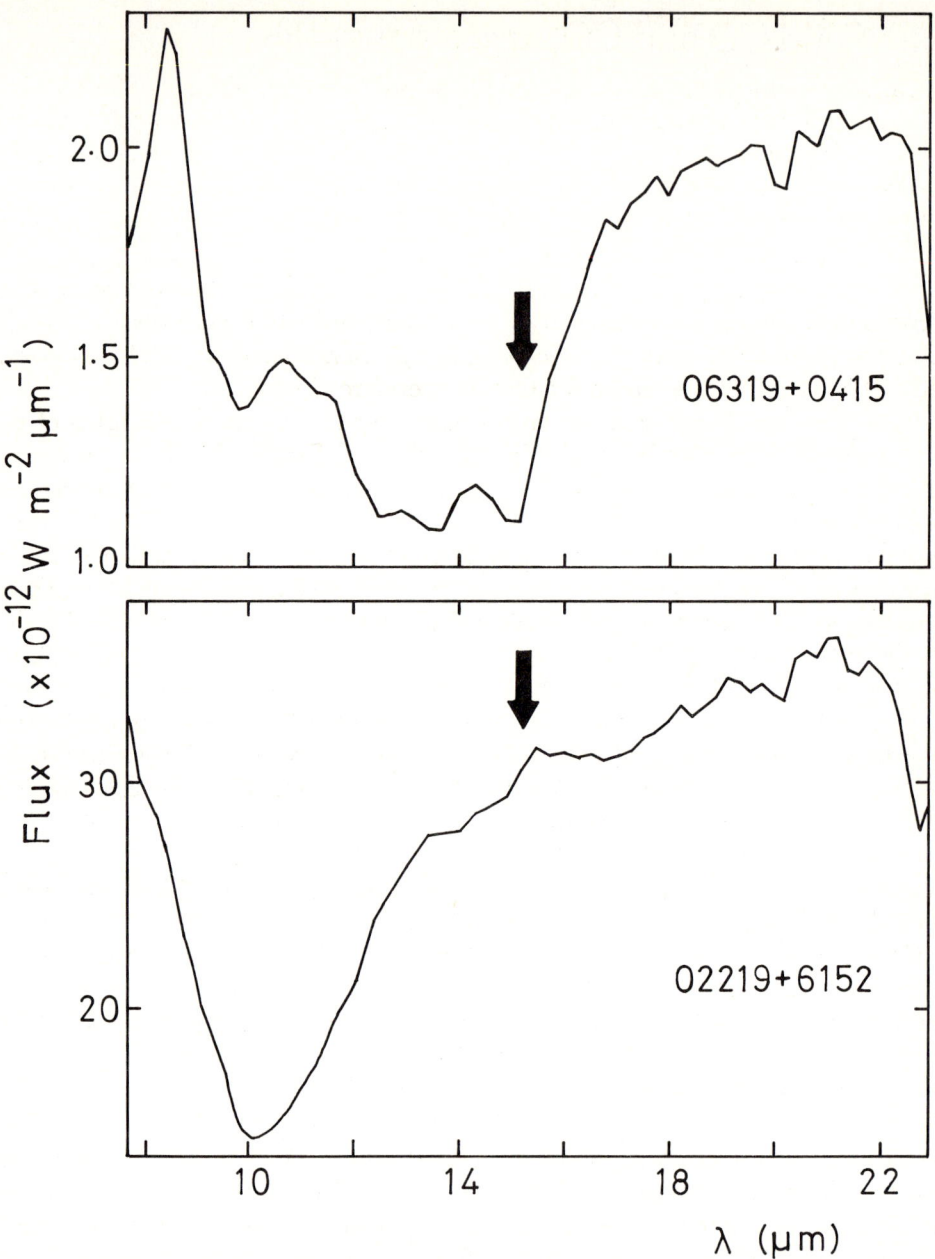

Figure 1: IRAS low-resolution spectra of sources 06319+0415 (GL961) and 02219+6152 (W3 IRS5). The position of the 15.3 μm feature associated with bending-mode vibrations in solid CO_2 is indicated by the arrows.

2 The search for CO_2

Carbon dioxide is not predicted to have appreciable abundance in the gas phase in molecular clouds, and so direct freeze-out should not lead to a significant presence of CO_2 in grain mantles. However, CO_2 may form by surface processing on the grains themselves, given the presence of CO, by means of the reaction $CO + O \rightarrow CO_2$. This reaction possesses an activation energy barrier (Grim and d'Hendecourt, 1986) and will not normally occur at low temperatures unless driven by ultraviolet photolysis or cosmic-ray processing. It has been demonstrated in the laboratory that CO_2 is produced readily in ice mixtures containing CO, when irradiated with ultraviolet light (e.g. Sandford et al., 1988). Following the detection of CO, a search for CO_2 is therefore timely.

Stretching and bending mode vibrations occur in CO_2 at wavelengths of 4.27 and 15.3 μm, respectively (e.g. Allamandola and Sandford, 1988). Neither of these spectral regions can be studied with ground-based or airborne instruments because of the strength of telluric CO_2 absorption; only the 15.3 μm region, covered by the IRAS low-resolution spectrometer (LRS), has so far been observed from space. The detection of CO_2 was reported recently by d'Hendecourt and de Muizon (1989), who found an absorption feature near 15.3 μm in LRS data for 06319+0415 (GL961) and also (more marginally) in 06384+0932 (GL989) and 06084−0611 (GL890). Of these, GL961 and GL989 are both known to be "protostellar" objects embedded in molecular clouds, displaying deep absorptions due to water-ice at 3 μm (Willner et al., 1982) and weaker absorptions due to solid CO at 4.67 μm (Geballe, 1986). Although it might appear reasonable to suppose that CO_2 formation might be activated in the vicinity of such sources, Smith et al. (1989) conclude on the basis of the shape of the 3 μm profile that the mantles are unannealed and maintained at temperatures < 30 K towards both GL961 and GL989.

If the feature discussed by d'Hendecourt and de Muizon (1989) is, indeed, identified with the bending mode of CO_2, it is reasonable to expect that it should be detectable towards other young stellar objects in molecular clouds which display comparable optical depths in the water-ice and solid CO features. Whittet and Walker (1990) carried out a systematic search of the LRS database for sources known from ground-based data to have 3 μm absorption with optical depth $\tau_{3.0} > 0.3$. A sample of 12 sources remain after rejection of poor-quality spectra. This includes two of the three sources discussed by d'Hendecourt and de Muizon, GL961 and GL989. The 15.3 μm feature may be weakly present in GL490 as well as in GL961 and GL989, but, remarkably, no other new detections were made: 9 of the 12 spectra show no discernible 15.3 μm absorption feature. Figure 1 compares our LRS spectrum of GL961 with that of W3 IRS5 (02219+6152). In GL961, the region from 12 to 16 μm is generally depressed, suggestive of blended absorption features at ∼ 13.5 μm and ∼ 15.3 μm. (The former may be due to the water-ice libration mode: see Cox, 1989). In contrast, the spectrum of W3 IRS5 is dominated by deep silicate absorption at 10 μm, and although weak absorption may be present at 14 μm, no feature is apparent at 15.3 μm. The spectrum of W3 IRS5 is fairly typical of young stellar objects in the LRS database, whereas that of GL961 is very *atypical*. The uniqueness of the GL961 spectrum in our sample, and the extreme weakness or absence of 15.3 μm absorption in all other spectra examined, casts doubt on the proposal that CO_2 is a significant constituent of dust in molecular clouds.

3 Future Observations

The uncertainty in the proposed identification of interstellar solid CO_2 would undoubtedly be resolved if data were available for the stretching-mode feature at 4.27 μm, as this is intrinsically an order of magnitude stronger than the bend feature at 15.3 μm. An analogy may be drawn with silicates: if the Earth's atmosphere were opaque in the region of the Si–O stretch feature at 10 μm, we would gain little impression of the ubiquity of silicate dust from the existing observations of the bend feature at 18 μm. The situation is actually somewhat worse for CO_2.

The spectral region from 4.1 to 4.5 μm represents one of the final frontiers of near infrared spectroscopy. Its exploration awaits the launch of ISO.

References

Allamandola, L.J. and Sandford, S.A., 1988, In: *Dust in the Universe*, ed. M.E. Bailey, and D.A. Williams, (Cambridge University Press), p. 229.
Cox, P., 1989, Astr. Astrophys., **225**, L1.
d'Hendecourt, L.B. and de Muizon, M.J., 1989, Astr. Astrophys., **223**, L5.
Duley, W.W., 1974, Astrophys. Space Sci., **26**, 199.
Geballe, T.R., 1986, Astr. Astrophys., **162**, 248.
Grim, R.J.A. and d'Hendecourt, L.B., 1986, Astr. Astrophys., **167**, 161.
Lacy, J.H., Baas, F., Allamandola, L.J., Persson, S.E., McGregor, P.J., Lonsdale, C.J., Geballe, T.R. and van de Bult, C.E.P., 1984, Astrophys. J., **276**, 533.
Mitchell, G.F., Allen, M. and Maillard, J.P., 1988, Astrophys. J., **333**, L55.
Sandford, S.A., Allamandola, L.J., Tielens, A.G.G.M. and Valero, G.J., 1988, Astrophys. J., **329**, 498.
Smith, R.G., Sellgren, K. and Tokunaga, A.T., 1989, Astrophys. J., **344**, 413.
Whittet, D.C.B. and Duley, W.W., 1990. Astr. Astrophys. Rev., (in press).
Whittet, D.C.B. and Walker, H.J., 1990. Mon. Not. Roy. Astron. Soc., (submitted).
Willner, S.P., Gillett, F.C., Herter, T.L., Jones, B., Krassner, J., Merrill, K.M., Pipher, J.L., Puetter, R.C., Rudy, R.J., Russell, R.W. and Soifer, B.T., 1982, Astrophys. J., **253**, 174.

Far-IR and submm studies of dust in molecular clouds

D. WARD-THOMPSON AND E. I. ROBSON

SCHOOL OF PHYSICS AND ASTRONOMY, LANCASHIRE POLYTECHNIC,
PRESTON, PR1 2TQ, U.K.

Abstract IRAS Calibrated Raw Detector Data and submm continuum data are presented of two molecular clouds, W49A and W75. The clouds' internal physical conditions are discussed and specific dust parameters calculated, such as mass, temperature, luminosity, optical depth, spectral index and cloud fragmentation. The parameters of W49A are compared with global values for clouds throughout the Galaxy, and preliminary evidence is presented of a possible old SNR around W75.

1 Data Reduction

IRAS Calibrated Raw Detector Data (CRDD) were obtained from the Rutherford Appleton Laboratory. The main problem in reduction of the IRAS raw data is to eliminate the characteristic striping by ensuring an accurate background subtraction. This was performed in the manner pioneered in association with the STARLINK IRAS Applications Programmer and described by Ward-Thompson et al. (1989 – Paper I), and Ward-Thompson and Robson (1990a, b – Papers II and III). Briefly, the procedure employed an iterative method to fit a polynomial to the background sky in each individual raw scan, which was then subtracted before the scans were co-added. No colour corrections (IRAS Explanatory Supplement, 1984) were applied to the maps. Submm data were obtained from UKIRT using the bolometer UKT14. Standard STARLINK EDRS and IRAS software was used, and all data reduction was carried out on the Lancashire Polytechnic MicroVAX minor node of STARLINK.

2 W49A

The W49A cloud lies at a distance of 13.8 ± 0.8 kpc and contains three separate far-infrared peaks, W49N, W49SE and W49SW (Harvey, et al., 1977). W49N contains a number of embedded IR sources, masers and HII regions. Figure 1a shows a 12 μm map of the whole region, and Figure 1b shows a 800 μm map of W49N. Figure 1c shows our fluxes of W49A, corrected for free-free emission, plotted along with data taken from the literature. The solid curve represents a greybody which is seen to be consistent with most of the available data. The parameters of the fit are $T = 50$ K, $\beta = 1.8$, $\nu_c = 2000$ GHz ($\lambda_c = 150\,\mu$m) and $\Omega = 1.5 \times 10^{-8}$ sr (Paper II).

The temperature and critical frequency agree with previous authors but the

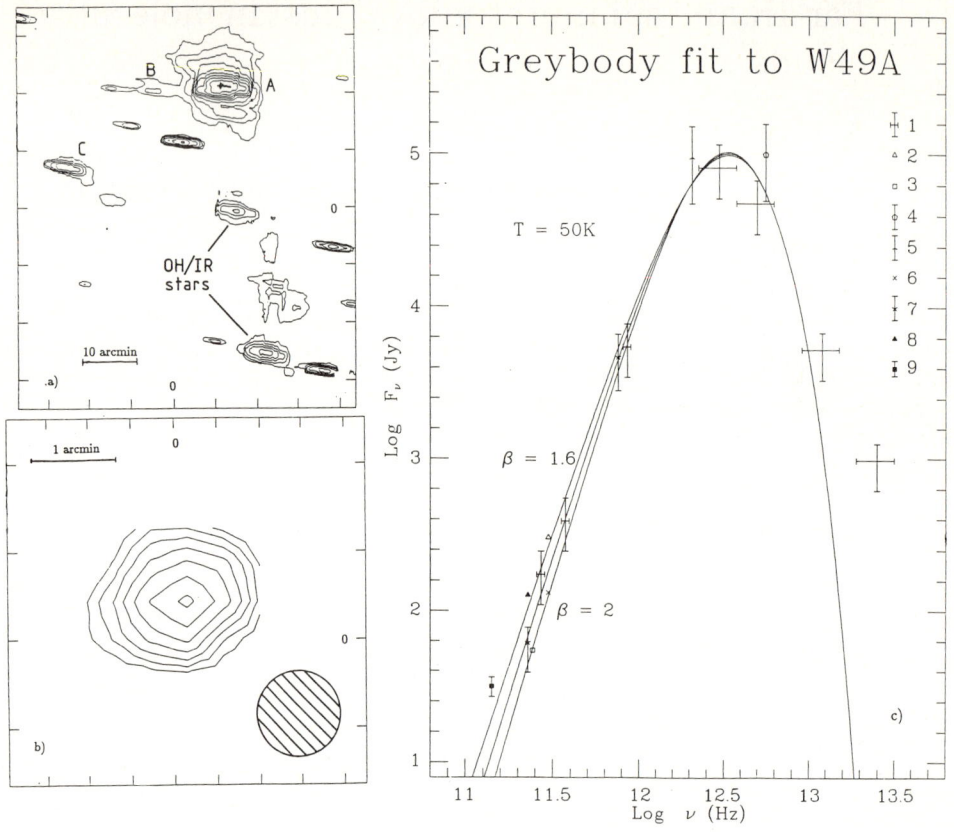

Figure 1: W49A at a) $12\,\mu m$, (0,0) is $19^h\,8^m\,30^s\,+8°40'\,0''$ and b) $800\,\mu m$, (0,0) is $19^h\,7^m\,50^s\,+\,9°\,1'\,0''$. c) Grey-body fit to the data (Paper II).

values of β and Ω are both somewhat less than was previously believed. The effects of varying β from 1.6 to 2.0 are illustrated by the two dashed lines. This range of values is typical of dust clouds containing unmantled grains. The critical frequency can be used to estimate A_v, which is found to be ~ 1000 mag.

The total far-infrared luminosity of W49A was calculated by integrating numerically under the solid curve in Figure 1c, and found to be $2.7 \times 10^7 L_\odot$, consistent with previous work. The value of Ω implies an effective total diameter of the emitting region of ~ 2 pc, suggesting that the cloud is highly fragmented. The mass of dust which is emitting in the submm was calculated to be $2400\,M_\odot \pm 20\%$ - in close agreement with Gordon (1987). The canonical value of 100 for the gas-to-dust mass ratio in dark clouds (Hildebrand, 1983) gives a cloud mass of $2.4 \times 10^5\,M_\odot$.

Comparison of our parameters of W49A with the global properties of molecular clouds (Scoville and Good, 1989) shows that, compared with the means, W49A has a dust temperature 15 – 20 K larger, and an order of magnitude greater luminosity and luminosity-to-mass ratio. These all confirm that W49A is one of the most intrinsically luminous star-forming regions in the Galaxy.

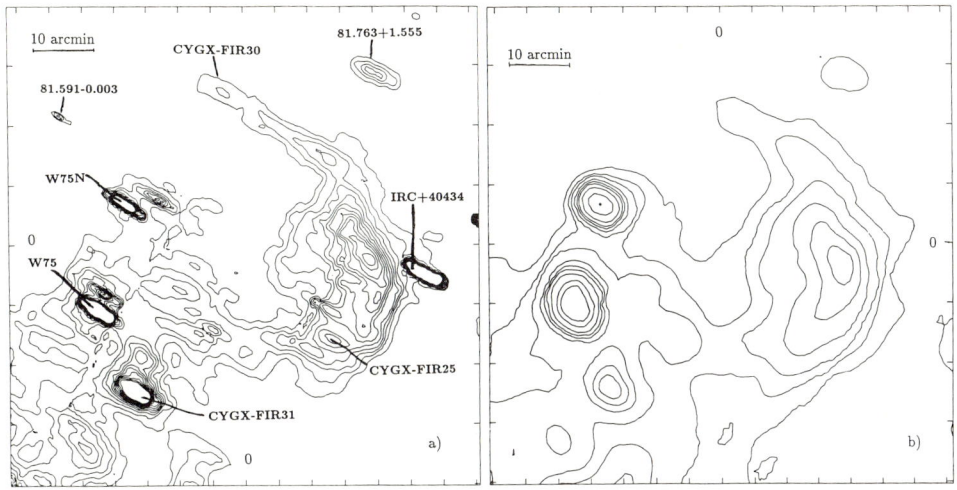

Figure 2: W75 region at (a) $12\,\mu m$ and (b) $100\,\mu m$, (0,0) is $20^h\,35^m\,+42°\,20'$.

3 W75

The W75 region lies on the north-eastern edge of the Cygnus-X complex, at an approximate distance of 2 kpc. It contains the molecular clouds W75 and W75N, and the bipolar outflow source DR21.

Figures 2a and 2b show isophotal contour maps of the region at $12\,\mu m$ and $100\,\mu m$. A ridge of emission stretches from the south-east of the maps and ends in a loop in the centre of the field. The previously known infrared sources are marked. The spatial correlation between the point sources and extended emission suggests embedded stars within extensive dust clouds. The brightest part of the loop is a knot at its western edge, centred roughly at $20^h\,33^m\,+42°\,15'$ and extending some $20'$ to both north and south, hereafter called K1.

The W75 loop does not correspond spatially with any previously known feature, although there is a high degree of correlation with the CO emission (Campbell et al., 1980). All of these observations can be interpreted in terms of a shell of gas and dust, whose edge can be seen by limb-brightening. We therefore suggest that the loop seen in the IRAS raw data in Figure 2 is an expanding shell-like shock front, which is sweeping up material in the arc of K1 (Paper III).

The $100\,\mu m$ flux was measured in a region 35 pc across and was found to be $\sim 3 \times 10^5$ Jy. This flux is clearly contaminated by emission from foreground dust, but the flux will underestimate the amount of dust if it is not optically thin at $100\,\mu m$. Previous experience (Papers I and II) shows that regions such as this only become optically thin at longer wavelengths.

The mass of dust was calculated from the $100\,\mu m$ flux to be $450\,M_\odot$. If the canonical gas-to-dust mass ratio of 100 is used, the mass of the cloud is $4.5 \times 10^4\,M_\odot$ (mean density $= 8 \times 10^{-20}\,\mathrm{kg\,m^{-3}}$). One possible origin of the shell is an old supernova remnant. If this were the cause of the shell, any optical remnant would be extinguished by the large A_v. The age can be calculated from the equation:

$$R = (2E/\rho_0)^{0.2} t^{0.4}$$

where R is the radius of the shell, E is the energy of the supernova explosion – typically 10^{51} erg – ρ_0 is the pre-explosion density of the cloud and t is the age of the SNR. The age was thus calculated to be of order 10^5 years, consistent with the age of W75N(B) calculated by Moore *et al.* (1988), and provides an explanation for their theory of coeval formation of the embedded stars. So it appears that a supernova remnant is consistent with all of the attributes of the dust loop in the W75 region.

An alternative explanation for the shell is a wind-blown bubble around an OB association. We do not favour this possibility, as is discussed in Ward-Thompson and Robson, 1990b – Paper III.

4 Conclusions

In our continuing work on star formation regions, we have presented in this paper IRAS raw data of two molecular clouds. We derive their specific parameters; W49A is seen to be one of the most luminous star-forming regions in the Galaxy; in the vicinity of W75 a loop of dust emission was detected, which we interpret in terms of a supernova explosion 10^5 years ago. The expanding shock front from the explosion has subsequently swept up the ambient dust and gas in its wake to a spherical shell of radius 17 pc.

References

Campbell, M.F., Hoffmann, W.F., Thronson, H.A. and Harvey, P.M., 1980, Astrophys. J., **238**, 122.
Gordon, M.A., 1987, Astrophys. J., **316**, 258.
Harvey, P.M., Campbell, M.F. and Hoffmann, W.F., 1977, Astrophys. J., **211**, 786.
Hildebrand, R.H., 1983, Q. J. Roy. Astron. Soc., **24**, 267.
Moore, T.J.T., Mountain, C.M., Yamashita, T. and Selby, M.J., Mon. Not. Roy. Astron. Soc., **234**, 95.
Scoville, N.Z. and Good, J.C., 1989, Astrophys. J., **339**, 149.
Ward-Thompson, D., Robson, E.I., Gordon, M.A., Whittet, D.C.B., Duncan, W.D. and Walther, D.M., 1989, (Paper I), Mon. Not. Roy. Astron. Soc., **241**, 119.
Ward-Thompson, D. and Robson, E.I., 1990a (Paper II), Mon. Not. Roy. Astron. Soc., **244**, 458.
Ward-Thompson, D. and Robson, E.I., 1990b (Paper III), Mon. Not. Roy. Astron. Soc., (in press).

Critical study of the assignment of PAHs as carriers of the diffuse interstellar bands

CLAUDINA COSSART-MAGOS[1] and SYDNEY LEACH[1,2]

Abstract Rotational band contours of vibronic transitions of selected polycyclic aromatic hydrocarbons (PAHs) are simulated and compared with the intrinsic profiles of 4 narrow diffuse interstellar bands (DIBs). The non-observation of the calculated multi-peak structures throws doubt on the assignment of PAHs as DIB carriers.

1 Introduction

About 80 diffuse interstellar absorption bands (DIBs) have been observed between 4400 and 8700 Å (Herbig, 1988, Cossart-Magos and Leach, 1990). Their assignment to specific carriers is an outstanding unresolved problem. Dust grain and molecular hypotheses as to their origin have been summarized by Bromage (1987). Arguments based on photostability and estimated abundances have been given in favour of PAHs as the DIB carriers (van der Zwet and Allamandola, 1985, Léger and d'Hendecourt, 1985); their spectroscopic properties have been little tested. The present study tests the PAH assignment by comparison of calculated rotational band contours of some typical polycyclic aromatic hydrocarbons, at temperatures in the 3–100 K range, with some narrow DIB profiles observed under specific spectral resolution and molecular cloud conditions.

The presence of several molecular clouds in the line of sight could give rise to broadened, distorted profiles if these clouds have different relative velocities and/or exhibit turbulent effects within themselves. Thus valid comparison between "laboratory" spectra and observed DIB profiles should be carried out on single cloud, non-turbulent material. "Intrinsic" DIB profiles have been obtained by deconvolution of multicomponent data (Westerlund and Krelowski, 1988a, b) as well as from observations on nearby stars which have relatively small colour excesses (Westerlund and Krelowski, 1988b).

We calculated the contours of vibronic bands in electronic transitions of typical polycyclic aromatic hydrocarbons under various conditions of temperature and spectral resolution (Cossart-Magos and Leach, 1990). These conditions were chosen in conformity with the expected range of interstellar temperatures, and the present, and future, observational resolution of DIBs. The results are used to compare with the "intrinsic" profiles derived from observed DIBs.

---

[1] LABORATOIRE DE PHOTOPHYSIQUE MOLECULAIRE DU CNRS, UNIVERSITÉ PARIS-SUD, 91405 ORSAY, FRANCE.
[2] DAMAP (URA 812), OBSERVATOIRE DE PARIS-MEUDON, 92195 MEUDON, FRANCE.

2 Rotational Contour Calculations and Results

The form of the integrated contour of a vibronic band of a polyatomic species depends on the following parameters: 1) $\Delta A = A' - A''$, $\Delta B = B' - B''$, $\Delta C = C' - C''$, where A, B and C are the rotational constants in the upper (') and lower('') states; 2) the direction of the transition moment, and effective Coriolis coefficients in the case of symmetric rotors; 3) the rotational temperature T_R; 4) the rotational linewidth LW.

Contour calculations were carried out on two classes of PAH with two molecules of different size being considered in each class: (i) symmetric rotors: benzene (C_6H_6) and coronene ($C_{24}H_{12}$) and (ii) asymmetric rotors: pyrene ($C_{16}H_{10}$) and ovalene ($C_{32}H_{14}$). The rotational constants A'', B'', C'' and A', B' and C' were calculated for reasonable geometrical models, where the bond-lengths were derived from the bond orders corresponding to the published molecular orbitals (MO) for the ground and first excited singlet states (Coulson et al., 1965). A classic bond-order–bond-length empirical relation was used (Coulson and Golebiewski, 1961).

As expected, sub-structure corresponding to P, Q, R branch features are present in many of the contours of coronene. These sub-structures are more evident as the temperature is increased, at a given linewidth (spectral resolution). The rotational contours calculated for benzene, pyrene and ovalene also show clearly visible substructures. We observe a shrinkage in the overall bandwidths and in the widths of individual structural features as PAH size increases.

3 Comparison of Intrinsic Profiles of Diffuse Interstellar Bands with computed Rotational contours of PAHs.

The observational material we use for comparison with contour calculations concerns the four DIB bands at 5780, 5797, 6196 and 6284 Å observed by Westerlund and Krelowski (1988a, b) at an instrumental resolution of 0.15 cm^{-1}.

The four DIB profiles were compared with the rotational band contours calculated for coronene at the same linewidth of 0.15 cm^{-1} at temperatures T = 3, 10, 50 and 100 K. It should be noted that the DIBs were observed with a Reticon detector, so that the interstellar absorption is recorded as relative transmission, whereas our simulated contours are calculated as absorbance, as is usual in absorption spectroscopy band contour studies (Ross, 1971). The number of subpeaks is identical in the two representations, but the FWHM of the band absorbance profile is greater than for the corresponding transmission profile. Thus the calculated absorbance FWHM represents an upper limit to that of a DIB profile measured by a Reticon detector.

Inspection of Figure 1 shows that multi-peak structures expected from our rotational band contour simulations for coronene are not observed in the DIB spectra. Multi-peak structures are also expected in the asymmetric top PAHs that we have studied as well as in still larger PAHs. Excitation of a single

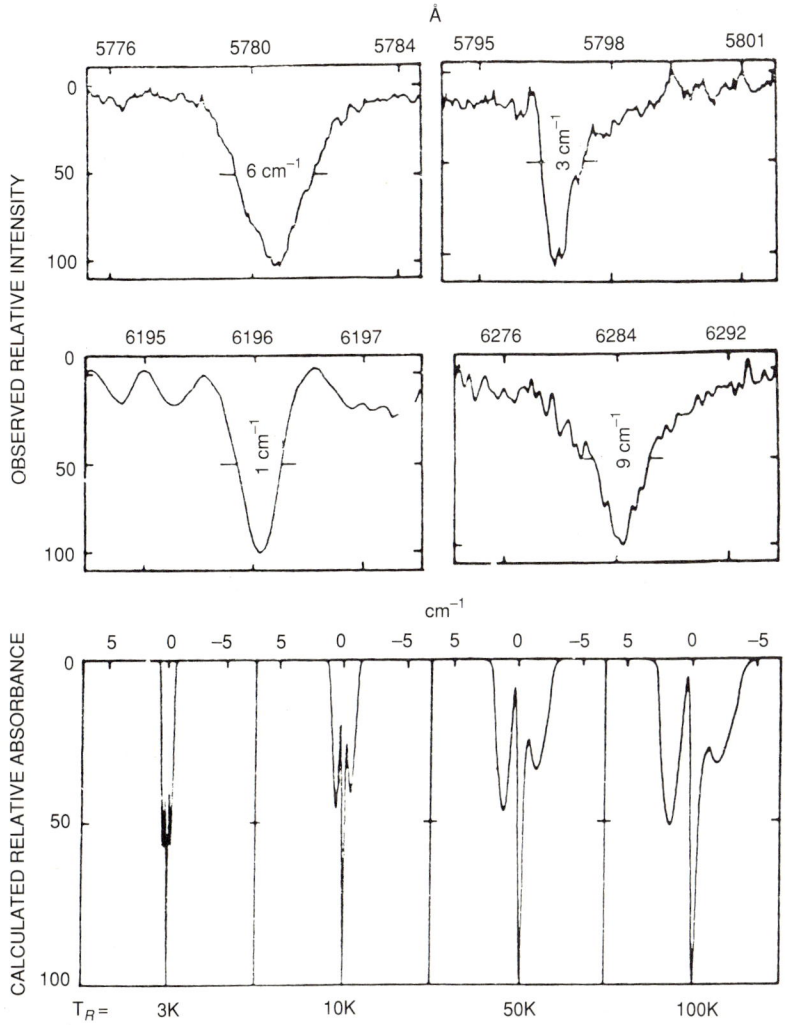

Figure 1: Intrinsic profiles (resolution $0.15\,\text{cm}^{-1}$) of four observed diffuse interstellar absorption bands (Westerlund and Krelowski, 1988b) compared with calculated rotational band contours for coronene (LW = $0.15\,\text{cm}^{-1}$) at T_R = 3, 10, 50 and 100 K.

π electron in analogous transitions in a series of bigger and bigger PAHs is predicted to have a decreasingly small effect on the moment of inertia changes. This should help sustain a multi-peak structure for the rotational band contour of the large PAHs. Furthermore, the larger the PAH, the smaller become the total bandwidths of the simulated contours for a given linewidth and temperature. We remark that the total bandwidths of our simulated band contours are compatible only with the narrowest DIB intrinsic profiles, e.g. that of the 6196 Å band.

4 Discussion and Conclusion

The simulated rotational band contours of vibronic transitions of polycyclic aromatic hydrocarbons lead us to expect multipeak structures in the narrow diffuse interstellar absorption bands assigned to these neutral or ionic species. Such fine structure components are not observed in the intrinsic profiles of four DIBs obtained by Westerlund and Krelowski (1988a, b). We have shown elsewhere (Cossart-Magos and Leach, 1990) that if, nevertheless, PAHs are the carriers of the DIBs, smoothed single-peak profiles could result from one or several sources: (i) inadequacies in the technique used for determining the intrinsic profiles of the observed DIBs, (ii) line broadening due to intramolecular relaxation in neutral or ionic PAHs, (iii) temperature variations within a single molecular cloud. Thus, although our results cast doubt on the PAH assignment of DIBs, more laboratory and observational studies are required in order to conclude definitely that the observed intrinsic profiles cannot be due to polycyclic aromatic hydrocarbons.

Acknowledgements

This research has benefitted from support by the C.N.R.S. Groupe de Recherche "Physico-Chimie des Molécules Interstellaires".

References

Bromage, G.E., 1987, Q. Jl. Roy. Astron. Soc., **28**, 294.
Cossart-Magos, C. and Leach, S., 1990, Astron. Astrophys., **233**, 559.
Coulson, C.A. and Golebiewski, A., 1961, Proc. Phys. Soc. (London), **78**, 1310.
Coulson, C.A., Streitweiser, A. Jr, Poole, M.D. and Brauman, J.I., 1965, *Dictionary of π-Electron Calculations*, (Pergamon, Oxford, London, New York).
Herbig, G.H., 1988, Astrophys. J., **331**, 999.
Léger, A. and d'Hendecourt, L., 1985, Astron. Astrophys., **146**, 81.
Ross, I.G., 1971, Adv. Chem. Phys., **20**, 341.
van der Zwet, G.P. and Allamandola, L., 1985, Astron. Astrophys., **146**, 76.
Westerlund, B.E. and Krelowski, J., 1988a, Astron. Astrophys., **189**, 221.
Westerlund, B.E. and Krelowski, J., 1988b, Astron. Astrophys., **203**, 134.

Ice in Barnard 5 IRS2

S. B. CHARNLEY[1], D. C. B. WHITTET[2] and D.A. WILLIAMS[3]

Barnard 5 (B5) is a nearby dark cloud in which low-mass star formation is occurring. It contains four IRAS protostellar objects and also several dense clumps embedded in a more diffuse interclump medium. The outflows associated with the protostellar objects disrupt the cloud structure and interact with the dense fragments. Following the model of Norman and Silk (1980), Goldsmith et al. (1986) have inferred from observations of B5 that continuous cycling of gas and dust is occurring between the clump and interclump phases, regulated by the effects of the stellar winds. They estimated that if clumps form on the order of a free-fall time ($\sim 4 \times 10^5$ years) then the maximum clump lifetime against dissipation is $\sim 6 \times 10^5$ years, much shorter than the time for a chemical steady-state to obtain.

Charnley et al. (1988) developed time-dependent models of the nonequlibrium chemistry in regions of low-mass star formation, appropriate for B5. Two scenarios were considered. One model (following Goldsmith et al., 1986) assumed that clumps form by free-fall collapse. The cycling time in this 'fast' model was $t_{fast} = 6 \times 10^5$ years. A second model assumed that the collapse time was about an order of magnitude longer, due to the assumed effects of magnetic field support. The cycling time in this 'slow' model was $t_{slow} = 6 \times 10^6$ years. A common feature of both dynamical-chemical models was the formation and removal of ice mantles during the clump-interclump cycling. When clumps are dispersed to form the interclump medium, the ice mantles are assumed to be completely removed, and hence the collapse to form clumps begins with 'clean' grain surfaces. (The mantles are presumed to be sputtered and thermally desorbed by the effects of nearby stars). The removal of accreted material during clump erosion is necessary to avoid complete loss of the entire heavy gas phase component over several cycle periods. The models predict spatial variations in the strength of the water ice feature within B5. In particular, the amount of ice observed in clumps should reflect the amount of accretion during clump formation, and hence may provide an estimate of the actual cycling timescale in B5.

The observations were made at the United Kingdom Infrared Telescope at Mauna Kea Observatory, Hawaii (Charnley, Whittet and Williams, 1990) and a substantial ice absorption feature is present in the spectrum. The profile is broad and asymmetrical in the same sense as those of other molecular cloud sources, with an extended wing to the red. It is not clear whether the detailed structure visible in the profile is real. A grating-resolution spectrum with high

---

[1] PHYSICS DEPARTMENT, RENSSELAER POLYTECHNIC INSTITUTE, TROY, NEW YORK 12180, U.S.A.
[2] SCHOOL OF PHYSICS AND ASTRONOMY, LANCASHIRE POLYTECHNIC, PRESTON, LANCASHIRE PR1 2TQ, U.K.
[3] MATHEMATICS DEPARTMENT, UMIST, PO BOX 88, MANCHESTER M60 1QD, U.K.

signal to noise ratio should be obtained to investigate this as the models of Charnley et al. (1988) also predict the mantle abundances of other molecules such as ammonia and methane. The optical depth of the feature is estimated to be $\tau_{3.0} = 1.20 \pm 0.15$. The column density of H_2O within the ice mantles is then found to be $N(H_2O) = 2 \times 10^{18}\,\text{cm}^{-2}$ with a probable error of $\sim 40\%$.

Charnley et al. (1990) have shown that simple estimates of the time required to obtain $N(H_2O)$ indicate that fast cycles in which some carbon is periodically released into the gas phase (due to grain erosion in shocks) are inconsistent with the observations, and also that, in both fast and slow cycling models, nearly all grains must accrete ice mantles. Furthermore it is possible to estimate the mean ice mantle thickness which should form over times t_{fast} and t_{slow} (e.g. Wickramasinghe, 1967). For the parameters quoted by Charnley et al. (1990) these are $0.016\,\mu\text{m}$ and $0.16\,\mu\text{m}$ respectively. Theoretical profiles have been calculated using a simple Mie theory analysis (assumed to be applicable when no hot dust is present, Charnley and Leung, 1990) which correspond to these mantle thicknesses. There is little difference in the shape of the profile; however it is clear that, as expected, cold amorphous ice provides the broad feature needed to fit the data around $3\,\mu\text{m}$. Information concerning the possible temperature history of the icy grains may be inferred. If thermal sublimation is important in removing mantles at some point in the cycle then the desorption of water must be highly efficient ($T_{grain} > 100\,\text{K}$). If this were not the case and the amorphous ice were heated to $\sim 80\,\text{K}$, followed by cooling to $\sim 10\text{--}20\,\text{K}$, the ice feature would become narrower and the absorption peak would be shifted to longer wavelengths (Hagen, Tielens and Greenberg, 1981). This is not apparent in the observed spectrum which suggests that the grains in B5 are periodically cleaned of their mantles, prior to the formation of clumps.

Acknowledgments

We are grateful to SERC for observing time and financial support. SBC was partially supported by the US Air Force under Grant AFOSR-89-0104.

References

Charnley, S.B., Dyson, J.E., Hartquist, T.W. and Williams, D.A., Mon. Not. Roy. Astron. Soc., 1988, **235**, 1257.

Charnley, S.B. and Leung, C.M., 1990, (in preparation).

Charnley, S.B., Whittet, D.C.B. and Williams, D.A., 1990, Mon. Not. Roy. Astron. Soc., 1990, **245**, 161.

Goldsmith, P.F., Langer, W.D. and Wilson, R.W., 1986, Astrophys. J., **303**, L11.

Hagen, W., Tielens, A.G.G.M. and Greenberg, J.M., 1981, J. Chem. Phys., **56**, 367.

Hollenbach, D. and Salpeter, E.E., 1970, J. Chem. Phys., **53**, 79.

Norman, C. and Silk, J., 1980, Astrophys. J., **238**, 138.

Wickramasinghe, N.C., 1967, *Interstellar Grains*, (Chapman and Hall, London).

Does dust mean gas?

S.J.CHAPMAN AND A.P.WHITWORTH

DEPARTMENT OF PHYSICS, UNIVERSITY OF WALES COLLEGE OF CARDIFF, P.O. BOX 913, CARDIFF CF1 3TH, WALES, U.K.

1 Introduction

It is normally assumed that where there is a high concentration of dust there is also a high concentration of gas, and *vice versa*. However, there are situations in which the forces on the dust act to separate it from the gas; there is clear evidence under our feet that the process can be very efficient. By modelling the motion of a dust grain as it orbits a grand-design two-arm spiral galaxy, we test whether significant fractionation of dust from gas occurs on a galactic scale.

2 Model

In our model, we map the orbit of a grain into a straight line. There is, therefore, a built-in balance between the centripetal acceleration and the gravitational acceleration of the galaxy. The grain motion is determined by the following forces: *(a) Gas Drag.* This is proportional to the gas density, the square of the relative velocity between the gas and grains, and the grain cross-section (modulated by a factor which takes into account the grain charge and ionization state of the gas). The gas density increases discontinuously (by a factor of between 2 and 10) at periodically spaced shocks (the spacing depends on the distance from the galactic centre). *(b) Radiation Pressure.* This is proportional to the integrated flux of visible and ultraviolet radiation, and the grain cross-section. The volume emissivity of the background star-field is assumed uniform, and the spiral arms are represented by a gaussian enhancement due to newly formed OB stars. The enhancement is 100 – 200 pc thick and located 50 – 100 pc downstream of the gas discontinuity.

We have evaluated the influence of *(c) grain growth and erosion*, and *(d) plasma drag and the Lorentz force* but have not included these effects because they do not seem to influence the overall dynamics of the dust significantly.

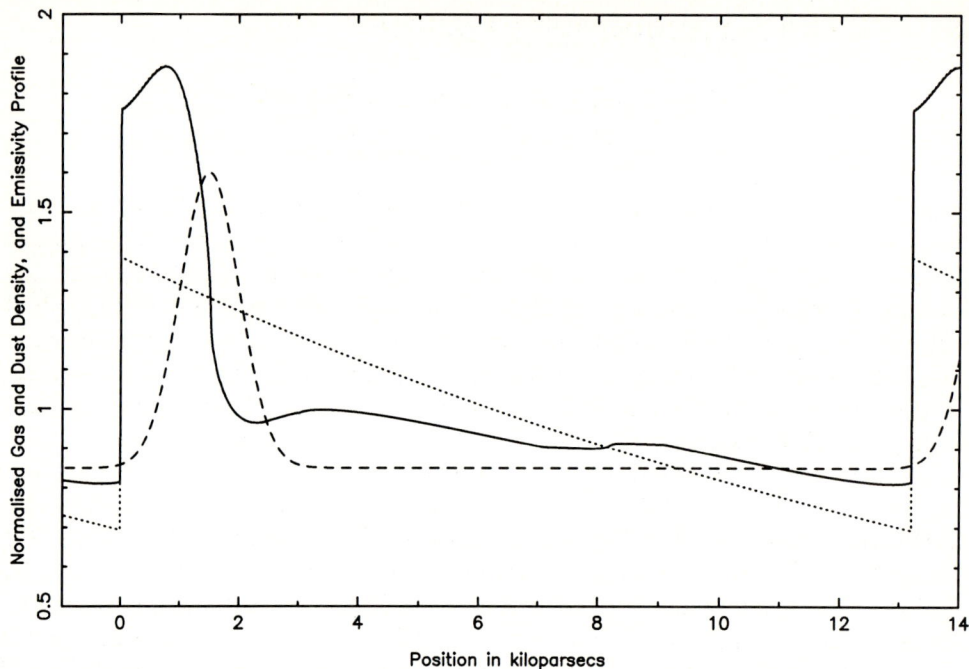

Figure 1. The gas density (dotted line), dust density (full line) and emissivity (dashed line) are plotted as a function of the orbital displacement in kpc. The ordinate-axis is dimensionless.

3 Results and Conclusions

Figure 1 shows a representative result of our calculations at R = 4.0 kpc. The model clearly predicts systematic fractionation of dust from gas; the dust lanes are offset from the maximum of the post-shock gas density, typically by 10 – 30 pc.

We emphasize that this model is only in an early stage of development. However, we feel that it strongly suggests a significant effect which should be evaluated properly. Inclusion of the following effects will improve our model: *(e) extinction; (f) motion of the gas and dust perpendicular to the galactic plane;* and *(g) stochastic processes such as star birth and death.*

AROME, a balloon borne experiment: detection of the 3.3 micrometre feature

N. SALES[1], M. GIARD[2], E. CAUX[1], J. M. LAMARRE[2],

F. PAJOT[2] and G. SERRA[1]

Introduction. Observations of the Orion nebulae and of the Galactic disk in the 3.3 μm feature are presented here. The 3.3 μm emission feature is assigned to one of the C–H stretching modes of polycyclic aromatic molecules (PAHs). Ground based observations are perturbed by the fluctuations of the atmospheric thermal emission. As a consequence, we built a balloon borne experiment, called AROME, designed to detect the 3.3 μm emission feature in extended sources. It uses two photometric bands, both centered at 3.3 μm. Two flights mapped a large part of the Galactic disk ($-50° < \ell < -5°, -5° < b < 5°$) (Giard et al., 1988, 1989) and particular areas such as Orion's clouds.

Detection of the emission feature at 3.3 μm in Orion's cloud. Sellgren (1981) has presented a 5'x 5' map of Orion Nebula with 30" resolution. The 3.3 μm feature flux, integrated over a $5' \times 5'$ area (Sellgren) is eight times lower than the flux measured in the 0.5° beam of AROME in Orion A. AROME results can be explained by the fact that our map covers a greater area than Sellgren's observations. The figure presents a map of the 3.3 μm emission around Orion A and Orion B and points out that the 3.3 μm emission feature is spread out in the interstellar medium, and not only located near the exciting stars. In Table 1, we compare our results to those of Sellgren. We have computed the feature to continuum ratio $R = (F(3.3)/\lambda\, I_\lambda(\mathrm{BB}))(\lambda/\delta\lambda)$, with $F(3.3)$ the total feature flux, $I_\lambda(\mathrm{BB})$ the measure flux intensity in the AROME broad band ($\Delta\lambda = 0.61\,\mu\mathrm{m}$) and $\delta\lambda = 0.05\,\mu\mathrm{m}$, the feature bandwidth (Tokunaga and Young, 1980). The lower value of the ratio R in $5' \times 5'$ compared with the one obtained in 0.5° (5.8 to 1.5), can be explained if R increases when we move away from the exciting stars.

Detection of the emission feature at 3.3 μm in the galactic disk. AROME led to the first detection of the 3.3 μm feature in the diffuse galactic emission. On one side, the maps of the 3.3 μm emission feature and those of IRAS in the 100 μm band present a very good similarity. On the other, the spatial distribution of the 3.3 μm continuum emission is very different from that of IRAS at 100 μm. As a consequence, we conclude that the 3.3 μm feature comes from the interstellar medium whereas the continuum at the same wavelength is of stellar origin.

Ubiquity of the 3.3 μm emission feature All the well-known HII Regions – Giant Molecular Cloud complexes which are in the galactic plane are strong emitters

---

[1] CENTRE D'ETUDE SPATIALE DES RAYONNEMENTS, 9, AVENUE DU COLONEL ROCHE, BP 4346, 31029 TOULOUSE CEDEX, FRANCE.
[2] LABORATOIRE DE PHYSIQUE STELLAIRE ET PLANETAIRE, BP 10, 91370 VERRIERE-LE-BUISSON CEDEX, FRANCE.

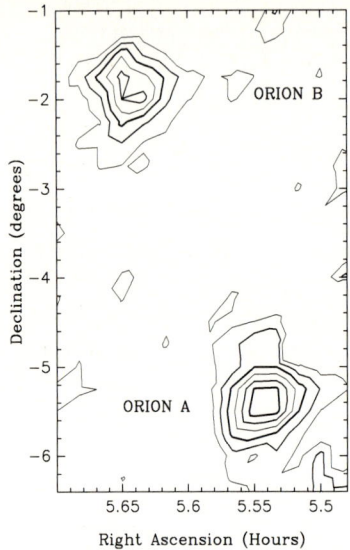

Figure 1. Orion's clouds narrow band ($\Delta\lambda = 0.17\,\mu$m)

Table 1 *Orion's clouds fluxes measured by AROME and Sellgren.*

	Area	Orion A	Orion B
F(3.3): Feature flux ($10^{-12}\,\mathrm{W\,m^{-2}}$)	0.5° (AROME)	5.8	4.3
	5'x 5' (Sellgren)	1.5	–
Feature to continuum ratio (R)	0.5°	48.3 ± 9	19.2 ± 3
	5'x 5'	6 ± 2	–

in the 3.3 μm feature. But the 3.3 μm feature is observed at large scale in the interstellar medium and therefore, is not specific of highly UV irradiated regions. The brightest sources are of course those associated with the ionized nebulae M17, NGC 6334, NGC 6357, M42 (Orion A) and NGC 2024 (Orion B), but a few tens of fainter objects can be isolated in our maps.

References

Giard, M., Pajot, F., Lamarre, J.M., Serra, G., Caux, E., Gispert, R., Lèger, A. and Rouan, D., 1988, Astron. Astrophys., **201**, L1.
Giard, M., Pajot, F., Lamarre, J.M., Serra, G. and Caux, E., 1989, Astron. Astrophys., **215**, 92.
Sellgren, K., 1981, Astrophys. J, **245**, 138.
Tokunaga, A.T. and Young, E.T., 1980, Astrophys. J. **237**, L93.

AUTHOR INDEX

The page number is emphasised for the first author of each paper

Anglada, G. **123**	Felli, M. 203, 207
Assendorp, R. 149	Genzel, R. 17, **75**, 183
Bailey, M.E. **273**	Giard, M. 325
Bajaja, E. 179	Gierens, K. 25
Bakes, E.L.O. **27**	Giovanardi, C. 203
Bally, J. 13	Graf, U.U. 17
Battinelli, P. **187**	Greaves, J. **125**
Beck, R. **179**	Gredel, R. **73**
Berkhuijsen, E.M. 179	Großman, V. 69
Black, J.H. 73, 255	Guélin, M. **97**
Blitz, L. **49**	Güsten, R. 107
Bloemen, H. **29**	Haikala, L. **25**
de Boisanger, C. 263	Harris, A.I. 17
Booth, R.S. **157**	Hasegawa, T. 115
Brand, J. **21**, 203, 207	Hayashi, S.S. 115
Brand, P.W.J.L. 243	Herbst, E. **209**
Bronfman, L. 23	Hills, R.E. 17
Brown, P.D. **253**	Hirano, N. 111, **115**
Bruni, G. 129	Hough, J.H. 153
Buj, J. 123	Jackson, J.M. 183
Cameron, M. 183	Johnston, K.J. 127
Campbell, B. 111	Just, A. **271**
Castets, A. 13	Kameya, O. **111**, 115
Catarzi, M. 203	Kasuga, T. 115
Caux, E. 325	Kawabe, R. 111
Cernicharo, J. 97	Kegel, W.H. 271
Cesaroni, R. 203, 207	Knacke, R.F. **131**
Chapman, S.J. **323**	Lamarre, J.M. 325
Charnley, S.B. **247**, 253, **321**	Langer, W.D. 13
Chièze, J.-P. **263**, 267	Larson, H.P. 131
Codella, C. **129**	Latter, W.B. **255**
Cohen, R.J. **189**	Leach, S. 317
Comoretto, G. 129, **203**, 207	Lioure, A. **267**
Cossart-Magos, C. **317**	López, R. 123
Deiss, B.M. 271	Maccaferri, G. 129
van Dishoeck, E.F. 73	Mantilla Ch., J. 269
Dutrey, A. **13**	Marquette, J.-B. **239**
Duvert, G. 13	Massi, M. 203
Eckart, A. **183**	Mauersberger, R. 107
Estelella, R. 123	Meyerdierks, H. **69**

Migenes, V.   **127**
Millar, T.J.   209
Miller, M.   25
Minchin, N.R.   **153**
Moorhouse, A.   243
Mundt, R.   145
Myers, P.C.   **133**
Nyman, L.-Å.   **23**
Pajot, F.   325
Palagi, F.   203
Palla, F.   203, **207**
Palumbo, G.G.C.   129
Parravano, A.   **269**
Pastor, J.   123
Pauls, T.A.   127
Planesas, P.   123
Poetzel, R.   145
Prusti, T.   149
Rawlings, J.M.C.   **257**
Ray, T.P.   **145**
Rist, C.   97
Roberge, W.G.   247
Robson, E.I.   313
Rothermel, H.   183
Russell, A.P.G.   17
Rydbeck, G.   183
Sales, N.   **325**
Scappini, F.   129
Scarrott, S.M.   **141**

Schilke, P.   **199**
Schmid-Burgk, J.   107
Schulz, A.   107
Serra, G.   325
Smith, M.D.   **243**
Sternberg, A.   **249**
Stutzki, J.   **17**, 183
Stüwe, J.A.   **155**
Thaddeus, P.   **1**, 23
Tofani, G.   203
Umemoto, T.   115
Vanysek, V.   **291**
de Vries, C.P.   73
Walker, H.J.   309
Ward-Thompson, D.   **313**
Watt, G.D.   **119**
Wesselius, P.R.   **149**
White, G.J.   27, 125
Whittet, D.C.B.   149, **309**, 321
Whitworth, A.P.   323
Wiklind, T.   183
Williams, D.A.   257, **295**, 321
Wilson, R.W.   13
Wilson, T.L.   **107**, 127
Winnewisser, G.   25
Wolf, M.   **259**
Wolfendale, A.W.   **41**
Wouterloot, J.G.A.   21
Ziurys, L.M.   107

OBJECT INDEX

AFGL 490   311
AFGL 890 (IRAS 06084-0611)   311
AFGL 961 (IRAS 06319+0415)   310-312
AFGL 989 (IRAS 06384+0932)   311
AFGL 2591   153, 154
AFGL 4029   145
AFGL 5142   123
AFGL 5157   123
Aquila Rift   4
Arp 220   171, 173, 174
Auriga   38, 261

B1   134, 135
B5   304, 305, 321, 322
B216   138
B217   138
B335   115, 117

3C 84 (Perseus A)   173
4C 12.50   173
Callisto   122
Carina   5
Cas A   180
Centaurus A (NGC 5128)   168, 169, 183-186
Cepheus   32, 38
Cepheus A   98, 145, 146, 194-196
Ceres   122
Chamaeleon   38, 149-152
Z CMa   145
Comet Brorsen-Metcalf   122
Comet Halley   291
Comet Okazaki-Levy-Rudenko   122
Crab Nebula   122
Cygnus Rift   4
Cygnus-X   315
V645 Cyg   145

30 Dor   6, 161
DR21   97, 315

Galactic Centre   1, 5, 12, 30, 32, 43, 122, 252, 282
Ganymede   122
GG30   141, 142
GG31   141
GG 38   141
GGD 37   145
G35.2-0.7N   194, 195
Gum Nebula   141

Haro 6-10 (L1524)   123, 124
HD210121   73, 74
HD211880   27
Heiles Cloud 2   129
HH 7-11   98, 148
HH 19-27 (NGC 2068)   123, 124
HH 120   141
HHL 73   123, 124
HLC 16   38
HLC 20   38
HLC 30   38

IC 342   160
IRAS 02219+6152 (W3 IRS5)   310
IRAS 06084-0611 (AFGL 890)   311
IRAS 06319+0415 (AFGL 961)   310-312
IRAS 06384+0932 (AFGL 989)   311
IRAS 12542-6115   141
IRAS 19345+0727   115-117
IRAS 20582+7724   25
IRC+10216   99, 103, 121

Juno   122

Lk Hα 198   145
Lk Hα 234   145
LMC   5-7, 9, 158, 160-163, 168, 227, 229
Lupus   38
L43 (RNO 91)   123, 124
L134N   99

Object index

L204   134, 135
L810   142-144
L1204   27
L1228   25, 26
L1524 (Haro 6-10)   123, 124
L1551   25, 26, 245
L1551 IRS1   26
L1641   135

M17   17, 76, 78-85, 91
M17SW   76-78, 80
M31   1, 7, 9-12, 179-181
M42   16, 97, 110
M43   16
M51   166, 167
M81   179, 181
M82   157, 160, 170, 172, 174
M101   170
MCW 1080   145
Mon R2   38
Mrk 1014   173

N159   161
NGC 253   157, 174
NGC 1977   16
NGC 2023   17, 20
NGC 2024 (Orion B)   17, 79, 80, 89-91, 93, 102
NGC 2068 (HH19-27)   123, 124
NGC 2264   99, 102
NGC 2688   99
NGC 3526   158, 171, 172
NGC 5128 (Centaurus A)   168, 169, 183-186
NGC 6946   167, 179, 181
NGC 7538   99, 102, 111-114, 199, 200, 202
North Celestial Pole Loop   69, 70, 135

Oort cloud   41, 273-278, 280-283, 286
Ophiuchus   2, 38, 138
ρ Oph   80, 83, 149-153
ζ Oph   70, 215

Orion A   1, 13, 75-77, 80, 88, 91, 97, 109, 1 07-110, 121, 125, 126, 135, 261
  Bar   17, 18, 91-93, 252
  BN-KL   13, 18, 77, 84, 85, 101, 131, 135, 194, 195, 232
  Compact Ridge   221, 232, 233
  Hot Core   77, 80, 232, 299
  H1C   17
  IRc2   17, 18, 80, 100-102, 107, 125, 232, 244
  KL   13, 77, 80, 84, 85, 98-101, 107, 109, 110, 194, 233
  Plateau 20, 101, 102, 232, 234
  Trapezium   91
OMC-1   15, 100, 102, 243-245
Orion B (NGC2024)   17-19, 80, 88-91, 93
Orion-Monoceros   1, 31, 38, 84
θ^1C Orionis   92

Pallas   122
Perseus A (3C 84)   173
Polaris Flare   5, 8, 70

Re 10   141, 142
RNO 91 (L43)   123, 124
Rosette Nebula   81-83
RX Boo   98

S106   17, 88, 91, 134
S140   27
S255   88
Sgr B2   121
SMC   187, 188, 227, 229
SN 1987A   6, 9, 257
Southern Coalsack   23, 24, 141, 142

Tapia 2   24
Taurus   1, 2, 6, 38, 76, 77, 80, 138, 261
Titan   122, 292
TMC-1   99, 129, 130, 225, 228, 230, 253, 299

Ursa Major   38

Object index

Vesta   122
Vulpecula Rift   4
VY CMa   98
W3   81-84, 98
W3(OH)   98, 102, 190, 192-194, 196
W33A   298
W49   98
W49A   313, 314, 316
W49N   98
W75   313, 315, 316
W75S   102

SUBJECT INDEX

Accretion   230, 232, 253, 254, 294, 295, 299, 321
Alfvén speed   65, 244, 244, 248
Ambipolar diffusion   84, 87, 244, 247, 297, 303, 304, 306
Ammonia, NH_3 80, 110, 123, 124, 127, 128, 133, 138, 174, 194, 199-202, 212, 217, 224, 225, 232, 247, 248, 299, 300, 302, 305, 322

Bipolar nebulae   141
Bipolar outflows   25, 27, 28, 101, 102, 109, 111, 115-118, 153, 154, 194, 197, 234, 243, 245, 315

[CI] observations   81, 91, 120, 121, 229
[CII] observations   80-82, 91, 229
Chemical models   214, 224-236, 247, 248, 253-255, 258, 294, 303-305, 309, 321, 322
Chemical processes   209-216
  cosmic-ray ionisation   216, 249, 250
  dissociative recombination   214, 215, 217-222, 225, 226, 231, 233, 255, 303
  gas-phase reactions   209-224, 230, 232, 249, 251
  ion-molecule reactions   103, 104, 211-213, 216, 217, 223, 224, 226, 231, 239-241, 249, 253
  ion-neutral reactions   103, 210, 219, 226, 232, 233, 291, 292
  neutral-neutral reactions   103, 211, 212
  radiative association   213, 214, 218, 220, 221, 231, 232, 239
Chemistry   41, 69, 126, 209-234, 247, 248, 263, 265, 286, 299
  comets   291-294
  cyclic models   230, 295, 296, 304, 305, 321, 322
  dark clouds   216-229, 247, 248, 253, 254, 299

Chemistry
  diffuse clouds   229
  high-latitude clouds   50
  hot molecular cores   232-234, 299
  shocks   71, 102, 104, 218, 230, 234, 247, 248
  star-forming regions   232-234, 249-252
  supernovae   257, 258
Clumps   6, 7, 12, 20, 25, 27, 49, 56, 57, 65, 66, 70, 73-87, 94, 115, 116, 125, 127, 128, 138, 142, 184, 186, 199, 210, 227-230, 233, 234, 244, 247, 258, 263-266, 268, 282, 284, 302, 304, 306, 321, 322
CO–H_2 conversion factor, (X)   6, 29-33, 37-39, 41-47, 54, 69, 157-162, 165, 166, 168, 170, 171, 175, 282
CO observations
  extragalactic objects   5-12, 157-175, 179-186, 257
  Galactic clouds   1-5, 13-33, 37, 38, 49, 51-55, 73-83, 88-94, 107-118, 120-122, 131, 133, 134, 153
  isotopes   13-24, 27, 46, 51, 52, 54, 69, 70, 73, 75-83, 88-94, 109, 111, 113, 115-117, 133
  masses of molecular clouds   5, 6, 12, 21-24, 27, 29, 41-47, 63, 91, 117, 157-162, 164, 170, 171, 175, 282
Comets   41, 122, 189, 273-287, 291
Cosmic rays   30, 31, 41, 43-45, 47, 87, 179, 225-227, 251, 269, 284, 297, 300, 301, 311
CS observations   16, 23, 24, 76, 77, 80, 88, 89, 101, 109, 112, 121, 123, 124, 133, 171, 173, 174, 183, 186

Deuterated molecules   97, 223, 224, 232, 233, 247-252, 255, 256, 292-294, 299, 304

Deuterium   97, 223, 247, 249, 250, 256, 292
Dust   27, 45, 50, 57, 60, 63, 64, 69, 75, 79, 100, 103, 121, 131, 153, 160, 165, 168, 199-202, 209, 210, 216, 219, 229, 245, 247, 257, 283-285, 293, 317-326
  emission   33, 88, 89, 122
  in molecular clouds   295-306, 313-316
  mantles   90, 131, 209, 210, 230, 232, 234, 293, 295-302, 309-312, 321, 322
  surface reactions   209, 210, 216, 221, 224, 249, 251, 253, 254, 292, 296, 299
  far-infrared emissivity   33-35, 37, 88, 90

Extinction   5, 29, 37, 39, 42, 43, 55, 60, 61, 69-71, 81, 133, 143, 155, 156, 159, 160, 229, 249, 266, 269, 295, 298, 324

Far-infrared observations   17, 45, 69, 81, 170, 179, 182, 313-316
Formaldehyde, H_2CO 54, 69-71, 109, 168, 174, 183, 225, 248, 299
Fractionation   97, 210, 223, 224, 232, 233, 247, 248, 256, 291-294
Fragmentation   83, 128, 259-262, 268, 313

Galactic Centre   1, 4, 5, 12, 30, 32, 43, 119, 122, 245, 252
Galactic cirrus   50, 57, 65, 69
Gamma rays   5, 30-32, 43-46, 69
Gamma ray emissivity   30-33
Gamma ray observations   29, 30, 160
Gas-grain interaction   253, 254, 299-302
Globules   23, 24, 63, 115, 141-144, 264

HII regions   13, 16, 21, 23, 28, 34, 35, 37, 44, 53, 76, 81, 88, 91, 109, 111-123, 185, 187, 192-194, 199, 202, 205-208, 227, 229, 313, 325, 326

HH objects   23, 107-110, 123, 124, 141, 145-148, 243
Heating
  by dust   20, 37, 91
  by interstellar radiation field   13, 20, 27, 29, 37, 249, 265, 266
  by photoelectric effect   27, 87, 263, 295, 297
  by shocks   20, 29
  by turbulence   271
High latitude clouds   29, 49-66, 69, 73, 74, 83

Infrared observations   153, 204
Initial mass function   149, 261, 262
Interfaces   34, 63, 76, 79, 80, 82, 91, 249-252, 305
Interstellar radiation field   35, 37, 57, 74, 76, 81, 85, 87, 91, 157, 160, 216, 226, 229, 249-252, 257, 258, 263-266, 323
Ionisation   187, 217, 243, 245, 248, 256, 296, 297, 303, 304, 306, 323
  by cosmic rays   216, 249-252, 263, 269
  by cosmic-ray-induced UV photons   216, 225-227
  by UV radiation   216, 217
Ionisation fronts   27
IRAS   5, 11, 22, 34, 57, 61, 69, 70, 158, 164, 168, 207, 295, 321, 325
IRAS observations   29, 33, 45, 51, 64, 149, 150, 164, 170, 192, 205, 309-316
IRAS sources   21, 23-26, 80, 115, 116, 123, 124, 141, 150, 203, 205, 310

Jeans length   272
Jeans mass   83, 260, 263, 265, 266
Jets   109, 110, 118, 145-148, 243

LTE   17, 19, 70, 116, 184, 185, 189, 190
LTE method   15
Luminosity function   149-152, 155

Subject index

LVG modelling   13-16, 27, 71, 112, 208
Magnetic fields   49, 50, 56, 64-66, 76, 84-86, 109, 133-139, 144, 164, 179-182, 189, 194-197, 243, 247, 260, 303, 304
Masers   98, 102, 127, 128, 189-208, 244, 313
  circumstellar   190, 204
  extragalactic   157, 170, 192
  interstellar   27, 111, 113, 114, 134, 135, 189-208
Metallicity   7, 12, 43, 47, 157, 158, 160, 161, 175, 229, 269
Molecular clouds   1-12, 21, 23, 34, 37, 38, 50, 51, 54, 60-63, 75-83, 85, 86, 127, 134, 182, 207, 209, 244, 259-263, 267-269, 273-277, 291-306, 317
  cores   13, 42, 57, 75, 79, 88, 111, 112, 121, 123-125, 127, 137, 174, 296, 302
  dark clouds   23, 49, 50, 76, 77, 80, 99, 104, 133-135, 138, 141-143, 155, 156, 185, 219, 224, 226, 248, 253, 254, 261, 293, 298, 303, 321, 322
  dense clouds   97, 100, 104, 160, 185, 207, 209, 216, 224, 225, 227, 229, 230, 232, 234, 239, 247, 250, 296
  diffuse clouds   49, 50, 64, 66, 70, 71, 87, 92, 133, 135, 137, 160, 210, 213, 215, 216, 229, 230, 249, 250, 295, 296, 298, 305
  giant   5, 37, 39, 41-47, 49, 50, 55, 76, 83, 133, 135, 162, 168, 216, 232, 273, 284, 325
  high-latitude   5, 8, 25, 49-66, 69-73, 83
  hot cores   77, 80, 102, 104, 232, 234
  in external galaxies   5-12, 157-165, 179-186
  structure   14, 70, 75-83, 94, 115-118, 155, 282
  translucent   70
  warm gas   17, 20, 75, 89, 91, 104, 160

Molecular outflows   23-25, 65, 84, 87, 90, 104, 107-110, 115-118, 121, 123, 125, 131, 145-148, 190, 193, 194, 205, 232, 244, 245
Molecular Ring   32

OB associations   34, 37, 76, 187, 188, 316
OB stars   76, 80, 81, 86, 90, 111, 134, 187, 323
Oort cloud   41, 273-278, 280-283, 286

PAHs   81, 214, 223, 232, 295, 317-320, 325
Photochemistry   298, 306
Photodesorption   297, 301, 302
Photodestruction   71, 74, 81, 215, 216, 226, 227, 229, 230
Photodissociation   103, 104, 162, 192, 215, 249, 250, 255, 263, 301
Photodissociation regions (PDRs)   20, 91-94, 228, 232, 249-252
Polarisation   76, 84, 133, 138, 139, 141-144, 153, 154, 179-182, 194-197, 295, 297
Protostars   75, 88, 112
Pseudo-time-dependent chemistry   224, 229, 230, 233, 305

Radiative transfer   17, 18, 37, 74, 185, 195
Reflection nebulae   5, 114, 141-143, 150, 153

Shocks   13, 16, 20, 62-64, 69, 70, 84, 92, 102, 104, 109, 112, 121, 145, 179, 187, 192, 229, 234, 243-245, 247, 248, 296, 297, 304, 305, 316, 322
Silicate features   295, 298, 311, 312
Starburst galaxies   91, 121, 160, 170-172, 174

Star formation   23, 24, 34, 37, 41, 47, 75,
    87, 107, 114, 124, 141-144, 164, 170,
    171, 173-179, 181, 182, 187, 189, 190,
    190, 197, 216, 247, 259, 261, 267-270,
    295, 297, 302, 304, 306, 321
  efficiency   84, 85, 157, 164, 168, 170,
    267, 269
  rate   84, 157, 164, 167, 168, 170, 179,
    261, 262, 268-270
Star-forming regions   17, 21, 49, 80,
    84, 87, 88, 90, 98, 99, 102, 104, 111,
    115, 123, 125, 149-152, 173, 179, 189,
    192, 195, 203-205, 209, 221, 224, 227,
    232-234, 244, 260, 269, 296, 313-316
Steady-state chemistry   70, 104, 224,
    233, 250, 300, 305, 321
Stellar winds   84, 147, 148, 181, 187, 190,
    234, 244, 295, 304, 306, 321

Submillimetre observations   17, 25, 75,
    79, 88, 121, 122, 313-316
Supernovae   5, 6, 44, 85, 179, 181, 187,
    250, 315, 316
T Tauri stars   23, 80, 134, 147, 150, 152
Turbulence   44, 57, 61, 79, 84, 127, 138,
    181, 259, 260, 263, 265, 284, 304

Virial equilibrium   78, 79, 83, 86, 87,
    127, 128, 137, 158, 159, 171
Virial masses   6, 21, 70, 160, 161, 163
Virial theorem   42, 45, 46, 137

Young stellar objects (YSOs)   145-148,
    153, 311

Zeeman effect   76, 133, 134, 195, 197